T0256970

Essential Radio Astronomy

PRINCETON SERIES IN MODERN OBSERVATIONAL ASTRONOMY

David N. Spergel, SERIES EDITOR

Written by some of the world's leading astronomers, the Princeton Series in Modern Observational Astronomy addresses the needs and interests of current and future professional astronomers. International in scope, the series includes cutting-edge monographs and textbooks on topics generally falling under the categories of wavelength, observational techniques and instrumentation, and observational objects from a multiwavelength perspective.

Statistics, Data Mining, and Machine Learning in Astronomy: A Practical Python Guide for the Analysis of Survey Data, by Željko Ivezić, Andrew J. Connolly, Jacob T. VanderPlas, and Alexander Gray

Essential Radio Astronomy

James J. Condon and Scott M. Ransom

PRINCETON UNIVERSITY PRESS • PRINCETON AND OXFORD

Published by Princeton University Press, 41 William Street,
Princeton, New Jersey 08540

In the United Kingdom: Princeton University Press, 6 Oxford Street,
Woodstock, Oxfordshire OX20 1TR

ISBN 978-0-691-13779-7

Library of Congress Control Number: 2016931166
British Library Cataloging-in-Publication Data is available

This book has been composed in Minion Pro w/ Universe light condensed for display
Printed on acid-free paper ∞

press.princeton.edu

Typeset by Nova Techset Private Limited, Bangalore, India
Printed in the United States of America

10   9   8   7   6   5   4   3   2   1

Contents

Preface

Essential Radio Astronomy (ERA) grew from lecture notes for the one-semester radio astronomy course taken by all astronomy graduate students at the University of Virginia. To attract advanced undergraduates with backgrounds in astronomy, physics, or engineering to radio astronomy, we limited the prerequisites to basic physics courses covering classical mechanics, macroscopic thermodynamics, electromagnetism, and elementary quantum mechanics. Prior courses covering electromagnetism with vector calculus, electrical engineering, special relativity, statistical thermodynamics, advanced quantum mechanics, Fourier transforms, or astrophysics were *not* required. Nearly everything in *ERA* has been derived from first principles in order to fill the gaps in students' backgrounds and make *ERA* a useful reference for practicing radio astronomers.

Classical radio astronomy textbooks such as *Radio Astronomy* by J. D. Kraus [62] emphasized radio engineering and were written for two-semester courses that prepared students for careers in radio astronomy. In this era of multiwavelength astronomy, most graduate students can afford to spend only one semester studying radio astronomy, during which only the most essential concepts can be presented. Our goal is to give them the background needed to read and understand the radio astronomy literature, to recognize when radio observations might help solve an astrophysical problem, and to design, propose, and analyze radio observations. *ERA* complements the longer and more technical *Tools of Radio Astronomy* [116] that "describes the tools radio astronomers need to pursue their goals" in greater detail. *ERA* was also shaped by our belief that radio astrophysics owes more to thermodynamics than to electromagnetism, so Kirchhoff's law appears more frequently than Maxwell's equations.

Originally, our brief lecture notes were printed and handed out to students in the traditional classroom environment: a chalk-dusted professor covered blackboards with equations which students faithfully copied into their notebooks for later study. To avoid this unnecessary work and free the students to concentrate on the ideas being presented, we expanded the abbreviated notes into complete texts with figures and full mathematical derivations, then converted them from TeX to html so they could be posted to the web and projected onto a screen in any classroom. Now the professor faces "heads up" students who can watch, listen, and ask questions without worrying about failing to write down a crucial step in some long derivation.

A broader goal of the National Radio Astronomy Observatory[1] is fostering the community of researchers using radio astronomy by attracting and training the most

---

[1] The National Radio Astronomy Observatory is a facility of the National Science Foundation operated under cooperative agreement by Associated Universities, Inc.

talented university students *anywhere in the world*. NRAO directors Paul Vanden Bout, Fred Lo, and Tony Beasley have generously supported our efforts to upgrade and expand all of the course materials (lectures, problem sets, exams) that are now available at http://www.cv.nrao.edu/course/astr534/ERA.shtml.

We hope that the combination of this book and its associated website will facilitate teaching radio astronomy at the university level, especially at the many colleges and universities lacking "black belt" radio astronomers. The website can display large galleries of color images, link to interactive demonstrations and relevant articles on the web, present problem sets and solutions, and be updated frequently to present new findings or report errata from the book. We thank Princeton University Press for agreeing to a nonexclusive copyright that allows the book and website to coexist.

We are extremely grateful to Jim Moran, who used a draft of *ERA* to teach his radio astronomy course at Harvard and gave us extremely valuable feedback. We also thank our outstanding copyeditor, Alison Durham, for significantly improving the style and accuracy of the text.

NRAO, Charlottesville
2015 October 15

James J. Condon   jcondon@nrao.edu
Scott M. Ransom   sransom@nrao.edu

Essential Radio Astronomy

1   Introduction

1.1 AN INTRODUCTION TO RADIO ASTRONOMY

1.1.1 What Is Radio Astronomy?

Radio astronomy is the study of natural radio emission from celestial sources. The range of **radio frequencies** or wavelengths is loosely defined by atmospheric opacity and by quantum noise in coherent amplifiers. Together they place the boundary between radio and far-infrared astronomy at frequency $\nu \sim 1\,\text{THz}$ ($1\,\text{THz} \equiv 10^{12}\,\text{Hz}$) or wavelength $\lambda = c/\nu \sim 0.3\,\text{mm}$, where $c \approx 3 \times 10^{10}\,\text{cm s}^{-1}$ is the vacuum speed of light. The Earth's **ionosphere** sets a low-frequency limit to ground-based radio astronomy by reflecting extraterrestrial radio waves with frequencies below $\nu \sim 10\,\text{MHz}$ ($\lambda \sim 30\,\text{m}$), and the ionized interstellar medium of our own Galaxy absorbs extragalactic radio signals below $\nu \sim 2\,\text{MHz}$.

The radio band is very broad logarithmically: it spans the five decades between $10\,\text{MHz}$ and $1\,\text{THz}$ at the low-frequency end of the electromagnetic spectrum. Nearly everything emits radio waves at some level, via a wide variety of emission mechanisms. Few astronomical radio sources are obscured because radio waves can penetrate interstellar dust clouds and Compton-thick layers of neutral gas. Because only optical and radio observations can be made from the ground, pioneering radio astronomers had the first opportunity to explore a "parallel universe" containing unexpected new objects such as radio galaxies, quasars, and pulsars, plus very cold sources such as interstellar molecular clouds and the cosmic microwave background radiation from the big bang itself.

Telescopes observing from above the atmosphere have since opened the entire electromagnetic spectrum to astronomers, but radio astronomy retains a unique observational advantage. **Coherent amplifiers**, which preserve phase information, allow the construction of sensitive multielement **aperture-synthesis interferometers** that can image complex sources with angular resolution and absolute astrometric accuracies approaching 10^{-4} arcsec. **Quantum noise** forever restricts sensitive coherent amplification to the low photon energies $E = h\nu$ (where h = Planck's constant $\approx 6.626 \times 10^{-27}$ erg s) of the radio band. Also, coherent signals can be shifted to lower frequencies and digitized, permitting the construction of radio spectrometers with extremely high spectral resolution and frequency accuracy.

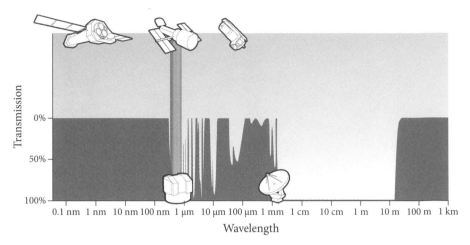

Figure 1.1. Ground-based astronomy is confined to the visible and radio **atmospheric windows**, wavelength ranges in which the atmosphere is nearly transparent. The radio window is much wider than the visible window when plotted on logarithmic wavelength or frequency scales, so it includes a wide range of astronomical sources and emission mechanisms. Radio astronomers usually measure (and think in terms of) frequencies $\nu = c/\lambda$ instead of wavelengths λ. Thus $\lambda = 0.3$ mm corresponds to $\nu = 1$ THz, the highest frequency accessible from the best terrestrial sites. The Earth's ionosphere reflects radio waves longer than $\lambda \sim 30$ m ($\nu \sim 10$ MHz). Abscissa: Wavelength. Ordinate: Atmospheric transmission. Image Credit: ESA/Hubble (F. Granato).

1.1.2 Atmospheric Windows

The Earth's atmosphere absorbs electromagnetic radiation at most infrared (IR), ultraviolet, X-ray, and gamma-ray wavelengths, so only optical/near-IR and radio observations can be made from the ground (Figure 1.1). The visible-light window is relatively narrow and spans the wavelengths of peak thermal emission from $T \sim 3000$ K to $T \sim 10,000$ K blackbodies. Early observational astronomy was limited to visible objects—hot thermal sources such as stars, clusters and galaxies of stars, and gas ionized by stars (e.g., the Orion Nebula in Orion's sword is visible as a fuzzy blob to the unaided eye on a dark night), and to cooler objects shining by reflected starlight (e.g., planets and moons). Knowing the spectrum of blackbody radiation, astronomers a century ago correctly deduced that stars having nearly blackbody spectra would be undetectably faint as radio sources, and incorrectly assumed that there would be no other celestial radio sources. Consequently they failed to develop radio astronomy until strong radio emission from our Galaxy was discovered accidentally in 1932 and followed up by radio engineers.

What physical processes limit the atmospheric windows? At the high-frequency end of the radio window, vibrational transitions of atmospheric molecules such as CO_2, O_2, and H_2O have energies $E = h\nu$ comparable with those of mid-infrared photons, so vibrating molecules absorb most extraterrestrial mid-infrared radiation. Lower-energy rotational transitions of atmospheric molecules define the fairly broad transition between the far-infrared band and the high-frequency limit of the radio window at $\nu \sim 1$ THz. Ground-based radio astronomy is increasingly degraded at frequencies $\nu < 300$ MHz (wavelengths $\lambda > 1$ m) by variable ionospheric refraction,

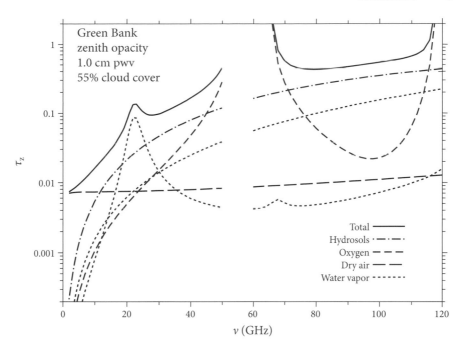

Figure 1.2. The atmospheric zenith opacity τ_z at Green Bank during a typical summer night. An opacity τ attenuates the power received from an astronomical source by the factor $\exp(-\tau)$. The oxygen and dry-air opacities are nearly constant, while the water vapor and hydrosol contributions vary significantly with weather.

and celestial radio waves having frequencies $\nu < 10\,\mathrm{MHz}$ (wavelengths $\lambda > 30\,\mathrm{m}$) are usually reflected back into space by the Earth's ionosphere. Total internal reflection in the ionosphere makes the Earth look like a mirror from space, like the glass face of an underwater wristwatch viewed obliquely.

Ultraviolet photons have energies close to the binding energies of the outer electrons in atoms, so electronic transitions in atoms account for the high ultraviolet opacity of the atmosphere. Higher-energy electronic and nuclear transitions produce X-ray and gamma-ray absorption. In addition, Rayleigh scattering of sunlight by atmospheric gas molecules and dust particles at visible and ultraviolet wavelengths brightens the sky enough to preclude daytime optical observations of faint objects. Radio wavelengths are much longer than atmospheric dust grains and the Sun is not an overwhelmingly bright radio source, so the radio sky is always dark and many radio observations can be made day or night.

The atmosphere is not perfectly transparent at any radio frequency. Figure 1.2 shows how its **zenith** (the zenith is the point directly overhead) opacity τ_z in Green Bank, WV varies with frequency during a typical summer night with a water-vapor column density of 1 cm, 55% cloud cover, and surface air temperature $T = 288\,\mathrm{K} = 15°\mathrm{C}$. The total **zenith opacity** (solid curve) is the sum of several components [65]:

1. The broadband or continuum opacity of dry air (long dashes) results from viscous damping of the free rotations of nonpolar molecules. It is relatively small ($\tau_z \approx 0.01$) and nearly independent of frequency.

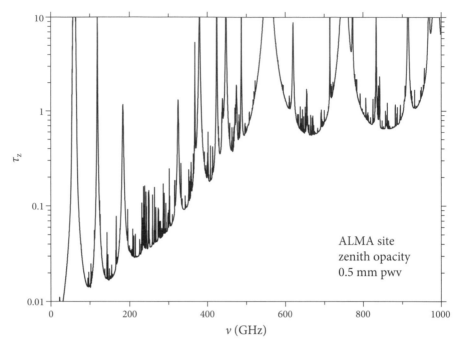

Figure 1.3. Zenith atmospheric opacity τ_z for 0.5 mm pwv at the Atacama Large Millimeter Array (ALMA) site. Water-vapor absorption is responsible for the broad opaque bands centered on 557 GHz, 752 GHz, and 970 GHz. The plotted data are from https://almascience.eso.org/about-alma/weather/atmosphere-model.

2. Molecular oxygen (O_2) has no permanent electric dipole moment, but it does have rotational transitions that can absorb radio waves because it has a permanent magnetic dipole moment. The atmospheric-pressure-broadened complex of oxygen spectral lines (short dashes) is quite opaque ($\tau_z \gg 1$) and precludes ground-based observations in the frequency range 52 GHz $< \nu <$ 68 GHz (1 GHz $\equiv 10^9$ Hz).

3. **Hydrosols** are liquid water droplets small enough (radius ≤ 0.1 mm) to remain suspended in clouds. They are much smaller than the wavelength even at 120 GHz ($\lambda \approx 2.5$ mm), so their emission and absorption obey the Rayleigh scattering approximation and their opacity (dot-dash curve) is proportional to λ^{-2} or ν^2.

4. The strong water-vapor spectral line at $\nu \approx 22.235$ GHz is pressure broadened to $\Delta \nu \approx 4$ GHz width. The so-called "continuum" opacity of water vapor at radio wavelengths is actually the sum of line-wing opacities from much stronger water lines at infrared wavelengths [106]. In the plotted frequency range, this continuum opacity is also proportional to ν^2. Both the line and continuum zenith opacities (dotted curves) are directly proportional to the column density of **precipitable water vapor (pwv)** along the vertical line of sight through the atmosphere. Conventionally pwv is expressed as a length (e.g., 1 cm) rather than as a true column density (e.g., 1 gm cm^{-2}), but the two forms are numerically equivalent because the mass density of water is unity in CGS units.

The partially absorbing atmosphere doesn't just attenuate incoming radio radiation; it also emits radio noise that can seriously degrade the sensitivity of ground-based radio observations. If the atmospheric opacity is τ, the atmospheric transparency is $\exp(-\tau)$ and emission from the atmosphere at kinetic temperature $T \sim 300\,\mathrm{K}$ adds $\Delta T_\mathrm{s} = T[1 - \exp(-\tau)]$ to the system noise temperature T_s. Radio astronomers use $T_\mathrm{s} \equiv P_\nu / k$, where $k =$ Boltzmann's constant $\approx 1.38 \times 10^{-16}\,\mathrm{erg\,K^{-1}}$, as a convenient measure of the noise power per unit bandwidth P_ν. The system noise temperature is normally much smaller than the atmospheric kinetic temperature, so the added noise from atmospheric emission degrades sensitivity even more than pure absorption alone. For example, emission by water vapor in the warm and humid atmosphere above Green Bank, WV precludes sensitive observations near the water-vapor line at $\nu \sim 22\,\mathrm{GHz}$ during the summer. Green Bank can be quite cold and dry in the winter, allowing observations at frequencies up to $\approx 115\,\mathrm{GHz}$. The very best sites for observing at higher frequencies are exceptionally high and dry. For example, the Atacama Large Millimeter Array (ALMA) shown in Color Plate 5 is located at 5000 m elevation on a desert plain near Cerro Chajnator in Chile, where the typical pwv is $< 1\,\mathrm{mm}$. Figure 1.3 shows the zenith atmospheric opacity at the ALMA site when pwv $= 0.5\,\mathrm{mm}$, for frequencies up to the $\nu \sim 1\,\mathrm{THz}$ atmospheric limit.

Finally, the refractive index of water vapor is about 20 times higher at radio than at optical wavelengths because the index of refraction at wavelength λ is proportional to the cumulative strength of the water-vapor absorption lines at *shorter* wavelengths, the strongest of which lie in the far-infrared range $0.03 < \lambda\,(\mathrm{mm}) < 0.6$. Water vapor is not well mixed in the atmosphere, so fluctuations in the column density of water vapor along the line of sight blur the image of a point radio source. The scale height of water vapor in the troposphere is $\sim 2\,\mathrm{km}$, so the largest fluctuations have transverse dimensions of several km. Consequently point sources seen by all radio telescopes or radio interferometers smaller than a few km in size are blurred by $\sim 0.5\,\mathrm{arcsec}$, and this blurring is nearly independent of wavelength for all $\lambda > 0.6\,\mathrm{mm}$. The angular size of the "seeing" disk for interferometers much larger than a few km is inversely proportional to the size of the interferometer. In contrast, optical seeing at wavelengths $\lambda \ll 0.03\,\mathrm{mm}$ is dominated by the much smaller ($\sim 10\,\mathrm{cm}$) turbulent density fluctuations of dry air. It is only a coincidence that the optical seeing disk is also $\sim 0.5\,\mathrm{arcsec}$ at the best terrestrial sites. For a thorough review of atmospheric and ionospheric propagation effects, see Thompson, Moran, & Swenson [106, Chapter 13].

1.1.3 Astronomy in the Radio Window

Because the radio window is so broad, (1) almost all types of astronomical sources, thermal and nonthermal radiation mechanisms, and propagation phenomena can be observed at radio wavelengths; and (2) a wide variety of radio telescopes and observing techniques are needed to cover the radio window effectively.

The radio window was explored before there were telescopes in space, so early radio astronomy was a science of discovery and serendipity. It revealed a "parallel universe" of unexpected sources not previously seen, or at least not recognized as being different from ordinary stars. Major discoveries of radio astronomy include

1. nonthermal radiation from our Galaxy [86] and many other astronomical sources;
2. The "violent universe" of powerful radio galaxies [4] and quasars (quasi-stellar radio sources) [48, 101] powered by supermassive black holes (SMBHs);

3. cosmological evolution of radio galaxies and quasars [98];
4. thermal spectral-line emission from cold interstellar gas atoms, ions, and molecules;
5. **maser** (the acronym for microwave amplification by stimulated emission of radiation) emission from interstellar molecules [114];
6. coherent continuum emission from stars and pulsars;
7. cosmic microwave background radiation from the hot big bang [80];
8. pulsars and neutron stars [50];
9. indirect but convincing evidence for gravitational radiation [104];
10. the supermassive black hole at the center of our Galaxy; [8]
11. evidence for dark matter in galaxies, deduced from their HI (neutral hydrogen) rotation curves [92];
12. extrasolar planets [117];
13. strong gravitational lensing [113].

The following items are some of the features of this parallel universe.

1. It is often violent, reflecting high-energy and explosive phenomena in radio galaxies, quasars, supernovae, pulsars, etc., in contrast to the steady light output of most visible stars.
2. Many radio sources are ultimately powered by gravity instead of by nuclear fusion, the principal energy source of visible stars.
3. It is cosmologically distant. Most continuum radio sources are extragalactic, and they have evolved so strongly over cosmic time that most are seen at lookback times comparable with the age of the universe.
4. It can be very cold. The cosmic microwave background dominates the electromagnetic energy of the universe, but its 2.7 K blackbody spectrum is confined to radio and far-infrared wavelengths. Cold interstellar gases emit spectral lines at radio wavelengths.

With the advent of telescopes in space, the entire electromagnetic spectrum has become accessible to astronomers. Many sources discovered by radio astronomers can now be studied in other wavebands, and new objects discovered in other wavebands (e.g., gamma-ray bursters) can now be followed up at radio wavelengths. Radio astronomy is no longer a separate and distinct field; it is one facet of multiwavelength astronomy. Even so, the radio band retains unique astronomical and technical features.

Most of the electromagnetic energy of the universe (Figure 1.4) is in the cosmic microwave background (CMB) radiation left over from the hot big bang. It has a nearly perfect 2.73 K blackbody spectrum peaking at $\nu \approx 220$ GHz. The strong optical/near-infrared (OIR) peak at $\nu \sim 3 \times 10^5$ GHz ($\lambda \sim 1\,\mu$m) is primarily thermal emission from stars plus a smaller contribution of thermal and nonthermal emission from the active galactic nuclei (AGN) in Seyfert galaxies and quasars. Most of the comparably strong cosmic far-infrared (FIR) background peaking at $\nu \sim 3 \times 10^3$ GHz ($\lambda \sim 100\,\mu$m) is thermal reemission from interstellar dust that was heated by absorbing about half of that OIR radiation. The cosmic X-ray and gamma-ray backgrounds are mixtures of nonthermal emission (e.g., synchrotron radiation or inverse-Compton scattering) from high-energy particles accelerated by AGN and thermal emission from very hot gas (e.g., gas in clusters of galaxies).

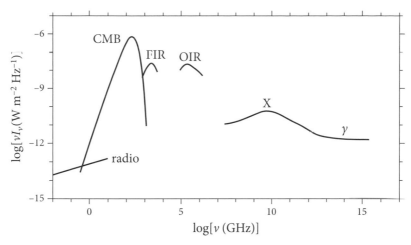

Figure 1.4. The electromagnetic spectrum of the universe at radio [29], far-infrared (FIR) [45], optical/near-infrared (OIR) [45, 46], and X-ray (X) and gamma-ray (γ) [42] frequencies. The extragalactic sky brightness νI_ν per logarithmic frequency interval is plotted as a function of the logarithm of frequency, so the highest peaks correspond to the most energetically important spectral ranges.

By comparison, the relatively faint extragalactic radio-source background (radio) is brighter than the CMB only for $\nu \leq 0.4\,\mathrm{GHz}$. Although many are energetically insignificant, radio sources do trace most phenomena that are detectable in other portions of the electromagnetic spectrum, and modern radio telescopes are sensitive enough to detect them.

1.1.4 What Is Special about Long Wavelengths and Low Frequencies?

Many unique scientific and technical features of radio astronomy result from radio waves occupying the long-wavelength end of the electromagnetic spectrum. At macroscopic wavelengths large groups of charged particles moving together in volumes $< \lambda^3$ may produce strong coherent emission, accounting for the astounding radio brightnesses of pulsars at $\lambda \sim 1\,\mathrm{m}$. Dust scattering is negligible because interstellar dust grains are much smaller than radio wavelengths, so the dusty interstellar medium (ISM) is nearly transparent. This allowed radio astronomers to see through the dusty disk of our Galaxy and discover the compact radio source Sgr A* [8] powered by the supermassive black hole at its center.

Low frequencies imply low photon energies $E = h\nu$. Thus radio spectral lines trace extremely low-energy transitions produced by atomic hyperfine splitting (the ubiquitous 21-cm line of neutral hydrogen at $\nu \approx 1.420$ GHz generates photons of energy $E \approx 6 \times 10^{-6}$ eV), the quantized rotation rates of polar molecules such as carbon monoxide in interstellar space, and high-level recombination lines from interstellar atoms.

At radio frequencies, the dimensionless ratio $h\nu/(kT)$ of photon energy to the mean kinetic energy of particles at temperature T is very small ($\ll 1$). In this limit, the brightness of a blackbody emitter is proportional to ν^2, ensuring that nearly every astronomical object is a thermal radio source at some low level.

On the negative side, the fact that nearly everything emits radio radiation means radio astronomers must deal with large and fluctuating natural foregrounds of emission from the ground, from the atmosphere, and even from their own antennas and receivers. Also, stimulated emission (negative absorption) becomes comparable with absorption when $h\nu/(kT) \ll 1$. This greatly lowers the opacities of radio spectral lines, makes their emission strengths nearly independent of the temperature of the emitting gas, and allows maser emission with only a small population inversion.

In contrast, $h\nu/(kT) \gg 1$ for cold sources at optical frequencies, where the exponential high-frequency cutoff of the blackbody radiation spectrum ensures that essentially no optical photons are emitted. Cold thermal emitters (e.g., the 2.73 K cosmic microwave background, or interstellar gas at temperatures below 100 K) are completely invisible. For example, a person can be approximated by a 300 K blackbody with a surface area $\sim 1\,\mathrm{m}^2$. Such a blackbody emits $\sim 10^{16}$ photons per second at radio frequencies below 10 GHz but only 0.01 photons per second at visible wavelengths $\lambda < 0.75\,\mu\mathrm{m}$.

Free electrons scatter electromagnetic radiation by a process called Thomson scattering or Compton scattering. The Thomson scattering cross section per electron is $\sigma_{\mathrm{T}} \approx 6.65 \times 10^{-25}\,\mathrm{cm}^2$ at all frequencies, and sources behind free-electron column densities $N_{\mathrm{e}} > \sigma_{\mathrm{T}}^{-1} \sim 10^{24}\,\mathrm{cm}^{-2}$ called **Compton thick** because they are obscured. Radio photons have energies much lower than the $\sim$eV (electron volt) binding energies of electrons in atoms, so those electrons are not "free" for radio photons, and radio waves can penetrate neutral Compton-thick sources. In contrast, electrons in atoms do not appear bound to X-ray photons with $\gg$ eV energies, and Compton-thick sources (e.g., "buried quasars" behind clouds of gas and dust) are hidden from X-ray observations.

Radio synchrotron sources live long after their emitting electrons were accelerated to relativistic energies, so they can provide long-lasting archaeological records of past energetic phenomena (e.g., see Color Plates 14 and 15). Likewise, neutral hydrogen stripped from colliding galaxies continues emitting at $\lambda = 21\,\mathrm{cm}$ for tens of millions of years (Color Plates 11 and 12).

Most plasma effects (scattering, dispersion, Faraday rotation, etc.) have strengths proportional to ν^{-2} and are strong enough at low radio frequencies to be useful tools for tracing interstellar electron densities and magnetic field strengths.

1.1.5 Radio Telescopes and Aperture-Synthesis Interferometers

Radio telescopes must have very large aperture diameters D to achieve good diffraction-limited angular resolution $\theta \approx \lambda/D$ radians at radio wavelengths. Even the biggest precision radio telescopes (e.g., telescopes with small rms reflector surface errors $\sigma < \lambda/16$) such as the Green Bank Telescope (GBT) (Color Plate 1) with $\sigma \approx 0.2\,\mathrm{mm}$ and $D = 100\,\mathrm{m}$ are limited to $\theta \gg 1$ arcsec. On the other hand, huge multielement interferometers spanning up to $D \sim 10^4$ km are practical (Figure 1.5). Paradoxically, the finest angular resolution for imaging faint and complex sources is obtainable at the long-wavelength (radio) end of the electromagnetic spectrum. Interferometers also yield extremely accurate astrometry (Figure 1.6) because interferometric positions depend on measuring time delays between telescopes rather than on the mechanical pointing errors of telescopes, and clocks are far more accurate than rulers.

Figure 1.5. The Very Long Baseline Array (VLBA) of 10 25-m telescopes extending 8000 km from St. Croix, VI to Mauna Kea, HI yields angular resolution as fine as $\theta = 0.00017$ arcsec, surpassing the resolution of the Hubble Space Telescope by two orders of magnitude. Image credit: NRAO/AUI/NSF.

Radio astronomers always measure frequencies directly, while wavelengths are usually measured in the rest of the electromagnetic spectrum. Measuring a frequency has two practical advantages: (1) frequency can be measured more accurately than wavelength because clocks are more accurate than rulers and (2) frequency doesn't change when radiation passes through a refractive medium, but wavelength does.

Coherent (phase-preserving) amplifiers are required for accurate interferometric imaging of faint extended sources because they allow the signal from each telescope in a multielement interferometer to be amplified before it is split and combined with the signals from the other telescopes, rather than just being divided among the other telescopes. The minimum possible noise temperature of a coherent receiver is $T \approx h\nu/k$ owing to quantum noise. Quantum noise is proportional to frequency, so even the best possible coherent amplifiers at visible-light frequencies must have noise temperatures $T > 10^4$ K. Aperture-synthesis interferometers at radio wavelengths provide unparalleled sensitivity, image fidelity, angular resolution, and absolute position accuracy.

1.2 THE DISCOVERY OF COSMIC RADIO NOISE

Natural radio emission from our Galaxy was detected serendipitously in 1932 by Karl Guthe Jansky, a physicist working as a radio engineer for Bell Telephone

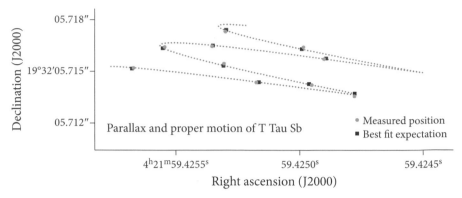

Figure 1.6. Multiepoch VLBA position measurements of T Tau Sb, a companion of the well-known young stellar object T Tauri, allowed Loinard et al. [67] to determine its parallax distance with unprecedented accuracy: $d = 146.7 \pm 0.6\,\mathrm{pc}$, a significant improvement over the Hipparcos distance $d = 177^{+68}_{-39}\,\mathrm{pc}$, and even to detect accelerated proper motion.

Laboratories. Why hadn't professional astronomers of that era vigorously pursued radio astronomy and made this discovery first? In part, because they knew too much. They knew that stars are nearly blackbody radiators at visible wavelengths. The spectral brightness B_ν at frequency ν of an ideal blackbody radiator is given by **Planck's law**

$$B_\nu(\nu, T) = \frac{2h\nu^3}{c^2} \frac{1}{\exp\left(\dfrac{h\nu}{kT}\right) - 1},$$ (1.1)

where B_ν is the power emitted per unit area per unit frequency per steradian of solid angle by a blackbody, $h \approx 6.63 \times 10^{-27}\,\mathrm{erg\,s} = 6.63 \times 10^{-34}\,\mathrm{joule\,s} =$ Planck's constant, $\nu =$ frequency in cycles per second, or hertz (so $\mathrm{Hz} = \mathrm{s}^{-1}$), $k \approx 1.38 \times 10^{-16}\,\mathrm{erg\,K}^{-1} = 1.38 \times 10^{-23}\,\mathrm{joule\,K}^{-1} =$ Boltzmann's constant, $c \approx 3.00 \times 10^{10}\,\mathrm{cm\,s}^{-1} = 3.00 \times 10^8\,\mathrm{m\,s}^{-1} =$ the speed of light, and T is the absolute temperature (K) of the blackbody.

The *subscript* ν in $B_\nu(\nu, T)$ denotes brightness *per unit frequency* and not brightness as a function of frequency. Likewise, the subscript λ in B_λ denotes brightness *per unit wavelength*, even if B_λ is written as a function of frequency, $B_\lambda(\nu, T)$. Thus $B_\nu(\nu, T) = B_\nu(\lambda, T)$ but $B_\nu(\nu, T) \neq B_\lambda(\nu, T)$. Both B_ν and B_λ appear in the astronomical literature, so you have to pay attention to which one is being used. Radio astronomers usually use B_ν because electronic spectrometers measure frequencies, but B_λ is often appropriate for mechanical spectrometers that measure wavelengths.

At radio frequencies, the dimensionless quantity $h\nu/(kT) \ll 1$ for most astronomical sources. For example, the temperature of the Sun's photosphere (the Sun's visible surface) is $T \approx 5800\,\mathrm{K}$. At $\nu = 1\,\mathrm{GHz} = 10^9\,\mathrm{Hz}$, which was near the high-frequency limit of radio technology in 1932,

$$\frac{h\nu}{kT} \approx \frac{6.63 \times 10^{-27}\,\mathrm{erg\,s} \cdot 10^9\,\mathrm{Hz}}{1.38 \times 10^{-16}\,\mathrm{erg\,K}^{-1} \cdot 5800\,\mathrm{K}} \approx 8 \times 10^{-6}.$$ (1.2)

Replacing the exponential denominator in Equation 1.1 by its Taylor-series approximation

$$\exp\left(\frac{h\nu}{kT}\right) - 1 \approx 1 + \frac{h\nu}{kT} + \cdots - 1 \approx \frac{h\nu}{kT} \tag{1.3}$$

yields the simple Rayleigh–Jeans approximation

$$B_\nu(\nu, T) \approx \frac{2h\nu^3}{c^2}\frac{kT}{h\nu} = \frac{2kT\nu^2}{c^2} \tag{1.4}$$

or

$$B_\nu(\lambda, T) = \frac{2kT}{\lambda^2} \tag{1.5}$$

to the blackbody spectrum valid at low frequencies or long wavelengths. The radio emission from a star, which subtends a very small solid angle, would have been too faint to detect. This argument is more-or-less correct; in fact, even the most sensitive modern radio telescopes could not detect the 1 GHz blackbody emission from the photosphere of a star like the Sun if it were moved to the distance $d > 1$ pc of the nearest stars (1 **parsec (pc)** $\approx 3.09 \times 10^{18}$ cm is defined as the distance at which the radius of the Earth's orbit subtends 1 arcsec $\approx 206265^{-1}$ rad).

Example. What would be the flux density S_ν at $\nu = 1$ GHz of a $T = 5800$ K blackbody the size of the Sun (radius $R_\odot \approx 7 \times 10^{10}$ cm) at the distance of the nearest star, about 1 parsec ($d \approx 3 \times 10^{18}$ cm)? This simple example illustrates conversions among CGS units (favored by astrophysicists), MKS or SI units (favored by engineers and radio observers), and "astronomical" units. See Appendix F for lists of useful constants and units, and a review of how to convert between CGS and MKS units.

Flux density S_ν is defined as the power received per unit detector area in a unit bandwidth ($\Delta\nu = 1$ Hz) at frequency ν, so the MKS units of S_ν are W m^{-2} Hz^{-1}. The flux density received from a compact source having brightness B_ν and subtending a small solid angle $\Omega \ll 1$ sr is

$$S_\nu = B_\nu\Omega.$$

For the Sun at 1 GHz,

$$B_\nu = \frac{2kT\nu^2}{c^2} \approx \frac{2 \cdot 1.38 \times 10^{-16} \text{ erg K}^{-1} \cdot 5800 \text{ K} \cdot (10^9 \text{ Hz})^2}{(3.00 \times 10^{10} \text{ cm s}^{-1})^2};$$

Hz = s^{-1} and sr is dimensionless, so

$$B_\nu \approx 1.78 \times 10^{-15} \text{ erg cm}^{-2} = 1.78 \times 10^{-15} \text{ erg s}^{-1} \text{ cm}^{-2} \text{ Hz}^{-1} \text{ sr}^{-1}.$$

Note that B_ν is a property of the source alone; it does not depend on the distance to the observer. The solid angle Ω subtended by the Sun does depend on the distance d to the observer:

$$\Omega = \frac{\pi R_\odot^2}{d^2} \approx \frac{\pi(7 \times 10^{10} \text{ cm})^2}{(3 \times 10^{18} \text{ cm})^2} \approx 1.71 \times 10^{-15} \text{ sr};$$

(Continued)

so the flux density of the Sun,

$$S_\nu = B_\nu \Omega \approx 3.0 \times 10^{-30} \text{ erg s}^{-1} \text{ cm}^{-2} \text{ Hz}^{-1},$$

also depends on the distance to the observer. The statement that "the flux density of the Sun is 3×10^{-30} erg s^{-1} cm^{-2} Hz^{-1}" misleadingly suggests that the flux density is an intrinsic property of the Sun; it makes sense only because the qualifier "for an observer located on the Earth" is implicitly assumed. In MKS units,

$$S_\nu \approx 3.0 \times 10^{-33} \text{ J s}^{-1} \text{ m}^{-2} \text{ Hz}^{-1} \approx 3.0 \times 10^{-33} \text{ W m}^{-2} \text{ Hz}^{-1}.$$

The flux densities of astronomical sources are so small in these units that astronomers introduced the unit "jansky" (honoring Karl Jansky, the first radio astronomer) defined by $1 \text{ Jy} \equiv 10^{-26} \text{ W m}^{-2} \text{ Hz}^{-1} = 10^{-23} \text{ erg s}^{-1} \text{ cm}^{-2} \text{ Hz}^{-1}$. The units mJy ($10^{-3}$ Jy), μJy (10^{-6} Jy), or even nJy (10^{-9} Jy) are used for the faintest sources. Thus

$$S_\nu \approx 0.3 \ \mu\text{Jy}.$$

This is too faint even for the most sensitive modern radio telescopes, which can barely detect continuum sources as faint as $S \sim 1 \ \mu$Jy.

Nonetheless, Professor Oliver Lodge at Liverpool University tried to detect "long wave" radiation from the Sun in 1894 by "filtering out the ordinary well-known waves by a blackboard" and using a "coherer" of metal filings to detect radio waves. Both "terrestrial sources of disturbance in a city like Liverpool" and an insufficiently sensitive detector foiled this effort [51].

In the 1920s, the Bell Telephone Company offered transatlantic telephone service based on "shortwave" ($\lambda \sim 15$ m) radio transmissions. Natural radio static was a serious source of interference, so Bell Telephone Laboratories asked their young electrical engineer Karl Jansky to determine its origin. Jansky built the antenna shown in Figure 1.7 to monitor radio static at 20.5 MHz ($\lambda \approx 15$ m). Its reception pattern was a fan beam (narrow horizontally and broader in the vertical plane) that pointed near the horizon and could be rotated in **azimuth**—the angle measured from north to east around the horizon. He found that most of the static is produced by lightning strokes in numerous tropical thunderstorms. In addition he discovered a steady "hiss" whose strength rose and fell daily, with a period of 23 hours and 56 minutes. He recognized that this is the length of the **sidereal day** (the time it takes the Earth to rotate once in the reference frame of the fixed stars), deduced that the hiss must originate somewhere outside the Solar System, and identified the direction to the Galactic center as the source of the strongest emission.

Jansky published his results in the paper "Electrical disturbances of apparently extraterrestrial origin" [56]. His discovery was even announced on the front page of the *New York Times*, but his employer had no practical interest in understanding the cosmic component of radio static and reassigned Jansky to other projects. Jansky himself believed that the cosmic noise was thermal emission because it produced a steady hiss in headphones that sounded like the hiss generated by hot electrons in vacuum-tube amplifiers. Skeptical astronomers couldn't understand how such strong

Figure 1.7. Karl Jansky and the antenna that discovered cosmic radio static. It rotated in azimuth on four wheels scavenged from a Ford Model T. An accurate replica of this antenna is located at the NRAO in Green Bank, WV. Image credit: NRAO/AUI/NSF.

(equivalent to the emission from a $T \sim 2 \times 10^5$ K blackbody covering most of the inner Galaxy) radio noise was produced, and generally ignored it.

The only person who took a serious interest in Jansky's discovery was the amateur radio operator and professional radio engineer Grote Reber. He later wrote,

> My interest in radio astronomy began after reading the original articles by Karl Jansky. For some years previous I had been an ardent radio amateur and considerable of a DX [long-distance communication] addict, holding the call sign W9GFZ. After contacting over sixty countries and making WAC [Worked All Continents, an amateur-radio award], there did not appear to be any more worlds to conquer. ([87])

Radio astronomy provided the new worlds to conquer, and radio astronomy became his obsession. He devoted years of his life to building the world's first radio antenna using a parabolic reflector (Figure 1.8) at his own expense in his back yard in Wheaton, IL and using it to map the Galaxy.

Because Reber also expected to find thermal emission with $B_\nu \propto \nu^2$, he started observing at $\nu = 3300$ MHz, the highest technically feasible observing frequency in 1937. When he failed to see anything, he concluded that the radio spectrum of the Galaxy was not Planckian. Next he tried 910 MHz, still with no luck, but "since I am a rather stubborn Dutchman, this had the effect of whetting my appetite for more." In 1938 he finally succeeded in detecting and mapping

Figure 1.8. Grote Reber's backyard radio telescope in Wheaton, IL. The parabolic reflector is about 10 m in diameter. His original telescope was dismantled and reassembled near the NRAO visitors science center in Green Bank, WV. Image credit: NRAO/AUI/NSF.

(with $\theta \approx 10°$ angular resolution) the Galaxy at 160 MHz, thereby confirming Jansky's discovery and demonstrating that the radio emission has a distinctly nonthermal spectrum. He observed only at night because automotive ignition interference in Wheaton, IL was too strong during the day. He patiently recorded meter readings by hand once per minute. His results were published in the *Astrophysical Journal* [86].

Then World War II intervened, hindering astronomical research but stimulating progress in radio and radar technology. Some of the engineers and physicists who developed and used this technology during the war led the rapid scientific development of radio astronomy immediately afterward.

If you are interested in learning more about the early history of radio astronomy, see the books *Cosmic Noise: A History of Early Radio Astronomy* by W. T. Sullivan III [102] and *The Evolution of Radio Astronomy* by J. S. Hey [51].

Figure 1.9. The radio sky is shown above an old photograph of the NRAO site in Green Bank, WV. The former 300-foot telescope (the largest dish) made this 4.85 GHz radio image [27], which is about 45 degrees across. Increasing radio brightness is indicated by lighter shades to indicate how the sky would appear to someone with a "radio eye" 300 feet (91 m) in diameter. Image credit: NRAO/AUI/NSF. Investigators: J. J. Condon, J. J. Broderick, and G. A. Seielstad.

1.3 A TOUR OF THE RADIO UNIVERSE

The visible and radio skies reveal distinct "parallel universes" sharing the same space. Most optically bright stars are undetectable at radio wavelengths, and many strong radio sources are optically faint or invisible. Familiar objects like the Sun and planets can appear quite different when seen through the radio and optical windows. The extended radio sources, spread along a band from the lower left to the upper right in Figure 1.9, lie in the outer regions of our Galaxy. The brightest irregularly shaped sources are clouds of hydrogen ionized by luminous young stars. Such stars quickly exhaust their nuclear fuel, collapse, and explode as supernovae; their supernova remnants appear as faint radio rings. Unlike the nearby ($d < 1000$ light years) stars visible to the human eye, almost none of the myriad radio "stars" (unresolved radio sources) scattered across the sky are actually stars. Most are extremely luminous radio galaxies or quasars, and their *average* distance is over 5×10^9 light years. Radio waves travel at the speed of light, so distant extragalactic sources appear to us today as they actually were billions (using the "short-scale" definition 1 billion $\equiv 10^9$) of years ago. Radio galaxies and quasars are beacons carrying information about galaxies and their

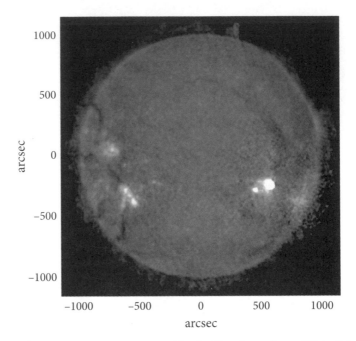

Figure 1.10. The Sun at $\nu = 4.6$ GHz imaged by the Very Large Array (VLA; Color Plate 4) with a resolution of 12 arcsec, which subtends about 8400 km at the surface of the Sun. The brightest features in this image have brightness temperatures $T_b \approx 10^6$ K and coincide with sunspots. At this frequency the radio-emitting surface of the Sun has an average temperature of 3×10^4 K, and the darker features are cooler yet. The radio Sun is somewhat bigger than the optical Sun: the solar limb (the edge of the disk) in this image is about 20000 km above the optical limb. Image credit: NRAO/AUI/NSF & S. M. White.

environs, everywhere in the observable universe and ever since the first galaxies were formed.

The brightest discrete radio source is the Sun (Figure 1.10), but the Sun is much less dominant than it is in visible light. The radio sky is dark even when the Sun is up because atmospheric molecules and dust particles don't scatter radio waves whose wavelengths are much larger than these particles. Most radio observations can be made day or night. Clouds are also nearly transparent at wavelengths $\lambda > 2$ cm, so long-wavelength radio observations can be made even when the sky is overcast.

The Moon and planets are not detectable by reflected solar radiation at radio wavelengths. However, they all emit thermal radiation, and Jupiter is a strong nonthermal source as well. If the Sun were suddenly switched off, the planets would remain radio sources for a long time, slowly fading as they cooled.

At first glance, the $\lambda = 0.85$ mm radio image of the Moon (Figure 1.11) looks familiar, but it is subtly different from the visible Moon. The darker right edge of the Moon is not being illuminated by the Sun, but it still emits radio waves because it does not cool to absolute zero during the lunar night. The radio emission is not produced at the visible surface; it emerges from a layer about 10 wavelengths thick. As a result, the monthly brightness variations of the Moon decrease as wavelength increases. These wavelength-dependent brightness variations encode information

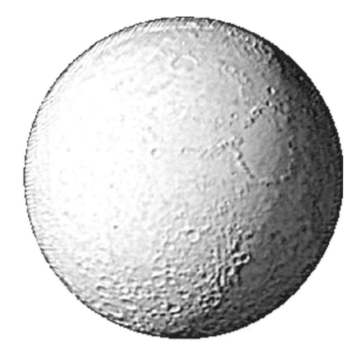

Figure 1.11. Thermal emission from the Moon at $\lambda = 850\,\mu$m. Image credit: http://www.eaobservatory.org/JCMT/publications/newsletter/jcmt-n15.pdf.

about the thermal conductivity and heat capacity of the rocky and dusty outer layers of the Moon. The craters stand out because their interiors are shielded from sunlight and hence cooler, and also because the steep angles of the crater walls reduce their emissivity (as later explained in Equation 2.47) owing to the Brewster angle effect for dielectric boundaries.

Radar studies of Solar-System objects are active experiments involving artificial radio signals reflected from targets, not just passive observations of natural emission. Planetary radar experiments first determined the rotation period of Venus by penetrating its optically opaque atmosphere, measured a more accurate value for the **astronomical unit** (the distance between the Earth and the Sun), imaged the topography of the solid planets and moons, and tracked asteroids and comets. Radar images like the one in Figure 1.12 were recently used to search for water ice trapped in cold craters near the lunar poles. For a good introduction to radar astronomy, see the Arecibo radar web page http://www.naic.edu/~pradar/radarpage.html. The principles of radar astronomy are covered in detail by the textbook [36] based on the 1960 MIT summer course in radar astronomy.

The cosmic static discovered by Karl Jansky is primarily diffuse emission originating in and near the disk of our Galaxy. The distribution of 408 MHz continuum emission displayed in Galactic coordinates (Figure 1.13) is strongest along the Galactic equator, or **Galactic latitude** $b = 0$, where b is the angle from the disk. The brightest radiation comes from near the center of our Galaxy (**Galactic longitude** $l = 0$), which is at the center of Figure 1.13.

Galactic interstellar gas emits spectral lines as well as broadband continuum noise. Neutral hydrogen (H$\textsc{i}$) gas is ubiquitous in the disk. The brightness of the $\lambda \approx 21$ cm

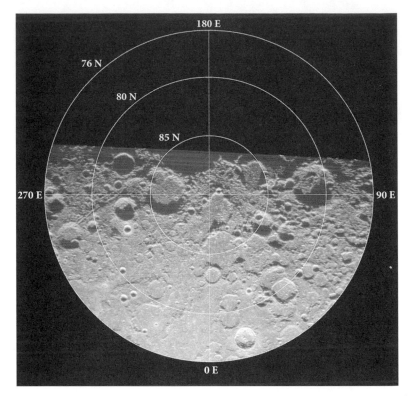

Figure 1.12. This Arecibo + GBT $\lambda = 70\,\text{cm}$ bistatic radar image of the lunar north pole did not find any water ice within a few meters of the lunar surface, even in cold polar craters. Image credit: B. Campbell, NAIC/NSF; NRAO/AUI/NSF.

Figure 1.13. This all-sky 408 MHz continuum image [43] is shown in Galactic coordinates, with the Galactic center in the middle and the Galactic disk extending horizontally from it. Most of the 408 MHz emission is synchrotron radiation from cosmic-ray electrons accelerated in supernova remnants. Image credit: HEASARC/LAMBDA/NASA/GSFC.

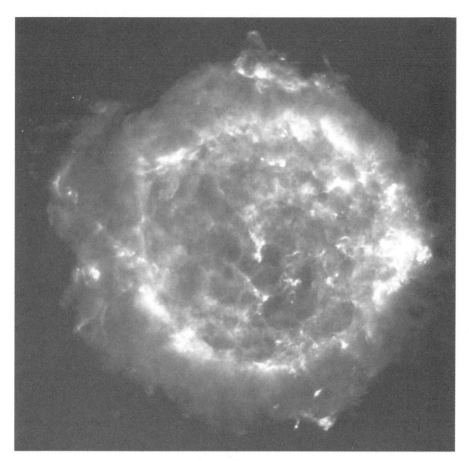

Figure 1.14. Cassiopeia A (usually shortened to Cas A) is the remnant of a supernova explosion that occurred over 300 years ago in our Galaxy, at a distance of about 11,000 light years. Its name is derived from the constellation in which it is seen: Cassiopeia, the Queen. A radio supernova is the explosion that occurs at the end of a massive star's life, and Cas A is the expanding shell of material that remains from such an explosion. This composite image is based on VLA data at three different frequencies: 1.4, 5.0, and 8.4 GHz. Image credit: NRAO/AUI/NSF. Investigators: L. Rudnick, T. DeLaney, J. Keohane, & B. Koralesky; image composite by T. Rector.

hyperfine line at $\nu \approx 1420.4$ MHz is proportional to the column density of HI along the line of sight and is nearly independent of the gas temperature. It is not attenuated by dust absorption, so HI can be seen throughout the Galaxy and nearby external galaxies (Color Plate 11).

Some of the diffuse continuum emission from our Galaxy can be resolved into discrete sources. Supernova remnants such as Cas A (Figure 1.14) and the Crab Nebula (Color Plate 10), and the relativistic electrons diffused throughout the Galaxy that were accelerated by them, account for about 90% of the $\nu \approx 1$ GHz continuum emission from our Galaxy. Most of the remaining continuum emission at 1 GHz is thermal emission from HII regions, hydrogen clouds ionized by UV radiation from extremely massive stars. The nearest large HII region is the Orion Nebula.

Orion's radio continuum is free–free thermal emission from the hot ionized hydrogen. The dusty nebula is transparent at high radio frequencies, so all of the ionized hydrogen contributes to the radio emission.

Thus massive, short-lived stars are responsible for nearly all of the radio continuum from our Galaxy, and the radio luminosities of most spiral galaxies are proportional to their recent star-formation rates. The nearby "starburst" galaxy M82 (Color Plate 13) has a star-formation rate about 10 times that of our Galaxy and is a correspondingly more luminous radio source. Most galaxies with little or no recent star formation (e.g., elliptical galaxies) are radio quiet. Star-forming galaxies are very common, but their radio sources are not especially luminous, so they account for < 1% of the strongest extragalactic radio sources and somewhat less than half of the cosmic radio-source background.

The strongest extragalactic radio source is the radio Cygnus A (usually shortened to Cyg A) shown in Figure 5.12. The identification of this source in 1954 with a distant (redshift $z \approx 0.057$, corresponding to a distance $d \sim 240$ Mpc and a lookback time of about 700 million years) galaxy stunned radio astronomers, who immediately recognized that such luminous radio sources (total radio luminosity $\approx 10^{45}$ erg s^{-1} = 10^{38} W) could be detected almost anywhere in the universe. The angular extent of Cyg A, about 100 arcsec, implies a linear extent $\sim$100 kpc, which is much larger than its host galaxy of stars. The energy source is clearly not stars. Gravitational energy released by matter accreting onto a supermassive ($M \sim 10^9 \, M_\odot$) black hole in the center of the host galaxy powers this and other luminous extragalactic radio sources. In Color Plate 14 the high-resolution (0.4 arcsec) radio (in red) and optical images of the radio galaxy 3C 348 are superimposed to illustrate their relative sizes.

The bright radio source 3C 273 (Figure 1.15) was identified with the first quasar at an even higher redshift, $z \approx 0.16$. Such quasars appear to be radio galaxies in an especially active state, when visible light from the region near the black hole overwhelms the starlight from the host galaxy and makes the quasar look like a bright star.

Some exotic phenomena are radio sources but were discovered in other wavelength ranges. Gamma-ray bursts (GRBs) are briefly the most luminous (up to 10^{53} erg s^{-1}) discrete sources in the universe, so bright that they were discovered in the 1960s by the VELA nuclear-test monitoring satellites. (For a good history, see the NASA/Swift GRB page.[1]) Their faint radio afterglows have proven very useful for constraining the energetics and parent populations of GRBs.

The final stop on any tour of the radio universe is the cosmic microwave background radiation (CMBR), which is thermal radiation from the hot big bang. It fills the universe and is the energetically dominant component of all electromagnetic radiation. We see the surface of last scattering beyond which the universe was ionized and opaque. No more distant radio sources, even if any exist, could be seen. The surface of last scattering is at redshift $z \approx 1100$, so the photons received today were emitted when the universe was only about 4×10^5 years old. The CMBR is very nearly isotropic and very nearly a perfect blackbody with $T \approx 2.73$ K. The Wilkinson Microwave Anisotropy Probe (WMAP), in orbit near the L2 Lagrange point, and the

---

[1] http://swift.sonoma.edu/about_swift/grbs.html.

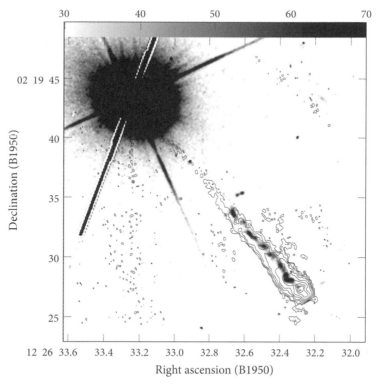

Figure 1.15. This Hubble Space Telescope (HST) gray-scale image of the quasar 3C 273 includes radio contours superimposed on the optical jet emission [7].

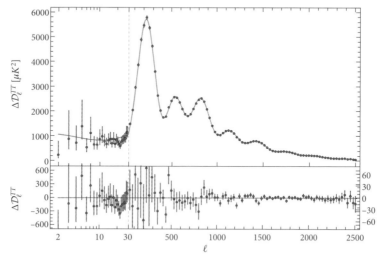

Figure 1.16. The upper panel shows the angular power spectrum of CMB brightness fluctuations measured by the Planck satellite [83]. The multipole number l corresponds to an angular separation $\theta \approx 180°/l$. The lower panel shows the offsets from the model and their errors. Abscissa: Multipole number l. Upper ordinate: Brightness power expressed in temperature units (μK^2). Lower ordinate: Offsets and their uncertainties (μK^2). Note the scale change at $l = 30$. Image credit: Planck Collaboration et al. [83], ESA/Planck, and the Planck Collaboration.

Planck satellite have made all-sky images of the tiny fluctuations in CMBR brightness (Color Plate 16). The angular power spectrum of these fluctuations (Figure 1.16) revealed by the Planck satellite of the European Space Agency (ESA) constrains a host of fundamental cosmological parameters. See the Planck[2] and WMAP[3] websites for the most recent results.

[2]http://www.cosmos.esa.int/web/planck.
[3]http://map.gsfc.nasa.gov/.

2 Radiation Fundamentals

2.1 BRIGHTNESS AND FLUX DENSITY

Astronomers study an astronomical source by measuring the strength of its radiation as a function of direction on the sky (by mapping or imaging) and frequency (spectroscopy), plus other quantities (time, polarization) to be considered later. Clear and quantitative definitions are needed to describe the strength of radiation and how it varies with direction, frequency, and distance between the source and the observer. The concepts of brightness and flux density are deceptively simple, but they regularly trip up even experienced astronomers. It is very important to understand them clearly because they are so fundamental.

Consider the simplest possible case of radiation traveling from a source through empty space, so there is no **absorption** (destruction of photons), **scattering** (changing the direction of photons), or **emission** (creation of new photons) along the path to an observer. In the **ray-optics approximation**, radiated energy flows in straight lines. This approximation is valid only for systems much larger than the wavelength λ of the radiation, a criterion easily met by astronomical sources. You may find it helpful to visualize electromagnetic radiation as a stream of light particles (photons), essentially bullets that travel in straight lines at the speed of light. To motivate the following mathematical definitions, imagine you are looking at the Sun. The "brightness" of the Sun appears to be about the same over most of the Sun's surface, which looks like a nearly uniform disk even though it is actually a sphere. This means that a photograph of the Sun would be uniformly exposed across the Sun's disk. It also turns out that the exposure would not change if photographs were made at different distances from the Sun, from points near Mars, the Earth, and Venus, for example (Figure 2.1).

The angular size of the Sun depends on the distance between the Sun and the camera, but the number of photons falling on the detector per unit area per unit time *per unit solid angle* does not. The photo taken from near Venus would not be overexposed, and the one from near Mars would not be underexposed. The total number of solar photons from all directions reaching the camera per unit area per unit time (or the total energy absorbed per unit area per unit time) does decrease with increasing distance, but *only* because the solid angle subtended by the Sun decreases. Thus we distinguish between the **brightness** or **intensity** of the Sun's radiation, which does not depend on distance, and the apparent **flux**,

Figure 2.1. The Sun as it would appear in three photos taken with the same camera from long (left), medium (center), and short distances (right) would have constant brightness but increasing angular size. Image credit: SOHO/EIT Consortium (ESA & NASA).

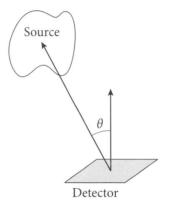

Figure 2.2. Specific intensity measured by a detector whose normal is at an angle θ from the line of sight.

which does. Some authors [16, 20] prefer to use *brightness* for the power per unit area per unit solid angle emitted at the source and *specific intensity* for the power per unit area per unit solid angle along the path to the detector; others [106, 116] do not distinguish between them. The two are identical if there is no absorption or emission between the source and the detector, and we will use these terms interchangeably.

Note also that the number of photons per unit area hitting the detector is proportional to $\cos\theta$ if the normal to the detector is tilted by an angle θ from the direction of the incoming rays (Figure 2.2). This is just the same projection effect that reduces the amount of water collected by a tilted rain gauge by $\cos\theta$. Likewise at the source, such as the spherical Sun, the projected area perpendicular to the line of sight is proportional to $\cos\theta$, where θ is the angle between the line of sight and the normal to the Sun's surface.

The total brightness is contributed by photons of all frequencies. The brightness per unit frequency is called the **specific intensity** (also **spectral intensity** or **spectral brightness**). The notation for specific intensity is I_ν, where the subscript ν is used to indicate "per unit frequency." In the ray-optics approximation, specific intensity can be defined quantitatively in terms of

- $d\sigma$ = an infinitesimal surface area (e.g., of a detector);
- θ = the angle between a "ray" of radiation and the normal to the surface;
- $d\Omega$ = an infinitesimal solid angle measured from the observer's location.

The surface containing $d\sigma$ can be any surface, real or imaginary; that is, it could be the physical surface of the detector, the source, or an imaginary surface anywhere along the ray. If energy dE from within the solid angle $d\Omega$ flows through the projected area $\cos\theta \, d\sigma$ in time dt and in a narrow frequency band of width $d\nu$, then

$$dE = I_\nu \cos\theta \, d\sigma \, d\Omega \, dt \, d\nu. \tag{2.1}$$

Power is defined as the flow of energy per unit time, so the corresponding power dP is

$$dP = \frac{dE}{dt} = I_\nu(\cos\theta \, d\sigma \, d\Omega \, d\nu).$$
$$\text{(watts)} \qquad (\text{m}^2 \ \text{sr} \ \text{Hz})$$

Thus the quantitative definition of **specific intensity** or **spectral brightness** is

$$\boxed{I_\nu \equiv \frac{dP}{(\cos\theta \, d\sigma) \, d\nu \, d\Omega}} \tag{2.2}$$

and the MKS units of I_ν are W m^{-2} Hz^{-1} sr^{-1}.

Radio astronomers almost always measure frequencies, but observers in other wavebands normally measure wavelengths rather than frequencies, so the brightness per unit wavelength

$$\boxed{I_\lambda \equiv \frac{dP}{(\cos\theta \, d\sigma) \, d\lambda \, d\Omega}} \tag{2.3}$$

is also widely used. The MKS units of I_λ are W m^{-3} sr^{-1}. The relation between the intensity *per unit frequency* I_ν and the intensity *per unit wavelength* I_λ can be derived from the requirement that the power dP in the frequency interval ν to $\nu + d\nu$ must equal the power in the corresponding wavelength interval λ to $\lambda + d\lambda$:

$$|I_\nu d\nu| = |I_\lambda d\lambda|. \tag{2.4}$$

Thus

$$\boxed{\frac{I_\lambda}{I_\nu} = \left|\frac{d\nu}{d\lambda}\right| = \frac{c}{\lambda^2} = \frac{\nu^2}{c}.} \tag{2.5}$$

The reasons for specifying the brightness in an infinitesimal frequency range $d\nu$ or wavelength range $d\lambda$ are (1) the detailed spectra of sources carry astrophysically important information, (2) source properties (e.g., opacity) may vary with frequency,

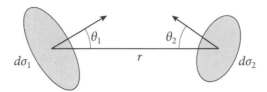

Figure 2.3. Specific intensity conserved along a ray of length r.

and (3) most general theorems about radiation are true for *all* narrow frequency ranges (e.g., specific intensity is conserved along a ray path in empty space), so they are also true for any wider frequency range. Thus the conservation of specific intensity (Equation 2.7) implies that **total intensity** defined as

$$I \equiv \int_0^\infty I_\nu(\nu)d\nu = \int_0^\infty I_\lambda(\lambda)d\lambda \tag{2.6}$$

is also conserved.

THEOREM. *Specific intensity is conserved (is constant) along any ray in empty space.*

This follows directly from geometry. Let $d\sigma_1$ and $d\sigma_2$ be two infinitesimal surfaces along a ray of length r, as shown in Figure 2.3. Let $d\Omega_1 \ll 1$ rad be the solid angle subtended by $d\sigma_2$ as seen from the center of the surface $d\sigma_1$ and $d\Omega_2 \ll 1$ rad be the solid angle subtended by $d\sigma_1$ as seen from the center of the surface $d\sigma_2$. Then

$$d\Omega_1 = \frac{\cos\theta_2 \, d\sigma_2}{r^2},$$

$$d\Omega_2 = \frac{\cos\theta_1 \, d\sigma_1}{r^2}.$$

The power dP_1 in the frequency range ν to $\nu + d\nu$ flowing through the area $d\sigma_1$ in solid angle $d\Omega_1$ is

$$dP_1 = \frac{dE_1}{dt} = (I_\nu)_1 \cos\theta_1 \, d\Omega_1 \, d\sigma_1 \, d\nu$$

$$= (I_\nu)_1 \cos\theta_1 \left(\frac{\cos\theta_2 \, d\sigma_2}{r^2} \right) d\sigma_1 \, d\nu$$

$$= (I_\nu)_1 \left(\frac{\cos\theta_1 \cos\theta_2 \, d\sigma_1 \, d\sigma_2}{r^2} \right) d\nu.$$

Likewise

$$dP_2 = \frac{dE_2}{dt} = (I_\nu)_2 \cos\theta_2 \left(\frac{\cos\theta_1 \, d\sigma_1}{r^2} \right) d\sigma_2 \, d\nu$$

$$= (I_\nu)_2 \left(\frac{\cos\theta_1 \cos\theta_2 \, d\sigma_1 \, d\sigma_2}{r^2} \right) d\nu.$$

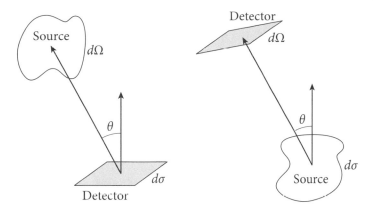

Figure 2.4. Two ways of looking at brightness.

Radiation energy is conserved in free space (where there is no absorption or emission), so $dE_1 = dE_2$ and

$$(I_\nu)_1 = (I_\nu)_2. \qquad \text{QED} \qquad (2.7)$$

The conservation of specific intensity has two important consequences:

1. **Brightness is independent of distance.** Thus the camera setting for a good exposure of the Sun would be the same, regardless of whether the photograph was taken close to the Sun (from near Venus, for example) or far away from the Sun (from near Mars, for example), so long as the Sun is resolved in the photograph.
2. **Brightness is the same at the source and at the detector.** Thus you can think of brightness in terms of energy flowing out of the source or as energy flowing into the detector (Figure 2.4).

The conservation of brightness applies to any lossless optical system (a system of lenses and mirrors, for example) that can change the direction of a ray. **No passive optical system can increase the specific intensity or total intensity of radiation**. If you look at the Moon through a large telescope, the Moon will appear bigger (in angular size) but not brighter. Many people are disappointed when they see a large galaxy through a telescope because it looks so dim; they expected to see a brilliantly glowing disk of stars, as in the photograph of Andromeda in Figure 2.5. The difference is not in the telescope; it is in the detector—the photograph appears brighter only because the photograph has accumulated more light over a long exposure time.

If a source is discrete, meaning that it subtends a well-defined solid angle, the spectral power received by a detector of unit projected area (Figure 2.6) is called the **flux density** S_ν of the source. Equation (2.2) implies

$$\frac{dP}{d\sigma\,d\nu} = I_\nu \cos\theta\,d\Omega, \qquad (2.8)$$

Figure 2.5. No passive optical system (e.g., a telescope) can increase the specific intensity of an extended source. The Andromeda galaxy (M31) appears much brighter in this photograph than it does to the eye, either with or without the aid of a telescope, only because a long photographic exposure accumulates more light. Image credit: Robert Gendler.

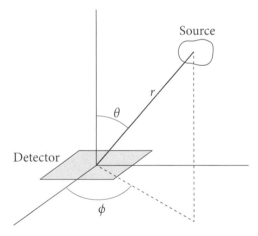

Figure 2.6. An illustration of the definition of flux density.

so integrating over the solid angle subtended by the source yields

$$S_\nu \equiv \int_{\text{source}} I_\nu(\theta, \phi) \cos \theta \, d\Omega. \tag{2.9}$$

If the source angular size is $\ll 1$ rad, $\cos\theta \approx 1$ and the expression for flux density is much simpler:

$$S_\nu \approx \int_{\text{source}} I_\nu(\theta, \phi)d\Omega. \qquad (2.10)$$

This is usually the case for astronomical sources, and astronomers rarely use flux densities to describe sources so extended that the $\cos\theta$ factor must be retained (e.g., the diffuse emission from our Galaxy).

In practice, when should spectral brightness and when should flux density be used to describe a source? If a source is **unresolved**, meaning that it is much smaller in angular size than the point-source response of the eye or telescope observing it, its flux density can be measured but its spectral brightness cannot. To the naked eye, the unresolved red giant star Betelgeuse appears to be one of the brightest stars in the sky. Yet calling it a "bright star" is misleading because the total intensity of this relatively cool star is lower than the total intensity of every hotter but more distant star that is scarcely visible to the eye. Betelgeuse appears "brighter" than most other stars only because it subtends a much larger solid angle and therefore its flux is higher. If a source is much larger than the point-source response, its spectral brightness at any position on the source can be measured directly, but its flux density must be calculated by integrating the observed spectral brightnesses over the source solid angle. Consequently, flux densities are normally used to describe only relatively compact sources.

The MKS units of flux density, W m^{-2} Hz^{-1}, are much too big for practical astronomical use, so astronomers use smaller ones:

$$1 \text{ jansky} = 1 \text{ Jy} \equiv 10^{-26} \text{ W m}^{-2} \text{ Hz}^{-1}, \qquad (2.11)$$

1 millijansky = 1 mJy $\equiv 10^{-3}$ Jy, and 1 microjansky = 1 μJy $\equiv 10^{-6}$ Jy. Optical astronomers often express flux densities as **AB magnitudes** defined in terms of Jy by

$$\text{AB magnitude} \equiv -2.5 \log_{10}\left(\frac{S_\nu}{3631 \text{ Jy}}\right). \qquad (2.12)$$

Unlike brightness, flux density depends on source distance d. Because $\int_{\text{source}} d\Omega \propto 1/d^2$ and brightness is conserved, Equation 2.10 implies the **inverse square law**:

$$S_\nu \propto d^{-2}. \qquad (2.13)$$

The specific intensity or brightness is an intrinsic property of a source, while the flux density of a source also depends on the distance between the source and the observer.

The **total flux** or **flux** S from a source is the integral over frequency of flux density:

$$S \equiv \int_0^\infty S_\nu d\nu. \qquad (2.14)$$

Its dimensions are power divided by area, so its MKS units are W m^{-2}. Total flux is a rarely used quantity in observational radio astronomy, so radio astronomers often

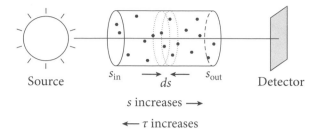

Source s_{in} $\xrightarrow{}$ ds $\xleftarrow{}$ s_{out} Detector

s increases $\longrightarrow$

$\longleftarrow \tau$ increases

Figure 2.7. Absorption between a source and a detector. The coordinate along the ray increases from s_{in} at the input end of the absorber to s_{out} at the output end of the absorber. The optical depth τ is measured in the opposite direction, starting from $\tau = 0$ at s_{out} and increasing as s decreases.

delete the subscript from S_ν and use the symbol S to indicate flux density. This is convenient but potentially confusing. Likewise, the word "flux" is sometimes used as a shorthand for "flux density" in the literature, even though it is formally incorrect.

The **spectral luminosity** L_ν of a source is defined as the total power per unit bandwidth radiated by the source at frequency ν; its MKS units are W Hz^{-1}. The area of a sphere of radius d is $4\pi d^2$, so the relation between the spectral luminosity and the flux density of an *isotropic* source radiating in free space is

$$\boxed{L_\nu = 4\pi d^2 S_\nu,}$$ (2.15)

where the distance d between the source and the observer is much larger than the dimensions of the source itself. Beware that some radio sources emit anisotropically, relativistically beamed quasars (Section 5.6.1) for example. Unfortunately, Equation 2.15 cannot be used to calculate the total (integrated over 4π sr) spectral luminosity of a beamed quasar from a flux-density measurement made from just one direction. Spectral luminosity is an intrinsic property of the source because it does not depend on the distance d between the source and the observer—the d^2 in Equation. 2.15 cancels the d^{-2} dependence of S_ν. The **luminosity** or **total luminosity** L of a source is defined as the integral over all frequencies of the spectral luminosity:

$$L \equiv \int_0^\infty L_\nu d\nu.$$ (2.16)

Astronomers sometimes call L the **bolometric luminosity** because a bolometer is a broadband detector that measures the heating power of radiation at all frequencies.

2.2 RADIATIVE TRANSFER

In free space, the specific intensity I_ν of radiation is conserved along a ray:

$$\frac{dI_\nu}{ds} = 0,$$ (2.17)

where s is the coordinate along the ray between the source and the detector. What happens if there is an intervening medium between s_{in} and s_{out} (Figure 2.7)?

2.2.1 Absorption

Think of the ray as a beam of photons, or light particles, some of which may be absorbed by the medium and vanish. The infinitesimal probability dP of a photon being absorbed (e.g., by hitting an absorbing particle) in a thin slab of thickness ds is directly proportional to ds: $dP = \kappa\, ds$, where the constant of proportionality

$$\kappa \equiv \frac{dP}{ds} \qquad (2.18)$$

is called the **linear absorption coefficient**, and its dimension is inverse length. A value $\kappa = 1\ \mathrm{m}^{-1}$ means that over some small distance $\Delta s \ll \kappa^{-1}$ along the ray, $\Delta s = 10^{-3}$ m for example, the small fraction $\kappa\,\Delta s = 1\ \mathrm{m}^{-1} \times 10^{-3}\ \mathrm{m} = 10^{-3}$ of the photons will be absorbed. Why consider such thin layers rather than the whole absorber at once? Only if the layer is thin enough that dP is infinitesimal does the probability of absorption within the layer increase linearly with thickness ds so that κ is a constant independent of ds. In a layer thick enough to absorb a significant fraction of the photons, fewer photons are absorbed near the far end of the layer simply because fewer photons have survived to be absorbed there, so the probability of absorption increases nonlinearly with thickness.

This simple *macroscopic* model doesn't depend on the microscopic physical processes by which photons are absorbed. It is like the ideal gas model, which relates the macroscopic properties of a gas (temperature, pressure, and density) without considering the microscopic details about how individual gas molecules behave.

The fraction of the specific intensity lost to absorption in the infinitesimal distance ds along the ray is

$$\frac{dI_\nu}{I_\nu} = -\kappa\, ds \quad \text{(absorption only)}. \qquad (2.19)$$

Integrating both sides of Equation 2.19 along the absorbing path gives the output specific intensity as a fraction of the input specific intensity:

$$\int_{s_{\mathrm{in}}}^{s_{\mathrm{out}}} \frac{dI_\nu}{I_\nu} = -\int_{s_{\mathrm{in}}}^{s_{\mathrm{out}}} \kappa(s')\, ds' = \ln I_\nu \Big|_{s_{\mathrm{in}}}^{s_{\mathrm{out}}}, \qquad (2.20)$$

$$\ln[I_\nu(s_{\mathrm{out}})] - \ln[I_\nu(s_{\mathrm{in}})] = -\int_{s_{\mathrm{in}}}^{s_{\mathrm{out}}} \kappa(s')\, ds', \qquad (2.21)$$

$$\frac{I_\nu(s_{\mathrm{out}})}{I_\nu(s_{\mathrm{in}})} = \exp\left[-\int_{s_{\mathrm{in}}}^{s_{\mathrm{out}}} \kappa(s')\, ds'\right]. \qquad (2.22)$$

The dimensionless quantity

$$\tau \equiv -\int_{s_{\mathrm{out}}}^{s_{\mathrm{in}}} \kappa(s')\, ds' \qquad (2.23)$$

is called the **optical depth** or **opacity** of the absorber. Note that $d\tau = -\kappa\, ds$. The "backward" direction of the path integration along the line of sight was chosen in Equation 2.23 to make $\tau > 0$ and increasing as you look deeper into an absorber.

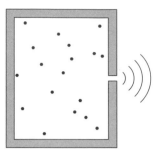

Figure 2.8. Thermal radiation from a cavity depends only on its temperature and is independent of both the material used to make the cavity walls and the contents of the cavity.

Thus

$$\frac{I_\nu(s_{\text{out}})}{I_\nu(s_{\text{in}})} = \exp(-\tau) \quad \text{(absorption only)}. \tag{2.24}$$

If $\tau \ll 1$, the absorber is said to be **optically thin**; if $\tau \gg 1$, it is **optically thick**.

2.2.2 Emission

The intervening medium may also emit photons, again by some unspecified microscopic process. In any infinitesimal volume ($ds\, d\sigma$) of thickness ds and cross section $d\sigma$, the probability per unit time that an *isotropic* source will emit a photon into the solid angle $d\Omega$ is directly proportional to the volume and solid angle:

$$\dot{P}_{\text{em}} \propto ds\, d\sigma\, d\Omega. \tag{2.25}$$

The **emission coefficient** j_ν is defined so that

$$\boxed{j_\nu \equiv \frac{dI_\nu}{ds}} \tag{2.26}$$

if there is no absorption. The dimensions of j_ν follow from its definition; they are power per unit volume per unit frequency per unit solid angle, and the corresponding MKS units are W m^{-3} Hz^{-1} sr^{-1}. Combining the effects of absorption (Equation 2.19) and emission (Equation 2.26) yields the **equation of radiative transfer**

$$\boxed{\frac{dI_\nu}{ds} = -\kappa I_\nu + j_\nu.} \tag{2.27}$$

Because the absorption and emission processes have not been specified, κ and j_ν seem to be independent. However, they are not independent in full **thermodynamic equilibrium (TE)**. In TE, matter and radiation are in equilibrium at the same temperature T. **Cavity radiation** from a container bounded by opaque walls pierced by a small hole (Figure 2.8) is equilibrium radiation. The intensity and spectrum of the radiation emerging from the hole is independent of the wall material (e.g., wood painted green, shiny copper, gray concrete, etc.) and any absorbing material (e.g., gas, dust, fog, etc.) that may be inside the cavity.

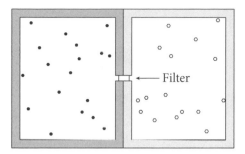

Figure 2.9. Kirchhoff's thought experiment invokes two cavities in thermodynamic equilibrium connected through a filter that passes radiation in the narrow frequency range v to $v + dv$. The cavities may be made of different materials and contain different emitting/absorbing particles.

Kirchhoff derived the relation between κ and j_v in TE using the thought experiment illustrated in Figure 2.9. Two cavities made of different materials and containing different absorbers are connected through a passive (meaning, no energy is needed for it to work) filter transparent only in the narrow frequency range v to $v + dv$. In equilibrium at any temperature T, radiation can transfer no *net* power from one cavity to the other, lest one cavity cool down and the other heat up. That would violate the second law of thermodynamics because the two cavities at different temperatures could be used to drive a heat engine.

In (full) thermodynamic equilibrium (TE) at temperature T,

$$\frac{dI_v}{ds} = 0 \quad \text{and} \quad I_v = B_v(T), \tag{2.28}$$

where $B_v(T)$ is the spectrum of equilibrium radiation at temperature T. The symbol $B_v(T)$ is used because equilibrium cavity radiation is the same as the **blackbody radiation** for a perfect absorber at temperature T, even if it is not in a cavity. Rearranging the equation of radiative transfer (Equation 2.27)

$$\frac{dI_v}{ds} = 0 = -\kappa B_v(T) + j_v \tag{2.29}$$

yields **Kirchhoff's law** for a system in TE:

$$\boxed{\frac{j_v(T)}{\kappa(T)} = B_v(T),} \tag{2.30}$$

which is valid at any frequency v. Equation 2.30 is remarkable because it connects the properties $j_v(T)$ and $\kappa(T)$ of any kind of *matter* to the single universal spectrum $B_v(T)$ of equilibrium *radiation*. Note also that Kirchhoff's argument is independent of the shapes of the cavities, so cavity radiation must be isotropic.

Although Kirchhoff's law was derived for a system in thermodynamic equilibrium, its applicability is not limited to radiation in full thermodynamic equilibrium with its material environment. Kirchhoff's law also applies whenever the radiating/absorbing *material* is in thermal equilibrium, in *any* radiation field. If the emitting/absorbing material is in thermal equilibrium at a well-defined temperature

T, it is said to be in **local thermodynamic equilibrium (LTE)** even if it is not in equilibrium with the radiation field. For example, gas molecules in the Earth's lower atmosphere have a Maxwellian speed distribution (Equation 4.34), so the gas is in LTE. The gas has a well-defined **kinetic temperature** $T \sim 300$ K measurable with an ordinary thermometer, but it is in a distinctly nonequilibrium radiation field (anisotropic $T \sim 5800$ K sunlight during the day, the cold dark sky at night, plus anisotropic emission from the ground).

Kirchhoff's law applies in LTE as well as in TE. To show this, recall that $B_\nu(T)$ is independent of the properties of the radiating/absorbing material. In contrast, both $j_\nu(T)$ and $\kappa(T)$ depend *only* on the materials in the cavity (e.g., whether the walls are made of copper or concrete) and on the temperature of that material; they do not depend on the ambient radiation field or its spectrum.

This generalized version of Kirchhoff's law is an exceptionally valuable tool for calculating the emission coefficient from the absorption coefficient or vice versa. For example, in Section 4.3 Kirchhoff's law gives $\kappa(\nu)$ immediately following a calculation of $j_\nu(\nu)$ for free–free emission by an HII region (a cloud of ionized hydrogen) in LTE.

The blackbody spectrum $B_\nu(T)$ (Equation 2.86) falls exponentially at frequencies above the peak, but only as a power law at lower frequencies (solid curve in Figure 2.20). Kirchhoff's law is not intuitively obvious from experience because room-temperature ($T \sim 300$ K) objects in our environment are much too cold ($h\nu/kT \gg 1$) to emit detectable amounts of visible light. A glass of water might absorb 10% of the sunlight passing through it, but we do not see any emitted light because 10% (or even 100%) of the blackbody radiation emitted by a room-temperature object is very nearly zero. One familiar example of Kirchhoff's law is a charcoal fire with flames and glowing coals. The infrared (heat) radiation from barely glowing black coals in LTE is much more intense than that from the hotter but nearly transparent visible flames, as you can verify by using a shield to cover either the coals or the flames. In contrast, the radio emission implied by Kirchhoff's law is always significant because $h\nu/kT \ll 1$ when $T \sim 300$ K and the blackbody spectrum below the peak falls off only as ν^{-2}. This point is illustrated by radio emission and absorption in the Earth's atmosphere (Section 2.2.3).

2.2.3 Emission and Absorption of Radio Waves in the Earth's Atmosphere

At radio frequencies higher than $\nu \sim 1$ GHz, absorption by the Earth's atmosphere may be large enough to affect the accuracy of flux-density measurements and atmospheric emission can increase noise errors. Radio astronomers can determine the amount of **atmospheric absorption** by measuring the amount of **atmospheric emission** as a function of **zenith angle**, or angle from the vertical, and using Kirchhoff's law to calculate the zenith opacity of the (roughly isothermal) atmosphere at frequency ν from its kinetic (thermometer) temperature $T_{\text{atm}} \sim 300$ K and its radio brightness. The celestial sky above the atmosphere is much colder ($T \sim 3$ K) at high frequencies, so the "background" emission above the atmosphere can usually be ignored.

This is done by tilting the radio telescope and measuring I_ν as a function of the zenith angle z as shown in Figure 2.10. There are two practical complications: (1) The atmospheric signal is noise indistinguishable from other noise sources,

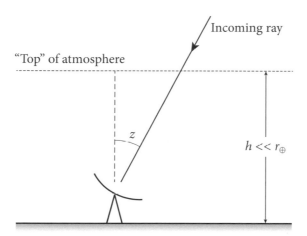

Figure 2.10. Most atmospheric emission and absorption occurs in a layer whose height h is only a few km, much smaller than the radius of the Earth $r_\oplus$, so the plane-parallel approximation for the spherical Earth shown here is accurate.

especially noise generated in the receiver itself. The output voltage of the receiver is proportional to the sum of these input noise powers. The amount of receiver noise power may not be well known, so only the *change* in I_ν with zenith angle z can be measured, not the absolute value of I_ν at any zenith angle. (2) The power gain of the receiver system is difficult to calculate from first principles, so most measurements are made relative to some calibration source. For example, the receiver might be calibrated by having the feed look alternately at two absorbing plates with different known temperatures.

Because B_ν is directly proportional to T in the Rayleigh–Jeans low-frequency approximation

$$B_\nu \approx \frac{2kT\nu^2}{c^2}, \tag{2.31}$$

radio astronomers often find it convenient to specify the spectral brightness I_ν, even if $I_\nu \neq B_\nu$, in terms of the equivalent Rayleigh–Jeans **brightness temperature** T_b *defined* by the equation

$$I_\nu = \frac{2kT_b\nu^2}{c^2}. \tag{2.32}$$

Thus for *any* I_ν,

$$\boxed{T_b(\nu) \equiv \frac{I_\nu c^2}{2k\nu^2}.} \tag{2.33}$$

Equation 2.33 explicitly indicates that T_b can vary with frequency. Brightness temperature is just another way to specify power per unit solid angle per unit bandwidth in terms of the Rayleigh–Jeans approximation. It is convenient because radio telescopes are often calibrated by absorbers or "loads" of known temperature and because the temperature of a radio source is frequently a quantity of physical

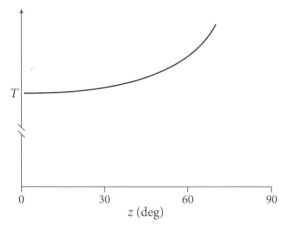

Figure 2.11. The receiver output from a tipping scan shows the increase in sky brightness with zenith angle, but the zero point of the sky brightness temperature is not well determined.

interest. Beware that brightness temperature is *not* the same as physical temperature. Nonthermal sources have frequency-dependent brightness temperatures but they do not have well-defined physical temperatures. Thermal sources have brightness temperatures lower than their physical temperatures if they are semitransparent (e.g., the atmosphere) or partially reflecting (e.g., the Moon). Even true blackbody radiators have $T_b = T$ only in the low-frequency limit $h\nu \ll kT$.

The observed output from the radio telescope might vary with z as shown in Figure 2.11. To understand and interpret these observations, start with the radiative transfer equation (2.27),

$$\frac{dI_\nu}{ds} = -\kappa I_\nu + j_\nu. \tag{2.34}$$

Kirchhoff's law (Equation 2.30) can eliminate the unknown j_ν because the atmosphere is in LTE:

$$j_\nu = \kappa B_\nu(T_{\text{atm}}), \tag{2.35}$$

where T_{atm} is the kinetic temperature of the atmosphere as measured by an ordinary thermometer. Dividing by κ_ν yields

$$\frac{1}{\kappa}\frac{dI_\nu}{ds} = \frac{-dI_\nu}{d\tau} = -I_\nu + B_\nu(T_{\text{atm}}). \tag{2.36}$$

Multiply both sides of this differential equation by $\exp(-\tau)$ and integrate along the ray in the telescope beam from the top of the atmosphere to the ground. Let $\tau_A(z)$ be the total optical depth of the atmosphere along the ray at zenith angle z. Then

$$\int_0^{\tau_A} e^{-\tau}\frac{dI_\nu}{d\tau}d\tau = \int_0^{\tau_A}[I_\nu - B_\nu(T_{\text{atm}})]e^{-\tau}d\tau. \tag{2.37}$$

Next integrate the left side by parts and take $B_\nu(T_{atm})$ outside the integral (i.e., make the approximation that the atmosphere below altitude h is nearly isothermal):

$$e^{-\tau} I_\nu \Big|_0^{\tau_A} - \int_0^{\tau_A} -e^{-\tau} I_\nu \, d\tau = \int_0^{\tau_A} I_\nu \, e^{-\tau} d\tau - B_\nu(T_{atm}) \int_0^{\tau_A} e^{-\tau} d\tau, \qquad (2.38)$$

$$I_\nu(\tau = \tau_A) e^{-\tau_A} - I_\nu(\tau = 0) = B_\nu(T_{atm})(e^{-\tau_A} - 1). \qquad (2.39)$$

The term $I_\nu(\tau = \tau_A) \approx 0$ can be neglected because the brightness of emission above the atmosphere is small and doesn't depend on zenith angle at high radio frequencies, so that

$$I_\nu(\tau = 0) = (1 - e^{-\tau_A}) B_\nu(T_{atm}). \qquad (2.40)$$

The path length through a plane-parallel atmosphere is proportional to $\sec z$ so at any zenith angle z,

$$\tau_A \approx \tau_Z \sec z, \qquad (2.41)$$

where $\tau_Z \equiv \tau_A(z = 0)$ is the zenith opacity of the atmosphere above the radio telescope. Inserting the Rayleigh–Jeans approximation for $B_\nu(T_{atm})$ gives

$$I_\nu = [1 - \exp(-\tau_Z \sec z)] \frac{2k T_{atm} \nu^2}{c^2}. \qquad (2.42)$$

Thus the brightness temperature of the atmospheric emission as a function of zenith angle is

$$T_b = \frac{I_\nu c^2}{2k\nu^2} = T_{atm}\big[1 - \exp(-\tau_Z \sec z)\big]. \qquad (2.43)$$

The variation of T_b with zenith angle is shown for different zenith opacities in Figure 2.12. By fitting observed data to such curves, radio astronomers can estimate the zenith opacity and correct the measured flux densities of celestial sources for atmospheric absorption.

If (1) the zenith opacity is low ($\tau_z \ll 1$), (2) the frequency is high enough ($\nu > 1$ GHz) that the radio-source background is weak, and (3) there is little pickup of "spillover" radiation from the ground, then a good approximation to the observed system noise temperature T_s (Equation 3.150) is

$$T_s \approx [T_r + T_{cmb}] + T_{atm}\tau_z \sec(z), \qquad (2.44)$$

where T_r is the noise temperature of the receiver alone and $T_{cmb} \approx 2.7$ K is the brightness temperature of the cosmic microwave background. The quantity in square brackets is independent of z and T_{atm} is close to the surface air temperature, which can easily be measured with a thermometer. Equation 2.44 implies that the zenith opacity

$$\tau_z \approx \frac{\Delta T_s / T_{atm}}{\Delta \sec(z)} \qquad (2.45)$$

can be read directly from the slope of the linear tipping-curve plot of T_s / T_{atm} versus $\sec(z)$. The solid line in Figure 2.13 shows an example in which

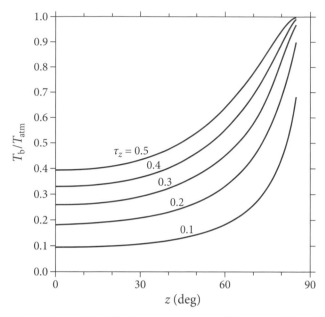

Figure 2.12. Ratios of atmospheric brightness temperature T_b to atmospheric kinetic temperature T_{atm} for different zenith opacities τ_z as functions of zenith angle z.

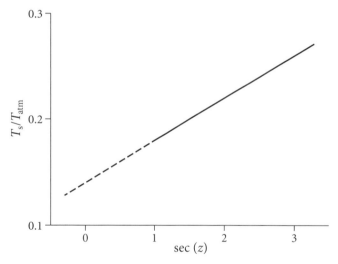

Figure 2.13. If the zenith opacity is low, a plot of the observed ratio (T_s/T_{atm}) versus $\sec(z)$ is nearly linear (solid line), and its slope equals the zenith opacity; $\tau_z \approx 0.04$ in this case. Extrapolating the solid line to $\sec(z) = 0$ (dashed line) yields the ratio $[T_r + T_{cmb}]/T_{atm} \approx 0.14$.

$\tau_z \approx 0.04$. Extrapolating the data to (the geometrically impossible) $\sec(z) = 0$ yields $[T_r + T_{cmb}]/T_{atm}$. If $T_{atm} = 280$ K, the example $[T_r + T_{cmb}]/T_{atm} = 0.14$ implies $[T_r + T_{cmb}] \approx 39.2$ K and thus $T_r \approx 36.5$ K. Alternatively, if T_r is measured in the laboratory, the tipping curve can be used to determine T_{cmb}, the temperature

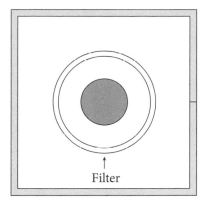

Figure 2.14. Thought experiment involving an opaque body surrounded by a filter passing frequencies ν to $\nu + d\nu$ inside a cavity in thermodynamic equilibrium.

of the cosmic microwave background, corrected for atmospheric emission. This is essentially how Penzias and Wilson [80] discovered the $T_{\mathrm{cmb}} \approx 3$ K CMB at $\nu \approx 4$ GHz using the low-spillover Bell Labs horn antenna (Figure 3.19).

2.2.4 Emission, Absorption, and Reflection from an Opaque Body

A body is **opaque** if its opacity is so high that photons cannot pass through it. The **absorption coefficient** $a(\nu)$ of an opaque body is the probability that an incident photon of frequency ν will be absorbed, and the **reflection coefficient** $r(\nu)$ is the probability that it will be reflected. Thus $a = 1$ for a blackbody and $r = 1$ for a perfect reflector. Each photon is either absorbed or reflected by any opaque body so

$$a(\nu) + r(\nu) = 1. \tag{2.46}$$

The **emission coefficient** $e(\nu)$ of an opaque body is defined as the ratio of the spectral power it emits per unit area at frequency ν to that of a blackbody of the same temperature.

The values of $a(\nu)$ and $e(\nu)$ for any opaque body in local thermodynamic equilibrium are not independent. Imagine an opaque body surrounded by a filter passing frequencies in the range ν to $\nu + d\nu$ inside a cavity (Figure 2.14) in thermodynamic equilibrium at some temperature T. In equilibrium, the spectral power absorbed by the opaque body must equal the spectral power emitted, so

$$\boxed{e(\nu) = a(\nu) = 1 - r(\nu).} \tag{2.47}$$

This is **Kirchhoff's law for opaque bodies**. It implies that the brightness temperature T_{b} of an opaque body in LTE at physical temperature T is given by

$$T_{\mathrm{b}}(\nu) = a(\nu)T = [1 - r(\nu)]T. \tag{2.48}$$

Because $a \leq 1$, then $T_{\mathrm{b}} \leq T$. A perfect reflector ($r = 1$) will always have $T_{\mathrm{b}} = 0$; that is, it does not emit.

Example. Temperature differences across the GBT reflector (Color Plate 1) caused by differential solar heating can deform the surface and degrade its performance. A special paint that is white at visible wavelengths, black in the mid-infrared, and transparent at radio wavelengths keeps the surface cool and does not harm performance at radio wavelengths by absorbing incoming radio waves or emitting radio noise. The special paint exploits Kirchhoff's law to perform three separate functions simultaneously: It is opaque and white in the visible portion of the spectrum to reflect sunlight ($T \approx 5800$ K). It is black in the mid-infrared so that the GBT ($T \approx 300$ K) can cool itself efficiently by reradiation. It is transparent at radio wavelengths so that it neither absorbs incoming radio waves nor emits thermal noise at radio wavelengths.

2.3 POLARIZATION

The instantaneous transverse electric field $\vec{E}$ of a monochromatic electromagnetic wave traveling in the $\hat{z}$-direction can be projected onto orthogonal $\hat{x}$- and $\hat{y}$- (e.g., horizontal and vertical) directions:

$$\vec{E} = [\hat{x} E_x \exp(i\phi_x) + \hat{y} E_y \exp(i\phi_y)] \exp[i(\vec{k} \cdot \hat{z} - \omega t)], \tag{2.49}$$

where $k \equiv 2\pi/\lambda$ is the magnitude of the **wave vector** $\vec{k}$ pointing in the direction of wave travel,

$$\omega \equiv 2\pi \nu \tag{2.50}$$

is the **angular frequency**,

$$\delta \equiv \phi_x - \phi_y \tag{2.51}$$

is the **phase difference** between the orthogonal fields E_x and E_y, and

$$E^2 = |\vec{E}|^2 = E_x^2 + E_y^2. \tag{2.52}$$

Any time-independent combination of phases and amplitudes yields an **elliptically polarized** wave (Figure 2.15) whose electric field vector traces out an ellipse in the (x, y) plane. If the phase difference δ is zero, the electric field vector does not rotate and the wave is **linearly polarized**. If $E_x = E_y$ and $|\delta| = \pi/2$, the electric field vector rotates with angular frequency ω and traces out a circle; such radiation is said to be **circularly polarized**. The **sign convention for circular polarization** used by the Institute of Electrical and Electronics Engineers (IEEE) and by the International Astronomical Union (IAU) is to call the polarization right-handed or left-handed depending on whether the rotation is clockwise ($\delta > 0$) or counterclockwise ($\delta < 0$), respectively, as viewed *from the source toward the observer*. An observer looking toward the source sees the electric-field vector from right-handed polarization rotating counterclockwise, as shown by the different time samples $\omega t = 0, \pi/4, \pi/2, \ldots$ in Figure 2.15. Beware that some optics textbooks use the opposite sign convention.

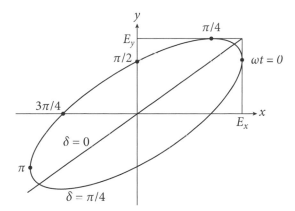

Figure 2.15. The electric field vector of any monochromatic wave traveling in the $\hat{z}$-direction pointing up out of the page traces an ellipse that can be written in the form $\vec{E} = [\hat{x} E_x \exp(i\phi_x) + \hat{y} E_y \exp(i\phi_y)] \exp[i(\vec{k} \cdot \hat{z} - \omega t)]$. The ellipse shown has $\delta = +\pi/4$ and $E_x/E_y = 1.4$. If $\delta \equiv \phi_x - \phi_y = 0$, the ellipse becomes a line, while $\delta = \pm\pi/2$, $E_x = E_y$ would make it a circle. When $\delta > 0$, the tip of the vector rotates clockwise as viewed from the source below the page and counterclockwise as seen by an observer above the page, as shown by the time samples at $\omega t = 0$, $\pi/4$, $\pi/2$, ...

The radiation from astronomical sources is wideband noise whose electric field vector varies rapidly and randomly in amplitude and direction. If radiation in a unit frequency range $\Delta\nu = \Delta\omega/(2\pi)$ is averaged over timescales $\tau \gg (\Delta\omega)^{-1}$, its average polarization can be characterized by the four **Stokes parameters**,

$$I = \langle E_x^2 + E_y^2 \rangle / R_0, \tag{2.53}$$

$$Q = \langle E_x^2 - E_y^2 \rangle / R_0, \tag{2.54}$$

$$U = \langle 2 E_x E_y \cos\delta \rangle / R_0, \tag{2.55}$$

$$V = \langle 2 E_x E_y \sin\delta \rangle / R_0, \tag{2.56}$$

where the brackets indicate time averages, R_0 is the radiation resistance of free space (Equation 3.29), and I is the total flux density, regardless of polarization. The **polarized flux density** is

$$I_\mathrm{p} = (Q^2 + U^2 + V^2)^{1/2} \tag{2.57}$$

and the **degree of polarization** is defined as

$$p \equiv \frac{I_\mathrm{p}}{I}. \tag{2.58}$$

If $\langle E_x \rangle$ and $\langle E_y \rangle$ have equal amplitudes and their phases are completely uncorrelated, then $Q = U = V = 0$, $I_\mathrm{p} = 0$, $p = 0$, and the wave is said to be **unpolarized**. For example, blackbody radiation is unpolarized. An antenna sensitive only to one polarization (e.g., a dipole antenna oriented so it is sensitive only to the $\hat{x}$ component of linear polarization or a helical antenna sensitive only to the left-handed component of circular polarization) will detect only half the power radiated by an unpolarized source. Two orthogonally polarized antennas (e.g., dipoles aligned in the $\hat{x}$- and $\hat{y}$-directions or left-handed and right-handed helical antennas) are needed to collect all of the unpolarized power.

Many astronomical sources are **partially polarized** with $0 < p < 1$. The quantity $(Q^2 + U^2)^{1/2}$ measures the linearly polarized component of flux, and the ratio Q/U depends on the linear polarization position angle. The circularly polarized flux is given by $|V|$, with $V > 0$ indicating right-handed and $V < 0$ indicating left-handed circular polarization.

2.4 BLACKBODY RADIATION

A perfect absorber has $a(\nu) = 1$ at all frequencies and is called a **blackbody**. Kirchhoff's law (Equation 2.47) requires that every blackbody has emission coefficient $e(\nu) = 1$ at all frequencies as well, so the radiated spectrum $B_\nu(T)$ of any blackbody at temperature T is the same as the spectrum of radiation in thermodynamic equilibrium inside a cavity of temperature T, even if the interior walls of the cavity are not black. Thus the intensity and spectrum of blackbody radiation depend *only* on the temperature of the blackbody or cavity. The most fundamental feature of blackbody radiation is that it is **equilibrium radiation**; the only reason for discussing **cavity radiation** is that the cavity traps radiation long enough for it to come into equilibrium. The radiation escaping from a small hole in the cavity is also **blackbody radiation** because radiation entering the hole from the outside has almost no chance of escaping. A small hole in the side of a very cold cavity looks black. The energy density and brightness spectrum of *equilibrium radiation* will be derived in Sections 2.4.1 and 2.4.2, even though it will be called *blackbody radiation*.

The same is true for the electrical noise generated by a warm **resistor**, a device that completely absorbs electrical energy, and which plays an important role in radio astronomy. The standard derivations of the Rayleigh–Jeans and Planck radiation equations are worth repeating because blackbody radiation is so fundamental and because their one-dimensional analogs yield the spectrum of electrical noise generated by a resistor in equilibrium at temperature T.

2.4.1 The Rayleigh–Jeans Approximation

Consider a large (side length $a \gg \lambda$, where λ is the longest wavelength of interest) cubical **cavity** filled with radiation in thermodynamic equilibrium. The purpose of the cavity is to generate radiation and to confine the radiation long enough for it to reach equilibrium. The radiation must be generated by thermal accelerations of charged particles in the walls of any cavity with $T > 0$. The walls have must have nonzero conductivity because walls of zero conductivity would be transparent— they would generate no currents in response to the electric fields of incoming radiation and the radiation would simply escape. The equilibrium radiation is otherwise independent of the wall material. For walls having nonzero conductivity, the transverse electric field strength at the walls is $E = 0$ in equilibrium because fields with $E \neq 0$ induce currents in the walls and lose energy. Only those standing waves with $E = 0$ at the walls will persist after some time $t \gg a/c$.

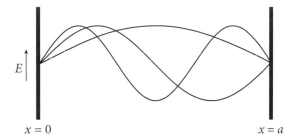

Figure 2.16. Standing waves corresponding to $n\lambda_x/2 = a$ for $n = 1, 2, 3$. Abscissa: x-axis of the cavity, bounded by walls at $x = 0$ and $x = a$. Ordinate: electric field strengths of the three longest standing waves satisfying the boundary condition $E = 0$ at the walls.

All possible standing-wave **modes** (a mode is a field pattern that oscillates sinusoidally with a fixed frequency and phase) in the cavity can be enumerated. For example, consider all standing waves whose wave vectors point in the x-direction (Figure 2.16). The boundary conditions $E = 0$ at $x = 0$ and at $x = a$ mean that only those waves having the discrete wavelengths

$$\frac{\lambda_x}{2} = a, \quad \frac{2\lambda_x}{2} = a, \quad \frac{3\lambda_x}{2} = a, \quad \ldots \tag{2.59}$$

can have nonzero amplitudes. They all satisfy

$$\frac{n_x \lambda_x}{2} = a, \tag{2.60}$$

where $n = 1, 2, 3, \ldots$. Likewise,

$$\frac{n_y \lambda_y}{2} = \frac{n_z \lambda_z}{2} = a \tag{2.61}$$

for wave vectors pointing in the y- and z-directions, respectively.

What about a wave vector pointing in some arbitrary direction? Let α, β, γ be the angles between the wave vector and the x-, y-, z-axes, respectively. From Figure 2.17 it is clear that

$$\lambda = \lambda_x \cos \alpha, \tag{2.62}$$

where λ is the wavelength measured in the direction of the wave vector and $\lambda_x \geq \lambda$ is the component of λ projected onto the x-axis. Thus

$$\lambda_x = \frac{\lambda}{\cos \alpha}. \tag{2.63}$$

Notice that $\lambda_x > \lambda$; it is λ divided by $\cos \alpha$, not multiplied by $\cos \alpha$. Simultaneously, the standing waves must also satisfy

$$\lambda_y = \frac{\lambda}{\cos \beta} \quad \text{and} \quad \lambda_z = \frac{\lambda}{\cos \gamma}. \tag{2.64}$$

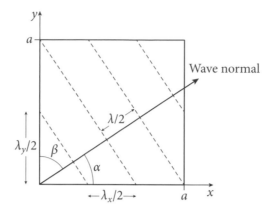

Figure 2.17. This two-dimensional figure illustrates standing waves propagating in a cavity with wave vectors at angles α and β from the x- and y-axes, respectively. Examples of wave nodes, where $|E| = 0$, are indicated by dashed lines for the case $n_x = 3$, $n_y = 2$.

Equations 2.60 through 2.64 imply that standing waves of *all* orientations must satisfy

$$n_x = \frac{2a}{\lambda_x}, \quad n_y = \frac{2a}{\lambda_y}, \quad n_z = \frac{2a}{\lambda_z}, \tag{2.65}$$

$$n_x = \frac{2a \cos \alpha}{\lambda}, \quad n_y = \frac{2a \cos \beta}{\lambda}, \quad n_z = \frac{2a \cos \gamma}{\lambda}. \tag{2.66}$$

Squaring and summing the three parts of Equation 2.66 gives

$$n_x^2 + n_y^2 + n_z^2 = \left(\frac{2a}{\lambda}\right)^2 (\cos^2 \alpha + \cos^2 \beta + \cos^2 \gamma). \tag{2.67}$$

The Pythagorean theorem implies $(\cos^2 \alpha + \cos^2 \beta + \cos^2 \gamma) = 1$ so all standing waves must satisfy

$$n_x^2 + n_y^2 + n_z^2 = \left(\frac{2a}{\lambda}\right)^2. \tag{2.68}$$

The permitted frequencies $\nu = c/\lambda$ of the standing waves are

$$\nu = \frac{c}{2a} \sqrt{n_x^2 + n_y^2 + n_z^2} \tag{2.69}$$

for all positive integers n_x, n_y, and n_z.

The permitted standing waves can be represented as a lattice in the positive octant of the space whose axes are $n_x > 0$, $n_y > 0$, and $n_z > 0$ (Figure 2.18). Each point of the lattice represents one possible mode of equilibrium cavity radiation. The space density of points in this lattice is unity, so the average number of points in any volume equals that volume.

Let ρ be the radial coordinate in (n_x, n_y, n_z)-space. Then

$$\rho^2 = n_x^2 + n_y^2 + n_z^2 \tag{2.70}$$

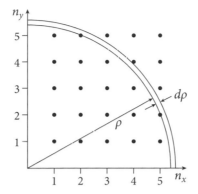

Figure 2.18. The (x, y) plane in the space whose axes are $(n_x > 0, n_y > 0, n_z > 0)$. Dots at all positive integers mark standing waves in the (n_x, n_y) plane. In this space, frequency is proportional to the radial distance ρ from the origin.

and

$$\nu = \frac{c}{\lambda} = \left(\frac{c}{2a}\right)\rho. \tag{2.71}$$

The number $N_\nu(\nu)d\nu$ of independent modes having frequencies in the range ν to $\nu + d\nu$ equals the volume of the spherical octant shell between ρ and $\rho + d\rho$, multiplied by 2 to account for the two independent orthogonal polarizations (Section 2.3) of electromagnetic waves:

$$N_\nu(\nu)d\nu = \frac{4\pi\rho^2 d\rho}{8} \times 2 = \pi\left(\frac{2a\nu}{c}\right)^2\frac{2a}{c}d\nu. \tag{2.72}$$

Classically, each radiation mode can have any energy $E > 0$. In thermodynamic equilibrium at temperature T, the probability $P(E)$ that any mode has energy E is given by the continuous **Boltzmann probability distribution**

$$P(E) \propto \exp\left(-\frac{E}{kT}\right). \tag{2.73}$$

Then the average energy per mode is

$$\langle E \rangle = \frac{\int_0^\infty E\,P(E)dE}{\int_0^\infty P(E)dE} = kT. \tag{2.74}$$

Thus each mode has the same average energy $\langle E \rangle = kT$. The spectral energy density $u_\nu(T)$ of cavity radiation at frequency ν is the total energy of all N_ν modes (Equation 2.72) at that frequency divided by the volume a^3 of the cavity:

$$u_\nu(T) = \frac{N_\nu(\nu)\,kT}{a^3} = \frac{8\pi a^3}{a^3}\frac{\nu^2}{c^3}kT = 8\pi kT\frac{\nu^2}{c^3}. \tag{2.75}$$

The **spectral energy density** of radiation (blackbody or not) is spectral energy per unit volume. It equals the total flow of spectral power per unit area divided by the flow speed c:

$$\boxed{u_\nu = \frac{1}{c} \int I_\nu \, d\Omega.}$$ (2.76)

Calling the specific intensity of blackbody radiation B_ν and making use of the fact the blackbody radiation is isotropic yields, for blackbody radiation,

$$u_\nu = \frac{1}{c} \int_{4\pi} B_\nu \, d\Omega = \frac{4\pi}{c} B_\nu.$$ (2.77)

Combining Equations 2.75 and 2.77 gives

$$\frac{4\pi}{c} B_\nu = \frac{8\pi k T \nu^2}{c^3}.$$ (2.78)

Solving for B_ν yields the **Rayleigh–Jeans approximation** for the spectral brightness of blackbody radiation

$$\boxed{B_\nu = \frac{2kT\nu^2}{c^2} = \frac{2kT}{\lambda^2}}$$ (2.79)

that is valid only in the low-frequency limit $h\nu \ll kT$.

Notice that

1. the spectral brightness B_ν is proportional to frequency squared because the volume of a spherical shell in three dimensions is proportional to frequency squared;
2. all modes have $\langle E \rangle = kT$, the classical assumption that breaks down at high frequencies;
3. the Rayleigh–Jeans approximation implies that B_ν diverges at high ν; this is called the **ultraviolet catastrophe**;
4. B_ν is independent of direction; blackbody radiation is **isotropic**;
5. blackbody radiation is **unpolarized**; you can show this with a thought experiment involving two cavities connected by a passive polarizing filter, as in Figure 2.9;
6. this result derived for a cubical cavity applies to equilibrium radiation in a cavity of any shape or (large) size, which you can demonstrate by a thought experiment in which a cubical cavity is connected through a small filter window to the other cavity.

2.4.2 The Planck Radiation Law

The only flaw in the derivation of the Rayleigh–Jeans approximation is the classical assumption that each radiation mode can have any energy $E > 0$. To eliminate the ultraviolet catastrophe, Planck postulated that the possible mode energies are not continuously distributed, but rather they are quantized and must satisfy the new constraint

$$E = nh\nu, \quad n = 1, 2, 3, \ldots,$$ (2.80)

where $h \approx 6.63 \times 10^{-27}$ erg s is Planck's constant and n is the number of **photons**, or discrete particles of light each having energy

$$\boxed{E = h\nu.}$$ (2.81)

Then

$$P(E) = P(nh\nu) \propto \exp\left(-\frac{nh\nu}{kT}\right)$$ (2.82)

and the average energy per mode is calculated by summing over only the discrete energies permitted instead of integrating over all energies. Planck's quantized replacement for Equation 2.74 is

$$\langle E \rangle = \frac{\sum\limits_{n=0}^{\infty} nh\nu \, P(nh\nu)}{\sum\limits_{n=0}^{\infty} P(nh\nu)} = \frac{\sum\limits_{n=0}^{\infty} nh\nu \exp\left(-\dfrac{nh\nu}{kT}\right)}{\sum\limits_{n=0}^{\infty} \exp\left(-\dfrac{nh\nu}{kT}\right)}.$$ (2.83)

Planck's sum is evaluated in Appendix B.1; it is

$$\langle E \rangle = \frac{h\nu}{\exp\left(\dfrac{h\nu}{kT}\right) - 1} = kT \left[\frac{\dfrac{h\nu}{kT}}{\exp\left(\dfrac{h\nu}{kT}\right) - 1}\right].$$ (2.84)

The quantity kT is the classical mean energy per mode, and the quantity in square brackets is the quantum correction factor. In the limit $h\nu \ll kT$ the quantum correction factor is unity, but in the limit $h\nu \gg kT$ it falls exponentially to zero. That is, the high-frequency or "ultraviolet" modes may contain few or even zero photons, thereby avoiding the "ultraviolet catastrophe."

Thus the correct blackbody radiation law becomes

$$B_\nu = \frac{2kT\nu^2}{c^2} \left[\frac{\dfrac{h\nu}{kT}}{\exp\left(\dfrac{h\nu}{kT}\right) - 1}\right],$$ (2.85)

where the first factor is the Rayleigh–Jeans approximation and the quantity in square brackets is the quantum correction factor. **Planck's equation** for the spectral brightness B_ν of blackbody radiation is usually written in the simpler form

$$\boxed{B_\nu(\nu, T) = \frac{2h\nu^3}{c^2} \frac{1}{\exp\left(\dfrac{h\nu}{kT}\right) - 1}.}$$ (2.86)

The corresponding brightness per unit wavelength B_λ follows from Equation 2.5; it can be written either as a function of frequency:

$$B_\lambda(\nu, T) = \frac{\nu^2}{c} B_\nu(\nu, T) = \frac{2h\nu^5}{c^3} \frac{1}{\exp\left(\dfrac{h\nu}{kT}\right) - 1}$$ (2.87)

or as a function of wavelength:

$$B_\lambda(\lambda, T) = \frac{2hc^2}{\lambda^5} \frac{1}{\exp\left(\frac{hc}{\lambda kT}\right) - 1}.$$ (2.88)

Integrating Planck's law over all frequencies (see Appendix B.2) gives the **Stefan–Boltzmann law** specifying a finite **integrated brightness** for a blackbody radiator at temperature T:

$$B(T) \equiv \int_0^\infty B_\nu(T)d\nu = \frac{\sigma T^4}{\pi}.$$ (2.89)

The quantity

$$\sigma \equiv \frac{2\pi^5 k^4}{15c^2 h^3} \approx 5.67 \times 10^{-5} \frac{\text{erg}}{\text{cm}^2 \text{ s K}^4 \text{ (sr)}}$$ (2.90)

is called the **Stefan–Boltzmann constant** and is also derived in Appendix B.2. The unit sr is dimensionless because solid angle has dimensions angle^2 = (length/length)^2. A dimensionless unit can be inserted or deleted without changing the dimensions of an equation—hence the parentheses around sr in Equation 2.90. For clarity it should be used where appropriate. For example, it is needed in the blackbody brightness (erg cm^{-2} s^{-1} K^{-4} sr^{-1}) Equation 2.89 but not in the radiation energy density (erg cm^{-3}) Equation 2.93 or in the flux (erg cm^{-2} s^{-1} K^{-4}) Equation 2.111.

For isotropic radiation, the spectral energy density (Equation 2.76) is simply

$$u_\nu = \frac{4\pi I_\nu}{c},$$ (2.91)

so the total **radiation energy density**

$$u \equiv \int_0^\infty u_\nu d\nu = \frac{4\pi I}{c}$$ (2.92)

of blackbody radiation is

$$u = \frac{4\sigma T^4}{c} = a T^4,$$ (2.93)

where the quantity

$$a \equiv 4\sigma/c \approx 7.56577 \times 10^{-15} \text{ erg cm}^{-3} \text{ K}^{-4}$$ (2.94)

is called the **radiation constant**.

In any narrow frequency range, the number density of photons is their spectral energy density divided by the energy per photon, $E = h\nu$. The total **photon number density** n_γ is

$$n_\gamma = \int_0^\infty \frac{u_\nu}{h\nu}d\nu.$$ (2.95)

The **photon number density of blackbody radiation** at temperature T is

$$n_\gamma = \frac{4\pi}{ch} \int_0^\infty \frac{B_\nu(T)}{\nu} d\nu \tag{2.96}$$

$$= \frac{8\pi}{c^3} \int_0^\infty \frac{\nu^2 d\nu}{\exp\left(\frac{h\nu}{kT}\right) - 1} \tag{2.97}$$

$$= \frac{8\pi}{c^3} \left(\frac{kT}{h}\right)^3 \int_0^\infty \frac{x^2 dx}{e^x - 1}, \tag{2.98}$$

where the integral

$$\int_0^\infty \frac{x^2 dx}{e^x - 1} \approx 2.404 \tag{2.99}$$

is evaluated in Appendix B.2. Numerically, the photon number density of blackbody radiation is

$$\boxed{\left(\frac{n_\gamma}{\text{cm}^{-3}}\right) \approx 20.3 \left(\frac{T}{\text{K}}\right)^3.} \tag{2.100}$$

The **mean photon energy of blackbody radiation** is

$$\boxed{\langle E_\gamma \rangle = \frac{u}{n_\gamma} = \frac{4\sigma T^4}{cn_\gamma} = \frac{\pi^4 kT}{15} \left(\int_0^\infty \frac{x^2 dx}{e^x - 1}\right)^{-1} \approx 2.70\, kT.} \tag{2.101}$$

The frequency $\langle \nu \rangle$ corresponding to this mean photon energy is

$$\left(\frac{\langle \nu \rangle}{\text{GHz}}\right) = \frac{\langle E_\gamma \rangle}{h} \approx 56 \left(\frac{T}{\text{K}}\right). \tag{2.102}$$

It is close to the frequency ν_{max} at which B_ν, the brightness *per unit frequency* of a blackbody, is maximum. Setting

$$\frac{\partial B_\nu(\nu)}{\partial \nu} = 0 \tag{2.103}$$

yields

$$\boxed{\left(\frac{\nu_{\text{max}}}{\text{GHz}}\right) \approx 59 \left(\frac{T}{\text{K}}\right).} \tag{2.104}$$

The wavelength λ_{max} at which

$$\frac{\partial B_\lambda(\lambda)}{\partial \lambda} = 0 \tag{2.105}$$

maximizes B_λ, the brightness *per unit wavelength* of a blackbody. It is

$$\boxed{\left(\frac{\lambda_{\text{max}}}{\text{cm}}\right) \approx 0.29 \left(\frac{T}{\text{K}}\right)^{-1}.} \tag{2.106}$$

Equation 2.106 is the familiar form of **Wien's displacement law** used by optical astronomers, whose spectrometers measure wavelengths instead of frequencies. Note that the peak frequency $c/\lambda_{max} \approx 103$ GHz T(K) is much higher than $\nu_{max} \approx 59$ GHz T(K).

The power emitted per unit area per unit frequency by any opaque isotropic radiator of brightness I_ν is the spectral flux density

$$F_\nu = \int I_\nu \cos\theta \, d\Omega, \qquad (2.107)$$

where the integration covers the hemisphere above the unit area:

$$F_\nu = \int_0^{\pi/2} I_\nu \cos\theta \left(\int_0^{2\pi} \sin\theta \, d\phi \right) d\theta = 2\pi I_\nu \int_0^{\pi/2} \cos\theta \sin\theta \, d\theta, \qquad (2.108)$$

$$\boxed{F_\nu = \pi I_\nu.} \qquad (2.109)$$

For a blackbody at temperature T, the spectral flux density is

$$F_\nu(T) = \pi B_\nu(T). \qquad (2.110)$$

The corresponding total power per unit area integrated over all frequencies is

$$\boxed{F(T) = \pi B(T) = \sigma T^4.} \qquad (2.111)$$

2.5 NOISE GENERATED BY A WARM RESISTOR

A **resistor** is a passive electronic component that absorbs the electrical power applied to it and converts that power into heat; it is the "blackbody" of electronic circuits. Just as motions of charged particles in the walls of a warm cavity generate photons, motions of charged particles in a resistor at any temperature $T > 0$ K generate electrical noise. The frequency spectrum of this noise depends only on the temperature of an ideal resistor and is independent of the material in the resistor.

An **antenna** is a passive device that converts electromagnetic radiation into electrical currents in wires (when it is used as a receiving antenna) or vice versa (when it is used as a transmitting antenna). The noise generated by a resistor is indistinguishable from the noise coming from a receiving antenna surrounded by blackbody radiation of the same temperature. Warm resistors are useful in radio astronomy as standards for calibrating antennas and receivers, so the noise power per unit bandwidth received by a radio telescope is often described in terms of the Rayleigh–Jeans **antenna temperature**. The antenna temperature of a receiving antenna is defined as the temperature of an ideal resistor that would generate the same Rayleigh–Jeans noise power per unit bandwidth as appears at the antenna output. Like brightness temperature, antenna temperature is not a physical temperature. The sensitivity and gain of a radio receiver can be calibrated by connecting its input alternately to hot and cold resistors (sometimes called hot and cold "loads") having known temperatures, and the amount of noise generated

Figure 2.19. Two resistors connected by a lossless transmission line of length $a \gg \lambda$, the longest wavelength of interest. In equilibrium, the transmission line can support only those standing waves having zero voltages at the ends; other modes are suppressed by the lossy resistors.

in a **radiometer** (a radio receiver that measures noise power) can be described by the **radiometer noise temperature**, the temperature of a resistor at the input of an imaginary noiseless radiometer having the same gain as the actual receiver that would generate the same noise power output. Like brightness temperature, radiometer noise temperature is *not* a physical temperature.

The derivation of the electrical power per unit bandwidth P_ν generated by current in a resistor is the one-dimensional analog of the the three-dimensional derivation of the blackbody spectrum [9, 74]. At low radio frequencies, $h\nu \ll kT$ and the Rayleigh–Jeans approximation is accurate. Recall that the Rayleigh–Jeans derivation of B_ν starts with a large cube of side length $a \gg \lambda$ containing standing waves of thermal radiation. The classical average energy in each standing-wave mode is $\langle E \rangle = kT$, and the number of modes with frequency ν to $\nu + d\nu$ is proportional to ν^2, so $B_\nu \propto \nu^2$.

Consider two identical resistors at temperature T connected by a lossless transmission line (e.g., a pair of parallel wires) of length a much larger than the longest wavelength of interest (Figure 2.19). Standing waves on the line must satisfy

$$a = \frac{n\lambda}{2}, \qquad n = 1, 2, 3, \ldots, \tag{2.112}$$

where λ is the wavelength. Electrical signals do not travel at exactly the speed of light on a transmission line, but at some slightly lower velocity $v < c$ so $\nu = v/\lambda$ and

$$n = \frac{2av}{v}. \tag{2.113}$$

For $a > \lambda$, the number of modes per unit frequency is

$$N_\nu = \frac{2a}{v}. \tag{2.114}$$

The classical Boltzmann law says that each mode has average energy $\langle E \rangle = kT$ in equilibrium, so the average energy per unit frequency E_ν in the transmission line is

$$E_\nu = N_\nu kT = \frac{2akT}{v}. \tag{2.115}$$

This energy takes a time $\Delta t = a/v$ to flow from one end of the transmission line to the other, so the classical power (energy per unit time) per unit frequency flowing on the transmission line is

$$P_\nu = \frac{E_\nu}{\Delta t} = 2kT, \tag{2.116}$$

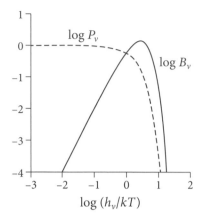

Figure 2.20. Comparison of the spectrum B_ν of blackbody radiation with the spectrum P_ν of noise generated by a resistor.

independent of ν. Thus the total spectral power P_ν generated by the two identical resistors must be $2kT$ and, by symmetry, the spectral power generated by each resistor is

$$P_\nu = kT \qquad (2.117)$$

in the limit $h\nu \ll kT$. This equation is called the **Nyquist approximation** and is the electrical equivalent of the Rayleigh–Jeans equation for radiation. Because the "space" of the transmission line has only one dimension instead of three, $P_\nu \propto \nu^0$ instead of $P_\nu \propto \nu^2$. The dimensions of P_ν are power per unit frequency (e.g., W Hz^{-1}) and the dimensions of kT are energy (e.g., joules), which appear to be different at first glance. However, Hz^{-1} = s, so power per unit frequency W Hz^{-1} = W s = joule also has dimensions of energy. The latter is simpler, but the former is used because it is conceptually more appropriate for noise power per unit bandwidth.

Like the Rayleigh–Jeans approximation, the Nyquist approximation implies an "ultraviolet catastrophe" that can be cured by Planck's quantization rule. If the electrical energy in each mode is restricted to an integer multiple of $h\nu$, the Nyquist formula becomes

$$P_\nu = kT \left[\frac{\dfrac{h\nu}{kT}}{\exp\left(\dfrac{h\nu}{kT}\right) - 1} \right], \qquad (2.118)$$

where the quantity in square brackets equals the quantization correction for black-body radiation (Equation 2.85). The exact **Nyquist formula** is usually written in the form

$$P_\nu = \frac{h\nu}{\exp\left(\dfrac{h\nu}{kT}\right) - 1}. \qquad (2.119)$$

Figure 2.20 compares the Nyquist and Planck spectra. At low frequencies $\nu \ll kT/h$, the specific intensity B_ν of blackbody radiation in three dimensions (solid curve) is proportional to ν^2. Its one-dimensional analog, the spectral power density of noise generated by a resistor (dashed curve), is proportional to ν^0. Quantization causes the sharp exponential cutoffs of both curves at high frequencies.

2.6 COSMIC MICROWAVE BACKGROUND RADIATION

The universe is filled with blackbody radiation whose temperature now is $T_0 = 2.725 \pm 0.002$ K. This **cosmic microwave background** (CMB) was produced by the hot "big bang" marking the birth of the universe. This section reviews the expanding universe, derives the evolution of blackbody radiation during the expansion, and indicates how the CMB carries cosmological information.

2.6.1 The Expanding Universe

The expanding universe is nearly homogeneous and isotropic on all scales much larger than clusters of galaxies. Consequently the recession velocity v of a nearby galaxy at rest in its local environment is proportional to its (independently measured) distance d. The constant of proportionality is called the **Hubble parameter**

$$H \equiv \lim_{d \to 0} \left(\frac{v}{d} \right). \tag{2.120}$$

$H(t)$ is everywhere the same at any particular time t after the big bang, but it changes with time. The present value of the Hubble parameter is called the **Hubble constant** H_0, and its measured value is [82]

$$H_0 \approx 67.8 \pm 0.9 \text{ km s}^{-1} \text{ Mpc}^{-1}. \tag{2.121}$$

Substituting 3.0856×10^{19} km ≈ 1 Mpc shows that the dimension of H is inverse time and the Hubble constant can be written as $H_0 \approx (4.55 \times 10^{17} \text{ s})^{-1}$. The **Hubble time** defined by

$$t_H \equiv H_0^{-1} \approx 4.55 \times 10^{17} \text{ s} / 10^{7.5} \text{ s yr}^{-1} \approx 1.44 \times 10^{10} \text{ yr} \tag{2.122}$$

would be the present age of the universe had it started as a point and expanded at a constant rate thereafter. (The useful approximation $10^{7.5}$ s ≈ 1 yr used in Equation 2.122 is easy to remember and accurate to 0.2%.)

The radius of the observable universe is roughly the **Hubble distance** at which the Hubble parameter indicates a recession velocity equal to the speed of light:

$$d_H \equiv \frac{c}{H}. \tag{2.123}$$

It is currently $d_H = c/H_0 \approx 1.36 \times 10^{28}$ cm $\approx 4.4 \times 10^9$ pc $\approx 1.4 \times 10^{10}$ light years.

Although general relativity is needed to describe the universe as a whole, the homogeneity of the universe means that many of its global properties can be derived by extrapolating Newtonian results obtained from regions of size $d \ll d_H$. For example, the **critical density** ρ_c is defined as the density at which the kinetic energy of expansion equals the gravitational energy decelerating the expansion. Consider a spherical region of radius d centered on an observer and a unit test mass on its surface, which is receding with radial velocity $v = Hd$. If the average density inside the sphere is ρ_c, the sum of the kinetic and gravitational potential energies per unit mass will be zero:

$$\frac{v^2}{2} - \frac{GM}{d} = 0, \tag{2.124}$$

$$\frac{H^2 d^2}{2} = \frac{G}{d} \left(\frac{4\pi d^3 \rho_c}{3} \right), \tag{2.125}$$

so the critical density today is

$$\boxed{\rho_c = \frac{3H_0^2}{8\pi G} \approx 8.6 \times 10^{-30} \text{ g cm}^{-3}} \tag{2.126}$$

for $H_0 = 67.8$ km s^{-1} Mpc^{-1}. Ordinary **baryonic matter** consists by mass primarily of protons and neutrons, whose masses are $m \approx 1.67 \times 10^{-24}$ g, so a mean number density $n_b \approx 5.2 \times 10^{-6}$ cm^{-3} of protons and neutrons would be needed to supply the critical density. However, in a matter-dominated universe, mutual gravitational attraction decelerates the expansion. The expansion age of a low-density ($\rho \ll \rho_c$) universe is nearly t_H, but a critical-density matter-dominated universe should be only $(2/3)H_0^{-1} \approx 9.6$ Gyr old, which is less than the ages of the oldest known stars. **Dark energy** actually dominates the mass-energy of the universe today, and it has the peculiar property that it *accelerates* the expansion enough that the actual age of the universe is closer to t_H, eliminating the problem of stars older than the universe.

2.6.2 Blackbody Radiation in the Expanding Universe

How does the blackbody CMB evolve as the universe expands with time t since the big bang? The universe is spatially homogeneous and isotropic on large scales, so the properties of the CMB can be calculated by considering what happens to blackbody radiation in a small imaginary cube whose side length slowly grows along with the universe. Here "small" only means much smaller than d_H so that relativistic effects can be ignored; the side length might be the distance between two galaxies that are currently separated by ~ 100 Mpc. Although the imaginary cube has no walls, CMB radiation escaping from the cube is balanced by CMB radiation entering the cube from similar adjacent cubes. Consequently the energy density and spectrum of CMB radiation within the cube is exactly the same as that in a cubical cavity having mirrored walls and undergoing a slow adiabatic (meaning, there is no heat transfer to or from the cube) expansion. The radiation remains in equilibrium during the slow expansion, so it is always blackbody radiation. The expansion of both the universe and the imaginary cube can be described by a single dimensionless **expansion scale factor** $a(t)$, where $a(t_0) \equiv 1$ today. The photon number density n_γ in the cube falls as a^{-3} because the total number of photons in the expanding cube doesn't change as a grows. The photon number density of blackbody radiation is proportional to T^3 (Equation 2.100), so the temperature of the CMB declines as $T = a^{-1}T_0$, where $T_0 \approx 2.725$ K. Also, the mean energy per photon of blackbody radiation is proportional to T. Thus the wavelength λ of any individual photon is proportional to a, even if the photon is from a source which does not have a blackbody spectrum.

Astronomers frequently use the term **redshift** defined by

$$z \equiv \frac{\lambda_o - \lambda_e}{\lambda_e} = \frac{\lambda_o}{\lambda_e} - 1 = \frac{\nu_e}{\nu_o} - 1, \qquad (2.127)$$

where λ_e and ν_e are the wavelength and frequency emitted by a source at redshift z, and λ_o and ν_o are the observed wavelength and frequency at $z = 0$. The redshift z and expansion factor a are related by

$$(1 + z) = a^{-1}. \qquad (2.128)$$

Thus the CMB temperature at redshift z is

$$T = T_0(1 + z). \qquad (2.129)$$

The CMB radiation is ubiquitous, so interstellar matter (e.g., interstellar dust or gas) is heated by the CMB to at least the temperature given by Equation 2.129. The interstellar dust temperature in nearby galaxies like our own is ~ 20 K, which is well above $T_0 \sim 3$ K but comparable with the CMB temperature felt by galaxies at $z \sim 6$. The CMB will influence the spectral line and continuum emission visible from high-redshift galaxies.

Prior to the time corresponding to the redshift $z_\star = 1091 \pm 1$, the CMB temperature was $T > 3000$ K and the radiation ionized enough of the hydrogen atoms filling the universe to keep the universe opaque. This "wall" beyond which photons cannot escape is called the **surface of last scattering**, and it defines the limit of the visible universe. During the **recombination era** when the age of the universe was only $t_\star = (379 \pm 5) \times 10^3$ yr, almost all of the free protons and electrons combined to form neutral hydrogen atoms and the universe became transparent to the CMB. (The term *re*combination is misleading because the protons and electrons were combining for the first time.) The CMB photons received today were last scattered at that time and have been traveling in straight lines ever since, so an image of the CMB (Color Plate 16) made now ($z = 0$) shows its temperature distribution as it was when $z_\star \approx 1091$.

2.6.3 Prediction and Discovery of the CMB

Well before the CMB was observed, George Gamow and his students calculated the relative hydrogen and helium abundances (Figure 2.21) produced by nucleosynthesis a few minutes after a hot big bang, when the temperature of the CMB was $T > 10^9$ K [2]. The abundance of deuterium in particular depends sensitively on the relative number densities n_b of baryons (primarily protons plus neutrons by mass) and of photons n_γ. The ratio $\eta \equiv n_b/n_\gamma \approx 6 \times 10^{-10}$ has remained constant since that time because both n_b and and n_γ are proportional to a^{-3}. Equation 2.100 gives the photon number density of the $T_0 \approx 2.725$ K CMB; it is $n_\gamma \approx 411$ cm^{-3}. If $\eta \approx 6 \times 10^{-10}$, then the baryon number density is $n_b \approx 2.5 \times 10^{-7}$ cm^{-3}, which is only about 5% of the density needed to close the universe (Equation 2.126).

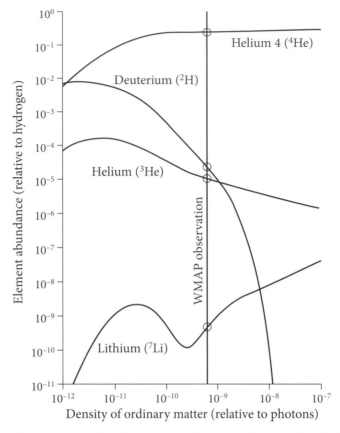

Figure 2.21. The energy density of the early universe was dominated by hot blackbody radiation. By $t \approx 1$ s, the temperature fell to $T \approx 10^{10}$ K, and neutrons decayed into protons and electrons or combined with protons to make deuterium nuclei, most of which combined to make helium nuclei during the first three minutes. The chemical composition of the early universe is a function of the baryon/photon ratio $\eta \approx 6 \times 10^{-10}$ by number. "Ordinary" baryonic matter now accounts for about 4.6% of the energy density of the universe. Image credit: NASA/WMAP Science Team.

Alpher and Herman [3] presented the first estimate $T_0 \approx 5$ K for the present temperature of the CMB based on the chemical composition of the universe. They also showed that only light elements can be formed by big-bang nucleosynthesis because the free neutrons needed to assemble heavier elements decay with a half life of about 10 minutes. Stars produce the heavier elements.

Although 5 K is a very strong signal by radio-astronomy standards, the CMB is hard to distinguish from other noise sources in receivers and radio telescopes because it is very nearly isotropic. Unlike emission from a compact radio source, the CMB cannot be isolated by a differential measurement comparing two adjacent patches of sky. It was accidentally discovered in 1964 by Arno Penzias and Robert Wilson [80], who reported painstaking measurements indicating an isotropic "excess antenna temperature" coming from the Bell Labs horn antenna at 4080 MHz. Subsequent measurements, especially by the COBE[1] (COsmic Background Explorer) satellite, have demonstrated that this "excess noise" matches a blackbody spectrum within 50 parts per million well into the far infrared.

2.6.4 Dipole Anisotropy

After contamination by foreground sources near the Galactic plane has been removed, the largest observed anisotropy in the CMB is the **dipole anisotropy** with amplitude

$$\Delta T \approx 3.37 \cos\theta \text{ mK}, \tag{2.130}$$

where θ is the angle from the direction (Galactic latitude $b = +48°$, Galactic longitude $l = 264°$) of the Earth's motion relative to the universe as a whole. The speed $v_\oplus$ of the Earth is much smaller than the speed of light, so the nonrelativistic Doppler equation can be used to estimate the redshift induced by the Earth's motion:

$$\frac{v_\oplus}{c} \approx \frac{\Delta\lambda}{\lambda} \approx z \approx \frac{\Delta T}{T} \approx \frac{3.37 \times 10^{-3} \text{ K}}{2.725 \text{ K}} \approx 1.24 \times 10^{-3}, \tag{2.131}$$

or $v_\oplus \approx 370$ km s^{-1}. This speed is consistent with the vector sum of independently measured motions of the Earth around the Sun, the Sun around the Galactic center, and the peculiar motion of our Galaxy in the local universe.

2.6.5 Intrinsic Anisotropy

The age $t_\star$ of the universe at recombination can be calculated because the physics of the homogeneous and isotropic early universe was so simple. During that time, small inhomogeneities in the density of dark matter were amplified by gravity. Baryons started falling into the gravitational potential wells, and the CMB was coupled to this flow by Thomson scattering (Equation 5.33) off the numerous free electrons present prior to $t_\star$. Radiation pressure resisted this infall, setting up **baryon acoustic oscillations** analogous to sound vibrations in air, but with a "sound" speed $c_s \approx c/\sqrt{3}$. The strongest oscillations in the CMB brightness had wavelengths equal to the diameter of the sound horizon at the time of recombination:

$$\lambda_\star \approx 2c_s t_\star \approx \frac{2ct_\star}{\sqrt{3}} \approx \frac{2 \cdot 379 \times 10^3}{\sqrt{3}} \text{ ly} \approx 438 \times 10^3 \text{ ly}, \tag{2.132}$$

which was about 134 kpc. After recombination, the photons decoupled and this "standard ruler" grew by a factor $(1 + z_\star) \approx 1092$ to its present size of 146 Mpc. Color Plate 16 shows these CMB brightness fluctuations, projected onto a sphere [105]. The angular power spectrum of CMB fluctuations (Figure 1.16) observed by the **Wilkinson Microwave Anisotropy Probe (WMAP)** and by the **Planck Mission** reveal that the angular size of this ruler is $\theta_\star = 0.010388 \pm 0.000027$ rad, so the comoving distance to the surface of last scattering is $D_A = (1 + z_\star)\lambda_\star/\theta_\star \approx 14.1 \pm 0.16$ Gpc. The **radius of the observable universe** (including gravitational radiation and neutrinos from redshifts $z \gg z_\star$) is only about 2% larger. If we live in a flat

---

[1]http://lambda.gsfc.nasa.gov/product/cobe/.

ΛCDM universe (Λ Cold Dark Matter universe, where Λ stands for the cosmological constant associated with dark energy), its present age is $t = 13.75 \pm 0.13$ Gyr.

The intrinsic rms of the CMB over the whole sky is $\sigma \approx 18\,\mu$K, which is only about $\sim 10^{-5}$ of the average $T \approx 2.725$ K. This leads to the **horizon problem**: Why should the CMB temperature be the same at points separated by more than $\theta_\star \approx 0.01$ rad, which were never in causal contact by the time of recombination? The simple big-bang model also has a **flatness problem**. The universe is nearly "flat" today so its density is close to the critical density, and it must have been *much* closer to the critical density when it was much smaller (e.g., during the nucleosynthesis era when $z \sim 10^9$). What "fine tuned" the density of the universe?

The preferred solution for both problems is that the early universe underwent a brief period ($\sim 10^{-32}$ s) of **inflation** during which the scale factor a grew exponentially by a factor $> 10^{26}$. The observable universe is only an infinitesimal and causal part of the whole universe, nearly all of which is outside our horizon. The **vacuum energy** associated with inflation ensures that the total density contributed by both matter and energy approaches the critical density.

2.7 RADIATION FROM AN ACCELERATED CHARGE

Maxwell's equations imply that *all classical electromagnetic radiation is generated by accelerating electrical charges.* It is possible to derive the intensity and angular distribution of the radiation from a point charge (a charged particle) subject to an arbitrary but small acceleration $\Delta v / \Delta t$ via Maxwell's equations, but the complicated math obscures the physical interpretation that remains clear in J. J. Thomson's illuminating derivation (see [68]).

If a particle with electrical charge q is at rest or moving with a constant velocity, its electric field lines are purely radial: $|\vec{E}| = E_r$. Suppose a charged particle initially at rest is accelerated to a small velocity $\Delta v \ll c$ in a short time Δt. This disturbs the lines of force, and the disturbance travels outward at the speed of light c. Figure 2.22 shows that at time t after the acceleration, the disturbance will have propagated to $r = ct$ and the perpendicular component of the electric field will have magnitude

$$\frac{E_\perp}{E_r} = \frac{\Delta v\, t\, \sin\theta}{c\,\Delta t}, \tag{2.133}$$

where θ is the angle between the acceleration vector and the line of sight connecting the charge to the observer. The web applet http://webphysics. davidson.edu/Applets/Retard/Retard.html provides a very clear and interactive demonstration of this effect.

Coulomb's law for the radial component E_r of the **electric field** (electric force per unit charge) a distance r from a stationary charge q is

$$\boxed{E_r = \frac{q}{r^2}} \tag{2.134}$$

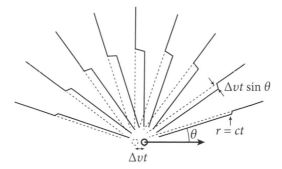

Figure 2.22. The electric field lines from an accelerated electron. The dotted circle shows the initial position of the electron, and the dotted lines are the radial lines of force emanating from that position. At time t after a small acceleration $\Delta v/\Delta t$, the electron position has moved by $\Delta v\, t$ and its lines of force have shifted transversely by $\Delta v\, t \sin\theta$.

in Gaussian CGS units. Substituting r/c for t in Equation 2.133 and using Coulomb's law to eliminate E_r gives

$$E_\perp = \frac{q}{r^2}\left(\frac{\Delta v}{\Delta t}\right)\frac{r \sin\theta}{c^2} \tag{2.135}$$

and

$$\boxed{E_\perp = \frac{q\dot{v}\sin\theta}{rc^2}}, \tag{2.136}$$

where

$$\dot{v} \equiv \lim_{\Delta t \to 0}\left(\frac{\Delta v}{\Delta t}\right). \tag{2.137}$$

Equation 2.136 is valid for *any* small acceleration, not just a sinusoidal acceleration at a single frequency. The transverse field $E_\perp$ at $r = ct$ mimics the acceleration at $t = 0$. Thus a sinusoidal acceleration would result in a sinusoidal variation of $E_\perp$ with the same frequency.

Notice that $E_\perp \propto r^{-1}$ (Equation 2.136) falls more slowly with distance r than $E_r \propto r^{-2}$ (Equation 2.134). Far from the charged particle, only $E_\perp$ will contribute significantly to the observed electric field. From a distant observer's point of view, only the "visible" component of acceleration perpendicular to the line of sight ($\dot{v}\sin\theta$) contributes to the radiated electric field; the "invisible" component of acceleration parallel to the line of sight ($\dot{v}\cos\theta$) does not appear to radiate—what you see is what you get. Likewise, the radiated electric field is linearly polarized in the direction parallel to the component of acceleration perpendicular to the line of sight.

How much power is radiated in each direction? The **Poynting flux**, or power per unit area (e.g., erg s^{-1} cm^{-2}), is

$$\vec{S} = \frac{c}{4\pi}\vec{E} \times \vec{B}. \tag{2.138}$$

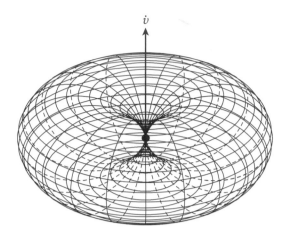

Figure 2.23. The power pattern of Larmor radiation from a charged particle shown for an acceleration vector $\dot{v}$ tilted 60 degrees from the line of sight. The power received in any direction is proportional to the component of $\dot{v}$ perpendicular to the line of sight.

In CGS units $|\vec{E}| = |\vec{B}|$ so

$$|\vec{S}| = \frac{c}{4\pi} E^2. \tag{2.139}$$

Inserting Equation 2.136 into Equation 2.139 gives

$$|\vec{S}| = \frac{c}{4\pi} \left(\frac{q\dot{v}\sin\theta}{rc^2} \right)^2 = \left(\frac{q^2\dot{v}^2}{4\pi c^3} \right) \frac{\sin^2\theta}{r^2} \tag{2.140}$$

at large distances r. The accelerated charge radiates with a dipolar power pattern $\propto \sin^2\theta$ shaped like a doughnut whose axis is parallel to the acceleration $\dot{v}$ (Figure 2.23).

The total power emitted is the integral of $|\vec{S}|$ over the spherical surface of radius r:

$$P = \int_{\text{sphere}} |\vec{S}|\, dA = \frac{q^2\dot{v}^2}{4\pi c^3} \int_{\phi=0}^{2\pi} \int_{\theta=0}^{\pi} \frac{\sin^2\theta}{r^2}\, r\sin\theta\, d\theta\, r\, d\phi \tag{2.141}$$

$$= \frac{q^2\dot{v}^2}{2c^3} \int_{\theta=0}^{\pi} \sin^3\theta\, d\theta. \tag{2.142}$$

The integral $\int_{\theta=0}^{\pi} \sin^3\theta\, d\theta = 4/3$, so the total power emitted by the accelerated charged particle is

$$P = \frac{2}{3} \frac{q^2\dot{v}^2}{c^3}. \tag{2.143}$$

This result is called **Larmor's equation**. It implies that any charged particle radiates when accelerated and that the total radiated power is proportional to the square of the acceleration. The largest astrophysical accelerations are usually produced by electromagnetic forces, so the acceleration is proportional to the charge/mass ratio of the particle. In celestial sources, radiation from electrons is typically $(m_p/m_e)^2 \approx 4 \times 10^6$ times stronger than radiation from protons.

Larmor's extremely useful equation will be the basis for our derivations of radiation from a short dipole antenna as well as for free–free and synchrotron emission from astrophysical sources. Beware that *Larmor's formula is nonrelativistic*; it is valid only in frames moving at velocities $v \ll c$ with respect to the radiating particle. To treat particles moving at nearly the speed of light in the observer's frame, we must first use Larmor's equation to calculate the radiation in the particle's rest frame and then transform the result to the observer's frame in a relativistically correct way. Also, *Larmor's formula does not incorporate the constraints of quantum mechanics*, so it should be applied only with great caution to microscopic systems such as atoms. For example, Larmor's equation incorrectly predicts that the electron orbiting in the lowest energy level of a hydrogen atom will quickly radiate away all of its kinetic energy and fall into the nucleus. On the other hand, it correctly predicts the radio power emitted by an electron orbiting in a very high energy level of a hydrogen atom.

2.8 DUST EMISSION AT RADIO WAVELENGTHS

All small solid particles in space are called **dust grains** by astronomers. **Interstellar dust** was first recognized because it scatters or absorbs ultraviolet, visible, and near-infrared photons from stars. **Extinction** is defined as the total dimming of the light coming directly from a point source caused by both photon scattering and photon absorption, and it is usually expressed in units of mag $= 0.4$ dex. The amount of extinction is inversely proportional to wavelength for $0.1\,\mu$m $\leq \lambda \leq 7\,\mu$m, indicating that the grains responsible have sizes $a \leq \lambda/(2\pi) \leq 1\,\mu$m. Interstellar dust grains are much smaller than common terrestrial dust particles visible to the naked eye, and they are much smaller than the wavelengths $\lambda > 300\,\mu$m (frequencies $\nu < 1000$ GHz) accessible to ground-based radio astronomy (Figure 1.3).

Dust scattering occurs because the oscillating electric field of incident radiation forces electrons within dust grains to oscillate and hence reradiate at the same frequency in all directions, as shown in Section 2.7. In the radio limit $\lambda \gg a$ these radiators are very inefficient, so their scattering cross section is proportional to λ^{-4} (Rayleigh's law). Rayleigh's law applies to all small grains, regardless of composition. Scattering in the limit $a \ll \lambda$ is called **Rayleigh scattering**, and Rayleigh scattering of sunlight by air molecules is why the sky is so blue and the setting Sun is so red. Although photon scattering by dust *in front of* a star makes the star appear both dimmer and redder in color, scattering by dust *within* an extended object such as an external galaxy affects neither its total brightness nor its color. Only the absorption component of its internal dust extinction can redden or dim an external galaxy.

Some of the incident photons are absorbed, and their energy heats the dust grains. The energy carried by a single ultraviolet photon can significantly raise the temperature of a very small ($a \ll 1\,\mu$m) dust grain, so the smallest grains are not in thermodynamic equilibrium with the local interstellar radiation field. Larger interstellar dust grains have enough heat capacity to come into equilibrium at well-defined temperatures $20 < T_d(\text{K}) < 200$ such that the power absorbed is balanced by

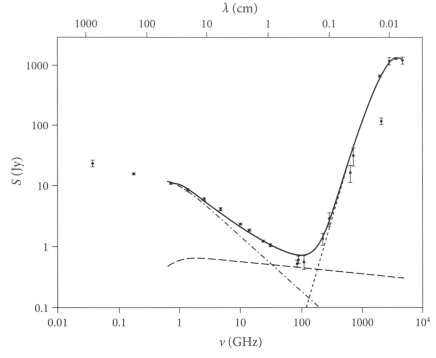

Figure 2.24. The radio and far-infrared spectrum of the nearby starburst galaxy M82 [24]. The contribution of free–free emission is indicated by the nearly horizontal dashed line. Synchrotron radiation (dot-dash line) and thermal dust emission (dots) dominate at low and high frequencies, respectively. Free–free absorption from HII regions distributed throughout the galaxy absorbs some of the synchrotron radiation and flattens the overall spectrum at the lowest frequencies. Color Plate 13 shows a radio image of M82.

power reemitted primarily at far-infrared (FIR) wavelengths $\lambda \sim 100\,\mu$m. The nearly equal UV/optical and FIR peaks in the cosmic background (Figure 1.4) imply that about half of all UV/optical starlight escapes and half is absorbed within its parent galaxy. This result is an average over all galaxies, but individual galaxies can range from nearly transparent (e.g., elliptical galaxies with little interstellar dust) to nearly opaque (e.g., the compact ultraluminous starburst galaxy Arp 220).

The dust absorption cross section varies as λ^{-2} when $a \ll \lambda$, so at radio wavelengths (1) dust absorption ($\propto \lambda^{-2}$) contributes much more than dust scattering ($\propto \lambda^{-4}$) to the total extinction, and (2) all galaxies are nearly transparent. Thus radio (and far-infrared $\lambda > 30\,\mu$m) observations can see into dusty starburst galaxies, and the radio/FIR luminosity is nearly proportional to the recent rate of star formation, unaffected by dust obscuration.

Obeying Kirchhoff's law for opaque bodies (Equation 2.47), small dust grains are also inefficient emitters. Their absorption and emission coefficients both scale with frequency as $\epsilon(\nu) \propto \nu^\beta$, where $1 < \beta < 2$ and $\beta \to 2$ in the Rayleigh limit. Thus dusty radio sources do not have Rayleigh–Jeans radio spectra $S_\nu \propto \nu^2$; they have relatively "blue" spectra rising as $S_\nu \propto \nu^{2+\beta}$. The dotted curve in Figure 2.24 showing dust emission from the galaxy M82 indicates $S_\nu \propto \nu^3$ to ν^4 for $\nu < 1000$ GHz.

A striking consequence of such a steeply rising spectrum is that the flux density of a galaxy observed through the $\lambda \sim 1$ mm ($\nu \sim 300$ GHz) atmospheric window (Figure 1.3) is nearly independent of its redshift in the range $1 < z < 10$. The strongest submm galaxies tend to be the most luminous, and radio surveys at wavelengths near $\lambda = 1$ mm are effective at tracing the rate of star formation over the entire redshift range during which most stars formed [10].

Astronomical dust is also present in **protoplanetary disks**. These "dust" particles are much larger than interstellar dust grains, so they can be strong radio emitters with nearly blackbody spectra at $\lambda \sim 1$ mm. The spectacular ALMA (Color Plate 5) image of the dust disk surrounding the young star HL Tau (Color Plate 9) has sufficient angular resolution (≈ 0.035 arcsec) to show dark rings carved out by young planets [1].

3 Radio Telescopes and Radiometers

3.1 ANTENNA FUNDAMENTALS

An **antenna** is a passive device that converts electromagnetic radiation in space into electrical currents in conductors or vice versa, depending on whether it is being used for receiving or for transmitting, respectively. Radio telescopes are receiving antennas, and radar telescopes are also transmitting antennas. It is often easier to calculate the properties of transmitting antennas and to measure the properties of receiving antennas. Fortunately, most characteristics of a transmitting antenna (e.g., its radiation pattern) are unchanged when that antenna is used for receiving, so any analysis of a transmitting antenna can be applied to a receiving antenna used in radio astronomy, and any measurement of a receiving antenna can be applied to that antenna when used for transmitting.

3.1.1 Radiation from a Short Dipole Antenna (Hertz Dipole)

The simplest antenna is a short (total length l much smaller than one wavelength λ) **dipole antenna**, which is shown in Figure 3.1 as two collinear conductors (e.g., wires or conducting rods). When they are driven at the small gap between them by an oscillating current source (a transmitter), the current going into the bottom conductor is 180 degrees out of phase with the current going into the top conductor. The radiation from a dipole depends on the transmitter frequency, so consider a sinusoidal driving current I with **angular frequency** $\omega \equiv 2\pi \nu$:

$$I = I_0 \cos(\omega t), \tag{3.1}$$

where I_0 is the peak current going into each half of the dipole. It is computationally convenient to replace the trigonometric function $\cos(\omega t)$ with its complex exponential equivalent (Appendix B.3), the real part of

$$\boxed{e^{-i\omega t} = \cos(\omega t) - i \sin(\omega t),} \tag{3.2}$$

so the driving current can be rewritten as

$$I = I_0 e^{-i\omega t} \tag{3.3}$$

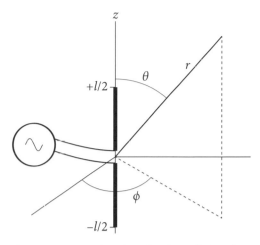

Figure 3.1. The coordinate system used to describe the radiation from a short (total length $l \ll \lambda$) dipole driven by a current source of frequency ν.

with the implicit understanding that only the real part of I represents the actual current. The driving current accelerates charges in the antenna wires, so Larmor's formula can be used to calculate the radiation from the antenna by converting from charges and accelerations to time-varying currents.

Example. Estimate the speed of electrons flowing through a copper wire of cross section $\sigma = 1 \, \text{mm}^2 = 10^{-6} \, \text{m}^2$ and carrying a current of 1 ampere.

The number density of free electrons is about equal to the number density of copper atoms in the wire, $n \approx 10^{29} \, \text{m}^{-3}$. In MKS units, the charge of an electron is

$$-e \approx 4.80 \times 10^{-10} \, \text{statcoul} \times \frac{1 \, \text{coul}}{3 \times 10^9 \, \text{statcoul}} \approx 1.60 \times 10^{-19} \, \text{coul.}$$

One ampere is defined as one coulomb per second, so the number of electrons flowing past any point along the wire in one second is

$$\dot{N} = \frac{I}{|e|} \approx \frac{1 \, \text{coul s}^{-1}}{1.60 \times 10^{-19} \, \text{coul}} \approx 6.25 \times 10^{18} \, \text{s}^{-1}.$$

The average electron velocity is only

$$v \approx \frac{\dot{N}}{\sigma n} \approx \frac{6.25 \times 10^{18} \, \text{s}^{-1}}{10^{-6} \, \text{m}^2 \cdot 10^{29} \, \text{m}^{-3}} \approx 6 \times 10^{-5} \, \text{m s}^{-1} \ll c.$$

Thus the nonrelativistic Larmor equation may safely be used to calculate the radiation from a wire.

The **electric current** in a wire is defined as the flow rate of electric charge along the wire:

$$I \equiv \frac{dq}{dt}. \tag{3.4}$$

For a wire on the z-axis,

$$I = \frac{dq}{dt} = \frac{dq}{dz}\frac{dz}{dt} = \frac{dq}{dz}v, \tag{3.5}$$

where v is the instantaneous flow velocity of the charges.

Many people incorrectly believe that the velocity v of individual electrons in a wire is comparable with the speed of light c because electrical *signals* do travel down wires at nearly the speed of light. However, a wire filled with electrons is like a garden hose already filled with water, a nearly incompressible fluid. When the faucet is turned on, water flows from the other end of a full hose almost immediately, even though individual water molecules have moved only a short distance along the hose. The example above shows that electrons move so slowly in a wire that Larmor's nonrelativistic equation accurately predicts the radiation from antennas.

Equation 2.136 from the derivation of Larmor's formula

$$E_\perp = \frac{q\dot{v}\sin\theta}{rc^2}$$

can be applied to yield the $dE_\perp$ contributed by each infinitesimal dipole segment of length dz. If the dipole is short ($l \ll \lambda$), all of these electric fields are in phase and add directly to give the total $E_\perp$ produced by the dipole:

$$E_\perp = \int_{z=-l/2}^{+l/2} \frac{dq}{dz}dz\frac{\dot{v}\sin\theta}{rc^2}. \tag{3.6}$$

At distances $r \gg l$, $(1/r)$ is nearly constant over the whole antenna and can be taken outside the integral. For a sinusoidal driving current, $\dot{v} = -i\omega v$ and

$$E_\perp = \frac{-i\omega\sin\theta}{rc^2}\int_{-l/2}^{+l/2}\frac{dq}{dz}v\,dz = \frac{-i\omega\sin\theta}{rc^2}\int_{-l/2}^{+l/2}I\,dz. \tag{3.7}$$

That is, the radiated electric field strength $E_\perp$ is proportional to the integral of the current distribution along the antenna. The current at the center is the driving current $I = I_0 e^{-i\omega t}$, and the current must drop to zero at the ends of the antenna, where the conductivity goes to zero. The current distribution along a short dipole is the tail end of a standing-wave sinusoid, which declines almost linearly from the driving current at the center to zero at the ends:

$$I(z) \approx I_0 e^{-i\omega t}\left[1 - \frac{|z|}{(l/2)}\right]. \tag{3.8}$$

Then

$$\int_{-l/2}^{+l/2} I\,dz \approx \frac{I_0 l}{2}e^{-i\omega t} \tag{3.9}$$

and

$$E_\perp \approx \frac{-i\omega \sin\theta}{rc^2} \frac{I_0 l}{2} e^{-i\omega t}. \tag{3.10}$$

Substituting $\omega = 2\pi c/\lambda$ gives

$$E_\perp \approx \frac{-i2\pi c \sin\theta}{\lambda rc^2} \frac{I_0 l}{2} e^{-i\omega t} = \frac{-i\pi \sin\theta}{c} \frac{I_0 l}{\lambda} \frac{e^{-i\omega t}}{r}. \tag{3.11}$$

The time-averaged Poynting flux (power per unit area) follows from Equation 2.139; it is

$$\langle S \rangle = \frac{c}{4\pi} \langle E_\perp^2 \rangle. \tag{3.12}$$

Thus

$$\langle S \rangle = \frac{c}{4\pi} \left(\frac{1}{2}\right) \left(\frac{I_0 l}{\lambda} \frac{\pi}{c}\right)^2 \frac{\sin^2\theta}{r^2}, \tag{3.13}$$

where the factor $(1/2)$ reflects the fact that $\langle \sin^2(\omega t) \rangle = \langle \cos^2(\omega t) \rangle = 1/2$. (This is a good relation to remember and an easy one to derive because $\sin^2(\omega t) + \cos^2(\omega t) = 1$ and $\langle \sin^2(\omega t) \rangle = \langle \cos^2(\omega t) \rangle$.)

The **power pattern** of a transmitting antenna is the angular distribution of its radiated power, often normalized to unity at the peak. From Equation 3.13 the normalized power pattern of a short dipole is

$$\boxed{P \propto \sin^2\theta.} \tag{3.14}$$

The radiation from a short dipole has the same polarization and the same doughnut-shaped power pattern as Larmor radiation from an accelerated charge because all of the charges in the short dipole are being accelerated along one line much shorter than one wavelength. From the observer's point of view, the power received depends only on the projected (perpendicular to the line of sight) length $l \sin\theta$ of the dipole. The electric field strength received is proportional to the *apparent* length of the dipole, and the radiation from the dipole is linearly polarized parallel to the projected dipole. The time-averaged total power emitted is obtained by integrating the Poynting flux over the surface area of a sphere of any radius $r \gg l$ centered on the antenna:

$$\langle P \rangle = \int \langle S \rangle dA = \frac{c}{4\pi} \left(\frac{1}{2}\right) \left(\frac{I_0 l}{\lambda} \frac{\pi}{c}\right)^2 \int_{\phi=0}^{2\pi} \int_{\theta=0}^{\pi} \frac{\sin^2\theta}{r^2} r \sin\theta \, d\phi \, r \, d\theta \tag{3.15}$$

$$= \frac{c}{4\pi} \left(\frac{1}{2}\right) \left(\frac{I_0 l}{\lambda} \frac{\pi}{c}\right)^2 2\pi \int_{\theta=0}^{\pi} \sin^3\theta \, d\theta. \tag{3.16}$$

Recall that $\int_0^\pi \sin^3\theta \, d\theta = 4/3$, so the time-averaged power radiated by a short dipole is

$$\boxed{\langle P \rangle = \frac{\pi^2}{3c} \left(\frac{I_0 l}{\lambda}\right)^2,} \tag{3.17}$$

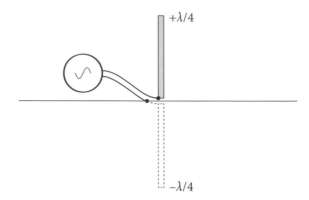

Figure 3.2. A ground-plane vertical antenna is just half of a dipole above a conducting plane. The lower half of the dipole is the reflection of the vertical in the mirror provided by the conducting "ground plane." The image vertical is 180 degrees out of phase with the real vertical. Above the ground plane, the radiation from the ground-plane vertical is exactly the same as the radiation from the dipole.

where $I_0 \cos(\omega t)$ is the driving current and l/λ is the total length of the dipole in wavelengths.

Most practical dipoles are **half-wave dipoles** ($l \approx \lambda/2$) because half-wave dipoles are **resonant**, meaning that they provide a nearly resistive load to the transmitter. When each half of the dipole is $\lambda/4$ long, the standing-wave current is highest at the center and naturally falls as $I = I_0 \cos(2\pi z/\lambda)$ to zero at the ends of the conductors.

The **ground-plane vertical** antenna shown in Figure 3.2 is very similar to the dipole. A ground-plane vertical is one half of a dipole above a conducting plane, which is called a "ground plane" because historically the conducting plane for vertical antennas was the surface of the Earth. The transmitter is connected between the base of the vertical, which is insulated from the ground, and the ground plane near the base. Many AM broadcast transmitting antennas are tall (at $\nu \sim 1\,\mathrm{MHz}$, $\lambda \sim 300\,\mathrm{m}$ and a $\lambda/4$ vertical antenna is about $75\,\mathrm{m}$ high), insulated towers acting as quarter-wave verticals. The conducting ground plane is a mirror that creates the lower half of the dipole as the mirror image of the upper half. Electric fields produced by the vertical antenna induce currents in the conducting plane to make the horizontal component of the electric field go to zero on the conductor. The virtual electric fields from the image vertical have the same amplitude but are 180 degrees out of phase, exactly as in a half-wave dipole. Consequently the radiation field from a ground-plane vertical is identical to that of a dipole in the half space above the ground plane and zero below the ground plane.

According to the strict definition of an antenna as a device for converting between electromagnetic waves in space and currents in conductors, the only antennas in most radio telescopes are half-wave dipoles and their relatives, quarter-wave ground-plane verticals. The large parabolic reflector of a radio telescope serves only to focus plane waves onto the **feed antenna**. (The term "feed" comes from radar antennas used for transmitting; the "feed" antenna feeds transmitter power to the main reflector. Receiving antennas used in radio astronomy work the other way around, and the "feed" actually collects radiation from the reflector.)

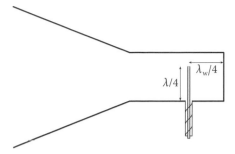

Figure 3.3. Most high-frequency feeds are quarter-wave ground-plane verticals inside waveguide horns. The only true antenna in this figure is the $\lambda/4$ ground-plane vertical, which converts electromagnetic waves in the waveguide to currents in the coaxial cable extending down from the waveguide.

Actual half-wave dipoles, backed by small reflectors about $\lambda/4$ behind them to focus the dipole pattern in the direction of the main dish, are normally used as feeds at low frequencies ($\nu < 1\,\mathrm{GHz}$) or long wavelengths ($\lambda > 0.3\,\mathrm{m}$) because of their relatively small size. However, the radiation patterns of half-wave dipoles backed by small reflectors are not well matched to most parabolic dishes, so their performance is less than optimum.

For shorter wavelengths, almost all radio-telescope feeds are quarter-wave ground-plane verticals inside **waveguide horns**. Radiation entering the relatively large (size $> \lambda$) rectangular or circular aperture of the tapered horn is concentrated into a rectangular or circular waveguide with parallel conducting walls. In the case of the rectangular waveguide whose cross section is shown in Figure 3.3, the side walls are separated by slightly over $\lambda/2$ so that vertical electric fields can travel down the waveguide with low loss. The top and bottom walls are separated by somewhat less than $\lambda/2$ so only the mode with vertical electric fields can propagate (Section 3.4). The $\lambda/4$ vertical antenna inserted through a small hole in the bottom wall collects most of this vertically polarized radiation and converts it into an electric current that travels down the coaxial cable to the receiver. The backshort wall about 1/4 of the guide wavelength λ_{w} (Equation 3.144) behind the dipole ensures that the dipole sees only radiation coming from the direction of the horn opening.

Both dipoles and quarter-wave verticals are **linearly polarized feeds**. The voltage response of a linearly polarized feed to a linearly polarized source is proportional to $\cos\Delta$, where Δ is the angle between the feed and the source electric field, and the power response is proportional to $\cos^2\Delta = [\cos(2\Delta) + 1]/2$. Consequently the degree of polarization and the polarization position angle of a partially linearly polarized radio source can be measured by rotating the linearly polarized feed of a radio telescope while tracking the source. The degree of polarization p of a partially linearly polarized source defined by Equation 2.58 is

$$p \equiv \frac{I_{\mathrm{p}}}{I} = \frac{I_{\mathrm{p}}}{I_{\mathrm{p}} + I_{\mathrm{u}}}, \tag{3.18}$$

where I_{p} is the polarized flux density and $I = I_{\mathrm{p}} + I_{\mathrm{u}}$ is the total flux density of the source. The power response of the feed $R(\Delta) \propto I_{\mathrm{p}} \cos^2\Delta + I_{\mathrm{u}}/2$ will be $R_{\parallel} \propto I_{\mathrm{p}} + I_{\mathrm{u}}/2$ when the feed and source polarizations are parallel ($\Delta = 0$) and $R_{\perp} = I_{\mathrm{u}}/2$

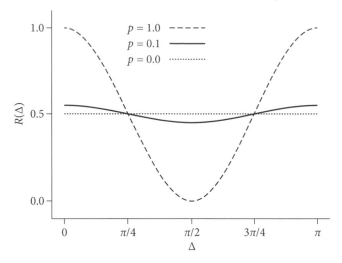

Figure 3.4. The relative power output from a linearly polarized antenna as a function of the polarization position angle difference between the source and the antenna for sources with fractional polarizations $p = 1.0$ (dashed curve), $p = 0.1$ (solid curve), and $p = 0.0$ (dotted line). Abscissa: Position angle difference Δ (rad). Ordinate: Relative power output R.

when they are perpendicular ($\Delta = \pi/2$). In terms of the observables $R_\parallel$ and $R_\perp$, the degree of polarization is

$$p = \frac{I_p}{I_p + I_u} = \frac{I_p + I_u/2 - I_u/2}{I_p + I_u/2 + I_u/2} = \frac{R_\parallel - R_\perp}{R_\parallel + R_\perp}. \tag{3.19}$$

Figure 3.4 shows how the relative power output $R(\Delta)$ of a linearly polarized feed varies as it is rotated through $\Delta = \pi$ radians relative to the polarization position angle of sources with fractional polarizations $p = 1.0, 0.1$, and 0.0.

To measure all four Stokes parameters of an arbitrarily polarized source, it is necessary to combine the voltage outputs of two orthogonally polarized feeds. For example, two orthogonal quarter-wave verticals can be inserted into a square waveguide to receive both the horizontally and the vertically polarized components simultaneously. If their output voltages are added in phase (phase difference $\delta = \phi_x - \phi_y = 0$ in Figure 2.15), the feed combination will respond to radiation linearly polarized in position angle $\pi/4 = 45°$. If a phase difference $\delta = \pi/2$ is inserted either mechanically (by moving one feed $\lambda/4$ behind the other) or electrically (by inserting a $\lambda/4$ longer cable between one feed and the point where the two outputs are added), then $\delta = \pi/2$ and the feed combination will respond to circular polarization.

3.1.2 Radiation Resistance

The power flowing through a circuit is

$$\boxed{P = VI,} \tag{3.20}$$

where V is the **voltage** (defined as energy per unit charge) and I is the **current** (defined as the charge flowing through the circuit per unit time), so P has dimensions of energy per unit time. The physicist George Simon Ohm observed that the current flowing through most (but not all) materials is proportional to the applied voltage, so most objects have a well-defined **resistance** R defined by **Ohm's law**,

$$R \equiv \frac{V}{I}. \tag{3.21}$$

When Ohm's law holds,

$$P = I^2 R = \frac{V^2}{R}. \tag{3.22}$$

The average power in a resistive circuit with time-varying currents is

$$\langle P \rangle = \langle I^2 \rangle R. \tag{3.23}$$

In the particular case of sinusoidal currents $I = I_0 \cos(\omega t)$ and $\langle I^2 \rangle = I_0^2/2$, so

$$\langle P \rangle = \frac{I_0^2 R}{2}. \tag{3.24}$$

Thus the (frequency-dependent) **radiation resistance** of an antenna is defined by

$$R \equiv \frac{2\langle P \rangle}{I_0^2}. \tag{3.25}$$

For a short dipole, the power emitted is given by Equation 3.17 and the radiation resistance is

$$R = \frac{2\pi^2}{3c} \left(\frac{l}{\lambda} \right)^2. \tag{3.26}$$

A ground-plane vertical of height $l/2$ emits exactly like a dipole of length l above the ground plane and nothing below the ground plane. Thus the total power emitted by the vertical is half the power emitted by the dipole, and the radiation resistance of the vertical is half the radiation resistance of the dipole.

Example. A "half-wave" (length $l = \lambda/2$) dipole is a resonant antenna. Resonant antennas are used in most real applications because the impedance of a resonant antenna is resistive; nonresonant antennas have large capacitive or inductive reactances as well. Most of the antenna current in a half-wave dipole is a standing wave with a current distribution $I = I_0 e^{-i\omega t} \cos(2\pi z/\lambda)$ that has a maximum $I = I_0 e^{-i\omega t}$ at the $z = 0$ feed point and declines co-sinusoidally to zero at the endpoints $z = \pm\lambda/4$. With the (no longer so accurate) assumption that the radiation from all parts of the dipole emit in phase, Equation 3.7 still holds:

$$E_\perp = \frac{-i\omega \sin\theta}{rc^2} \int_{-l/2}^{+l/2} I \, dz.$$

(Continued)

The current distribution of the half-wave dipole is

$$I = I_0 e^{-i\omega t} \cos\left(\frac{2\pi z}{\lambda}\right)$$

so

$$\int_{-l/2}^{+l/2} I\, dz \approx I_0 e^{-iwt} \int_{-\lambda/4}^{+\lambda/4} \cos\left(\frac{2\pi z}{\lambda}\right) dz = \frac{I_0 \lambda}{\pi} e^{-i\omega t},$$

which is a factor of $2\lambda/(\pi l)$ larger than the comparable integral for a short dipole (Equation 3.8):

$$\int_{-l/2}^{+l/2} I\, dz \approx \frac{I_0 l}{2} e^{-iwt}.$$

The average power $\langle P \rangle$ radiated by a given I_0 is proportional to the square of this factor (see Equations 3.10 through 3.17), or

$$\left(\frac{2\lambda}{\pi l}\right)^2.$$

Thus the radiation resistance R of a half-wave dipole is the radiation resistance of a short dipole (Equation 3.26) multiplied by the factor squared:

$$R = \left[\frac{2\pi^2}{3c}\left(\frac{l}{\lambda}\right)^2\right]\left(\frac{2\lambda}{\pi l}\right)^2 = \frac{8}{3c}$$

$$= \frac{8}{3 \cdot 3 \times 10^{10}\ \text{cm s}^{-1}} \approx \frac{8}{9} \times 10^{-10}\ \text{s cm}^{-1}.$$

Engineers and real test instruments use the MKS "ohm" (symbol Ω) as the unit of resistance. The conversion factor is $1\ \Omega = (10^{-11}/9)\ \text{s cm}^{-1}$, so

$$R = \left(\frac{8}{9} \times 10^{-10}\ \text{s cm}^{-1}\right) \Big/ \left(\frac{1}{9} \times 10^{-11}\ \text{s cm}^{-1}\ \text{ohm}^{-1}\right) = 80\ \Omega.$$

This is pretty close to the $R \approx 73\ \Omega$ result from an exact calculation that doesn't use the $l \ll \lambda$ approximation.

The **radiation resistance R_0 of free space** (sometimes called the impedance Z_0 of free space) can be obtained from the relations

$$|\vec{S}| = \frac{c}{4\pi} E^2 \quad \text{and} \quad P = \frac{V^2}{R}. \tag{3.27}$$

The electric field E is just the voltage per unit length V/l and the flux is the power per unit area l^2, so

$$|\vec{S}| = \frac{c V^2}{4\pi l^2} = \frac{V^2}{R_0 l^2} \tag{3.28}$$

and

$$R_0 = \frac{4\pi}{c} = \frac{4\pi}{3 \times 10^{10} \text{ cm s}^{-1}} = 4.19 \times 10^{-10} \text{ s cm}^{-1}. \tag{3.29}$$

Converting from CGS to MKS units yields the radiation resistance of space in ohms:

$$R_0 = \frac{4\pi}{3 \times 10^{10} \text{ cm s}^{-1} \cdot 1/9 \times 10^{-11} \text{ s cm}^{-1} \, \Omega^{-1}} = 120\pi \ \Omega \approx 377 \ \Omega. \tag{3.30}$$

The tapered opening of a waveguide horn feed (Figure 3.3) acts as an impedance transformer to match the impedance of the waveguide to the impedance of free space to minimize standing waves and couple power efficiently between the waveguide and space, just as the bell of a trombone is an acoustic transformer matching sound vibrations of air in the trombone to the outside environment.

A black hole is a perfect absorber of radiation, so its resistance must also be $120\pi \ \Omega$ to match that of free space. A black hole spinning in an external magnetic field can generate electrical power with a voltage/current ratio of $120\pi \ \Omega$, and this process may be important in powering quasar jets [12].

3.1.3 The Power Gain of a Transmitting Antenna

The **power gain** $G(\theta, \phi)$ of a transmitting antenna is defined as the power transmitted per unit solid angle in direction (θ, ϕ) relative to an **isotropic antenna**, which has the same gain in all directions. Frequently, the value of G is expressed logarithmically in units of **decibels (dB)**:

$$\boxed{G(\text{dB}) \equiv 10 \log_{10}(G).} \tag{3.31}$$

For any lossless antenna, energy conservation requires that the gain averaged over all directions be

$$\boxed{\langle G \rangle = 1.} \tag{3.32}$$

Consequently, all lossless antennas obey

$$\boxed{\int_{\text{sphere}} G \, d\Omega = 4\pi.} \tag{3.33}$$

Different lossless antennas may radiate with different directional patterns, but they do not alter the total amount of power radiated. Consequently, the gain of a lossless antenna depends only on the angular distribution of radiation from that antenna. In general, an antenna having peak gain G_0 must beam most of its power into a solid angle $\Delta\Omega$ such that $\Delta\Omega \approx 4\pi/G_0$. This motivates the definition of the **beam solid angle** Ω_A:

$$\Omega_A \equiv \frac{4\pi}{G_0}. \tag{3.34}$$

Thus the higher the gain, the smaller the beam solid angle.

Example. What is the power gain of a lossless short dipole? It is sufficient to recall only the angular dependence of the short-dipole power pattern (Equation 3.14)

$$P \propto \sin^2 \theta,$$

where θ is the angle from the dipole axis. Thus

$$G \propto \sin^2 \theta = G_0 \sin^2 \theta.$$

The maximum gain G_0 is determined by energy conservation:

$$\int_{\text{sphere}} G \, d\Omega = \int_{\phi=0}^{2\pi} \int_{\theta=0}^{\pi} G_0 \sin^2 \theta \, d\theta \, \sin\theta \, d\phi = 4\pi,$$

$$2\pi G_0 \int_0^{\pi} \sin^3 \theta \, d\theta = 4\pi.$$

Recall that $\int_0^{\pi} \sin^3 \theta \, d\theta = 4/3$ so

$$G_0 = \frac{4\pi}{2\pi} \frac{3}{4} = \frac{3}{2}$$

and

$$G(\theta, \phi) = \frac{3 \sin^2 \theta}{2}.$$

Expressed in dB, the maximum gain G_0 of a short dipole is

$$G_0 = 10 \log_{10}(3/2) \approx 1.76 \, \text{dB}.$$

Note that $G(\theta, \phi)$ is nearly independent of the antenna length so long as $l \ll \lambda$ because the power pattern of a short dipole is nearly independent of l. Varying $l \ll \lambda$ affects only the radiation resistance.

The **antenna efficiency** η is defined as the ratio of input power to radiated power. If ohmic losses reduce η, then the gain G in Equations 3.31 through 3.34 should be replaced by the **directivity** defined by $D \equiv G/\eta$.

3.1.4 The Effective Area of a Receiving Antenna

The receiving counterpart of transmitting power gain is the **effective area** or **effective collecting area** of a receiving antenna. Imagine an ideal antenna of geometric area A that could collect all of the radiation falling on it from a distant point source and convert it to electrical power—a "rain gauge" for collecting photons. The total spectral power incident on the antenna is the product of its geometric area and the incident spectral power per unit area, or flux density S_ν. However, any single antenna can respond to only one polarization, so its output P_ν can equal all of the input spectral power ($P_\nu = A S_\nu$) from a fully polarized source whose polarization matches that of the antenna, but only half of the incident power ($P_\nu = A S_\nu / 2$) from

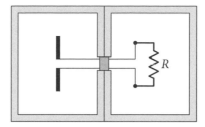

Figure 3.5. A cavity in thermodynamic equilibrium at temperature T containing a resistor R is coupled to an antenna, also at temperature T, through a filter blocking electromagnetic radiation but passing currents having frequencies in the range ν to $\nu + d\nu$.

an unpolarized source and nothing at all from an orthogonally polarized source. The output of a real antenna is always smaller than this and most radio sources are nearly unpolarized, so radio astronomers find it useful to define the *effective* collecting area A_e of an antenna whose output spectral power is P_ν in response to an *unpolarized* point source of total flux density S_ν by

$$A_e \equiv \frac{2 P_\nu}{S_\nu}. \tag{3.35}$$

The **average collecting area**

$$\langle A_e \rangle \equiv \frac{\int_{4\pi} A_e \, d\Omega}{\int_{4\pi} d\Omega} = \frac{1}{4\pi} \int_{4\pi} A_e \, d\Omega \tag{3.36}$$

of any lossless antenna can be calculated via another thermodynamic thought experiment. Imagine an antenna inside a cavity in full thermodynamic equilibrium at temperature T connected through a transmission line to a matched resistor (whose resistance equals the radiation resistance of the antenna) in a second cavity at the same temperature (Figure 3.5). A filter between the cavities passes only currents in a narrow range of frequencies between ν and $\nu + d\nu$. Because this entire system is in thermodynamic equilibrium, no net power can flow through the wires connecting the antenna and the resistor. Otherwise, one cavity would heat up and the other would cool down, in violation of the second law of thermodynamics. The total spectral power P_ν from all directions collected in one polarization is half the total spectral power in the unpolarized blackbody radiation, so

$$P_\nu = \frac{1}{2} \int_{4\pi} A_e(\theta, \phi) B_\nu \, d\Omega \tag{3.37}$$

must equal the Nyquist spectral power P_ν produced by the resistor. Inserting the Nyquist formula from Equation 2.119 and Planck's law from Equation 2.85,

$$P_\nu = kT \left[\frac{\dfrac{h\nu}{kT}}{\exp\left(\dfrac{h\nu}{kT}\right) - 1} \right] \quad \text{and} \quad B_\nu = \frac{2kT}{\lambda^2} \left[\frac{\dfrac{h\nu}{kT}}{\exp\left(\dfrac{h\nu}{kT}\right) - 1} \right] \tag{3.38}$$

leads to

$$kT = \frac{2kT}{2\lambda^2} \int_{4\pi} A_e(\theta, \phi) \, d\Omega, \tag{3.39}$$

and finally,

$$\int_{4\pi} A_e(\theta, \phi) \, d\Omega = 4\pi \langle A_e \rangle = \lambda^2. \tag{3.40}$$

Without using Maxwell's equations we have obtained the remarkable result

$$\boxed{\langle A_e \rangle = \frac{\lambda^2}{4\pi}} \tag{3.41}$$

which implies that all lossless antennas, from tiny dipoles to the 100-m diameter Green Bank Telescope (GBT), have the same average collecting area. $\langle A_e \rangle$ is proportional to λ^2 because space has two more dimensions than a transmission line has.

The collecting area of an isotropic receiving antenna is proportional to λ^2, so most satellite broadcast services, GPS (Global Positioning System) or satellite FM radio for example, operate at relatively long wavelengths (10 to 20 cm). Likewise, practical radio telescopes constructed from arrays of dipoles are reasonably sensitive only at long wavelengths.

By analogy with Equation 3.34, the **beam solid angle** of a lossless receiving antenna is defined as

$$\Omega_A \equiv \int_{4\pi} \frac{A_e(\theta, \phi)}{A_0} \, d\Omega, \tag{3.42}$$

where A_0 is the maximum effective collecting area, so

$$\boxed{A_0 \, \Omega_A = \lambda^2.} \tag{3.43}$$

The much larger peak collecting area of the GBT implies it has a much smaller beam solid angle Ω_A.

3.1.5 Reciprocity Theorems

Many antenna properties are the same for both transmitting and receiving. It is often easier to *calculate* the gain of a transmitting antenna than the collecting area of a receiving antenna, and it is often easier to *measure* the receiving power pattern of a large radio telescope than to measure its transmitting power pattern. Thus this receiving/transmitting "reciprocity" greatly simplifies antenna calculations and measurements. Reciprocity can be understood via Maxwell's equations or by thermodynamic arguments.

Burke and Graham-Smith [20] state the electromagnetic case for **reciprocity** clearly: "An antenna can be treated either as a receiving device, gathering the incoming radiation field and conducting electrical signals to the output terminals, or as a transmitting system, launching electromagnetic waves outward. These two cases are equivalent because of time reversibility: the solutions of Maxwell's equations are valid when time is reversed."

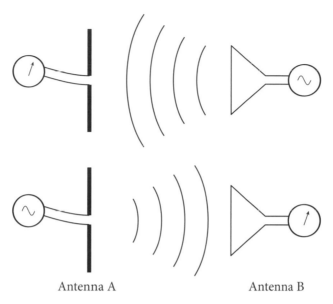

Figure 3.6. The strong reciprocity theorem implies that the transmitter voltages V_A and V_B are related to the receiver currents I_A and I_B by $I_B^{-1} V_A = I_A^{-1} V_B$ for any pair of antennas A and B.

The **strong reciprocity theorem** states,

If a voltage is applied to the terminals of an antenna A and the current is measured at the terminals of another antenna B, then an equal current (in both amplitude and phase) will appear at the terminals of A if the same voltage is applied to B. (Figure 3.6)

It can be formally derived from Maxwell's equations (see a partial derivation in Wilson et al. [116, Appendix D]) or by network analysis (see Kraus et al. [63, "Antennas", p. 252]).

Most radio astronomical applications do not depend on the detailed phase relationships of voltages and currents, so it is sufficient to use a **weak reciprocity theorem** that relates the angular dependences of the transmitting power pattern and the receiving collecting area of any antenna: "The power pattern of an antenna is the same for transmitting and receiving"; that is,

$$G(\theta, \phi) \propto A_e(\theta, \phi). \tag{3.44}$$

The weak reciprocity theorem can be proven by another simple thermodynamic thought experiment: An antenna is connected to a matched load inside a cavity initially in equilibrium at temperature T. The antenna simultaneously receives power from the cavity walls and transmits power generated by the resistor. The total power transmitted in all directions must equal the total power received from all directions because no net power can be transferred between the antenna and the resistor; otherwise the resistor would not remain at temperature T. Moreover, in any direction, the power received and transmitted by the antenna must be the same, else the cavity wall in directions where the transmitted power was greater than the

Example. Use the transmitting power pattern of a short dipole (Equation 3.14) to calculate the effective collecting area of a short dipole used as a receiving antenna:

$$A_e(\theta, \phi) = \frac{\lambda^2 G(\theta, \phi)}{4\pi} = \frac{\lambda^2}{4\pi} \frac{3\sin^2\theta}{2},$$

$$A_e = \frac{3\lambda^2 \sin^2\theta}{8\pi}.$$

The effective collecting area of a short receiving dipole does not depend on the length l of the dipole itself because the transmitting power pattern of a short ($l \ll \lambda$) dipole is independent of l.

received power would rise in temperature and the cavity wall in directions of lower transmitted/received power ratio would cool, leading to a violation of the second law of thermodynamics.

The constant of proportionality relating G and A_e can be derived from Equations 3.32 and 3.41:

$$\langle A_e \rangle = \frac{\lambda^2}{4\pi} \qquad \text{and} \qquad \langle G \rangle = 1. \tag{3.45}$$

Thus energy conservation and the weak reciprocity theorem imply

$$\boxed{A_e(\theta, \phi) = \frac{\lambda^2 G(\theta, \phi)}{4\pi}} \tag{3.46}$$

for any antenna. This extremely useful equation shows how to compute the receiving power pattern from the transmitting power pattern and vice versa.

3.1.6 Antenna Temperature

A convenient practical unit for the power output per unit frequency from a receiving antenna is the **antenna temperature** T_A. Antenna temperature has nothing to do with the physical temperature of the antenna as measured by a thermometer; it is only the temperature of a matched resistor whose thermally generated power per unit frequency *in the low-frequency Nyquist approximation* (Equation 2.117) equals that produced by the antenna:

$$\boxed{T_A \equiv \frac{P_\nu}{k}.} \tag{3.47}$$

It is widely used for the following reasons:

1. 1 K of antenna temperature is a conveniently small power per unit bandwidth. $T_A = 1$ K corresponds to $P_\nu = kT_A = 1.38 \times 10^{-23}$ J K$^{-1} \cdot 1$ K $= 1.38 \times 10^{-23}$ W Hz^{-1}.
2. It can be calibrated by a direct comparison with hot and cold **loads** (another word for matched resistors) connected to the receiver input.

3. The units of receiver noise are also K, so comparing the signal in K with the receiver noise in K makes it easy to compare the signal and noise powers.

Combining Equations 3.35 and 3.47 shows that an unpolarized point source of flux density S increases the antenna temperature by

$$T_A = \frac{P_\nu}{k} = \frac{A_e S}{2k},$$

(3.48)

where A_e is the effective collecting area. It is often convenient to express the point-source sensitivity of a radio telescope in units of "kelvins per jansky" rather than in units of effective collecting area (m^2). The effective collecting area corresponding to a sensitivity of $1 \, \mathrm{K \, Jy^{-1}}$ is

$$A_e = \frac{2k T_A}{S} = \frac{2 \cdot 1.38065 \times 10^{-23} \, \mathrm{J \, K^{-1} \cdot 1 \, K}}{10^{-26} \, \mathrm{W \, m^{-2} \, Hz^{-1}}} = 2761 \, \mathrm{m^2}.$$

(3.49)

In an arbitrary radiation field $I_\nu(\theta, \phi)$, Equation 3.37 becomes

$$P_\nu = \frac{1}{2} \int_{4\pi} A_e(\theta, \phi) \, I_\nu(\theta, \phi) \, d\Omega.$$

(3.50)

Replacing P_ν by antenna temperature T_A using Equation 3.47 and inserting $I_\nu(\theta, \phi) = 2k T_b(\theta, \phi)/\lambda^2$ (Equation 2.33) gives

$$k T_A = \frac{1}{2} \int_{4\pi} A_e(\theta, \phi) \, \frac{2k}{\lambda^2} T_b(\theta, \phi) \, d\Omega,$$

(3.51)

$$T_A = \frac{1}{\lambda^2} \int_{4\pi} A_e(\theta, \phi) \, T_b(\theta, \phi) \, d\Omega.$$

(3.52)

In the limit of a very extended source having nearly constant T_b across the entire beam,

$$T_A = \frac{T_b}{\lambda^2} \int_{4\pi} A_e(\theta, \phi) \, d\Omega$$

(3.53)

so

$$\boxed{T_A = T_b.}$$

(3.54)

In words, *the antenna temperature produced by a smooth source much larger than the antenna beam equals the source brightness temperature.*

If a lossless antenna is pointed at a compact source covering a solid angle Ω_s much smaller than the beam and having uniform brightness temperature T_b, then

$$T_A = \frac{A_0 T_b \Omega_s}{\lambda^2},$$

(3.55)

where A_0 is the on-axis effective collecting area. Substituting $A_0 \Omega_A = \lambda^2$ (Equation 3.43) gives the result:

$$\boxed{\frac{T_A}{T_b} = \frac{\Omega_s}{\Omega_A}.}$$

(3.56)

Stated in words, *the antenna temperature equals the source brightness temperature multiplied by the fraction of the beam solid angle filled by the source.* A $T_b = 10^4$ K source covering 1% of the beam solid angle will add 100 K to the antenna temperature. The ratio (Ω_s / Ω_A) is called the **beam filling factor**.

The **main beam** of an antenna is defined as the region containing the principal response out to the first zero; responses outside this region are called **sidelobes** or, very far from the main beam, **stray radiation**. The **main beam solid angle** Ω_{MB} is defined as

$$\Omega_{MB} \equiv \frac{1}{G_0} \int_{MB} G(\theta, \phi) \, d\Omega. \tag{3.57}$$

The fraction of the total beam solid angle lying inside the main beam is called the **main beam efficiency** or, loosely, the **beam efficiency**:

$$\eta_B \equiv \frac{\Omega_{MB}}{\Omega_A}. \tag{3.58}$$

3.2 REFLECTOR ANTENNAS

3.2.1 Paraboloidal Reflectors

Antennas useful for radio astronomy at short wavelengths must have collecting areas much larger than the $\lambda^2/(4\pi)$ collecting area of an isotropic antenna and much higher angular resolution than a short dipole provides. Because arrays of dipoles are impractical at wavelengths $\lambda < 1$ m or so, most radio telescopes use large **reflectors** to collect and focus power onto their small feed antennas, such as waveguide horns or dipoles backed by small reflectors, that are connected to receivers. The most common reflector shape is a paraboloid of revolution because it can focus the plane wave from a distant point source onto a single **focal point**.

To focus plane waves onto a single point, the reflector must keep all parts of an on-axis plane wavefront in phase at its focal point. Thus the total path lengths to the focus must all be the same, and this requirement is sufficient to determine the shape of the desired reflecting surface. Clearly the surface must be rotationally symmetric about its axis. In any plane containing the axis, the surface looks like the curve in Figure 3.7.

The requirement of constant path length can be written by equating the on-axis path length $(f + h)$ from any height h to the reflector and then back to the **prime focus** at height f with the off-axis path length:

$$(f + h) = \sqrt{r^2 + (f - z)^2} + (h - z). \tag{3.59}$$

This yields the reflector height z as a function of radius r:

$$\sqrt{r^2 + (f - z)^2} = f + z,$$

$$r^2 + f^2 + z^2 - 2fz = f^2 + z^2 + 2fz;$$

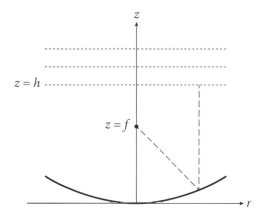

Figure 3.7. A plane containing the axis of a **paraboloidal reflector** with focal length f. Plane wave fronts from a distant point source are shown as dotted lines perpendicular to the z-axis. From a wavefront at height h above the **vertex** (the point $r = 0$, $z = 0$) of the paraboloid, the ray path (dashed line) lengths at all radial offsets r down to the reflector and up to the prime focus at $z = f$ must be equal.

the result is

$$z = \frac{r^2}{4f}. \tag{3.60}$$

This is the equation of a paraboloid with **focal length** f.

The ratio of the focal length f to the diameter D of the reflector is called the **f/D ratio** or **focal ratio**. Note that the gain, collecting area, and beamwidth of a reflector antenna depend only weakly and indirectly on f/D, via the effect of f/D on illumination taper. In principle, f/D is a free parameter for the telescope designer, but in practice it is constrained. If the reflector f/D is too high, the support structure needed to hold the feed or subreflector at the focus of a large radio telescope becomes very long and unwieldy. Consequently large radio telescopes usually have $f/D \approx 0.4$, an order-of-magnitude lower than the typical focal ratio of an optical telescope. The drawback of a low f/D is a small field of view. The **focal ellipsoid** is the volume around the exact focal point that remains in reasonably good focus, and the **focal circle** is defined by the intersection of the focal ellipse and the transverse plane at $z = f$. Ruze [95] showed that the angular radius of the focal circle is proportional to $(f/D)^2$, and only a small number (about seven) of discrete feeds can fit inside the focal circle of an $f/D \approx 0.4$ paraboloid. Large arrays of feeds or imaging cameras require larger f/D ratios, obtained either by using a shallower paraboloid or by using magnifying subreflectors to increase the effective focal length.

The primary mirrors of most radio telescopes are circular paraboloids or sections thereof for the following reasons:

1. The effective collecting area A_e of a reflector antenna can approach its projected geometric area $A = \pi D^2/4$.
2. They are electrically simple (compared with a phased array of dipoles, for example).

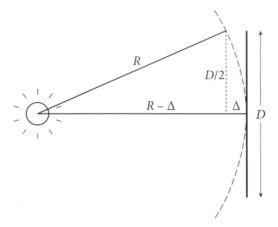

Figure 3.8. The spherical wavefront (dashed circle) emitted by a point source at distance R deviates from a plane by Δ at the edge of an aperture of diameter D.

3. A single reflector can work over a wide range of frequencies. Changing frequencies only requires changing the feed antenna and receiver located at the focal point, not building a whole new radio telescope.

3.2.2 The Far-Field Distance

How far away must a point source be for the received waves to satisfy the assumption that they are nearly planar across the reflector? The answer depends on both the wavelength λ and the reflector diameter D. Figure 3.8 shows the spherical wave emitted by a point source a finite distance R from a flat **aperture**, an imaginary circular hole that covers the reflector. It could be located at the plane $z = h$ shown in Figure 3.7, for example. The maximum departure Δ from a plane wave occurs at the edge of the aperture. The **far-field distance** R_{ff} is somewhat arbitrarily defined by requiring that $\Delta < \lambda/16$. At the aperture edge, the Pythagorean theorem gives

$$R^2 = (R - \Delta)^2 + \left(\frac{D}{2}\right)^2. \tag{3.61}$$

Thus

$$R = \frac{\Delta}{2} + \frac{D^2}{8\Delta}. \tag{3.62}$$

In the limit $\Delta \ll D$, we have $\Delta/2 \ll D^2/(8\Delta)$ and

$$R \approx \frac{D^2}{8\Delta}. \tag{3.63}$$

Given the $\Delta = \lambda/16$ criterion, the far-field distance is

$$\boxed{R_{ff} \approx \frac{2D^2}{\lambda}.} \tag{3.64}$$

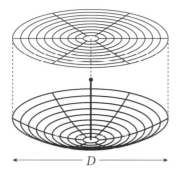

Figure 3.9. The aperture plane associated with a paraboloidal dish of diameter D.

Example. What is the far-field distance of the Green Bank Telescope ($D = 100$ m) observing at $\lambda = 1$ cm?

Equation 3.64 gives

$$R_{\mathrm{ff}} = \frac{2(100\,\mathrm{m})^2}{1\,\mathrm{cm}} = \frac{2 \times 10^4\,\mathrm{m}^2}{0.01\,\mathrm{m}} = 2 \times 10^6\,\mathrm{m} = 2000\,\mathrm{km}.$$

Such a large far-field distance makes ground-based measurements of the GBT antenna pattern impractical. To measure small errors in the GBT reflector surface using radio holography, it is necessary to observe a geostationary satellite having an orbital altitude $R \sim 36{,}000 \gg 2000$ km. Similarly, the easiest way to determine the transmitting power pattern for a large radar antenna such as the $D = 305$-m Arecibo reflector is to scan across a celestial point source in the far field and use the reciprocity theorem to equate the transmitting and receiving patterns.

If $R < R_{\mathrm{ff}}$, the path-length errors will introduce significant phase errors in the waves coming from the off-axis portions of the reflector, reducing the effective collecting area and degrading the antenna pattern.

3.2.3 Patterns of Aperture Antennas

In optics, the term **aperture** refers to the opening through which all rays pass. For example, the aperture of a paraboloidal reflector antenna would be the plane circle, normal to the rays from a distant point source, that just covers the paraboloid (Figure 3.9). The phase of the plane wave from a distant point source would be constant across the aperture plane when the aperture is perpendicular to the line of sight. Another example of an aperture is the mouth of a waveguide horn antenna (Figure 3.10). How can the **beam pattern**, or power gain as a function of direction, of an aperture antenna be calculated? For simplicity, first consider a one-dimensional aperture of width D (Figure 3.11) and calculate the **electric field pattern** at a distant ($R \gg R_{\mathrm{ff}}$) point.

When used in a transmitting antenna, the feed can **illuminate** the aperture antenna with a sine wave of fixed frequency $\nu = \omega/(2\pi)$ and electric field strength

Figure 3.10. "Doc" Ewen looking into the rectangular aperture of the horn antenna used to discover the λ = 21-cm line of neutral hydrogen. Image credit: NRAO/AUI/NSF.

$g(x)$ that varies across the aperture. The illumination induces currents in the reflector. The currents will vary with both position and time:

$$I \propto g(x) \exp(-i\omega t). \tag{3.65}$$

The constant of proportionality doesn't matter yet; it can be calculated later from energy conservation. **Huygens's principle** asserts that the aperture can be treated as a collection of small elements which act individually as small antennas. Huygens's principle actually applies to waves of any type, sound waves for example. The electric field produced by the whole aperture at large distances is just the vector sum of the elemental electric fields from these small antennas. The field from each element extending from x to $x + dx$ is

$$df \propto g(x) \frac{\exp(-i2\pi r(x)/\lambda)}{r(x)} dx, \tag{3.66}$$

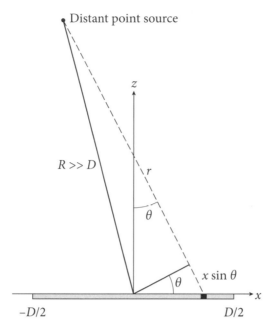

Figure 3.11. Coordinate system for a one-dimensional linear aperture spanning $-D/2 <$ $x < +D/2$. For a source distant from the center of the aperture ($R \gg D$), the *fractional* change in the distance r between the source and any aperture element at a distance $|x|$ from the aperture center is small, so the variable $r(x)$ in the denominator of Equation 3.66 can be replaced by the constant R. However, the variation of $r(x)$ across the aperture can be much larger than the wavelength λ, so the oscillating numerator of Equation 3.66 cannot be replaced by a constant.

where $r(x)$ is the distance between the source and the aperture element at position x (Figure 3.11). In the far field (Equation 3.64), the **Fraunhofer approximation**

$$r \approx R + x \sin \theta \qquad (3.67)$$

is valid. This equation is usually written in the form

$$r \approx R + xl, \qquad (3.68)$$

where

$$\boxed{l \equiv \sin \theta.} \qquad (3.69)$$

For the small angles $\theta \ll 1$ rad relevant to large $D \gg \lambda$ apertures, $l = \sin \theta \approx \theta$.

At large distances, the quantity

$$\frac{1}{r} \approx \frac{1}{R} \qquad (3.70)$$

is nearly constant across the aperture and can be absorbed by the constant of proportionality in Equation 3.66. Although $\exp(-i2\pi R/\lambda)$ is a constant because R is fixed, the variable part xl of $r = R + xl$ in the numerator of Equation 3.66 cannot be ignored at any distance:

$$df \propto g(x) \exp(-i2\pi xl/\lambda)dx. \tag{3.71}$$

When $\theta \neq 0$ the phase $xl/\lambda \approx \sin\theta/\lambda$ varies linearly across the aperture, and different parts of the aperture add constructively or destructively to the total electric field $f(l)$. Defining

$$\boxed{u \equiv \frac{x}{\lambda}} \tag{3.72}$$

to express position along the aperture in units of wavelength yields

$$\boxed{f(l) = \int_{\text{aperture}} g(u)e^{-i2\pi lu}du.} \tag{3.73}$$

In words, this very important equation says that *in the far field, the electric-field pattern $f(l)$ of an aperture antenna is the **Fourier transform** (Appendix A.1) of the electric field distribution $g(u)$ illuminating that aperture.*

3.2.4 The Electric-Field and Power Patterns of a Uniformly Illuminated Aperture

What is the electric-field pattern of a uniformly illuminated one-dimensional aperture of width D at wavelength λ? **Uniform illumination** means that the strength of the illumination is constant over the aperture:

$$g(u) = \text{constant}, \qquad -\frac{D}{2\lambda} < u < +\frac{D}{2\lambda}.$$

This question is best answered in two steps: first find the far-field pattern of a unit aperture ($D = \lambda$) and then use the **similarity theorem** (Equation A.11) for Fourier transforms to scale the first result to an aperture of any size.

The **unit rectangle function** is defined as

$$\Pi(u) \equiv 1, \quad -1/2 < u < +1/2, \tag{3.74}$$

and $\Pi(u) = 0$ otherwise. The function symbol (an uppercase pi) is easy to remember because it looks like the function graph shown in the top panel of Figure 3.12.

Inserting $\Pi(u)$ into Equation 3.73 gives the field pattern $f(l)$ of the uniformly illuminated unit aperture:

$$f(l) = \int_{-\infty}^{\infty} \Pi(u)e^{-i2\pi lu}du. \tag{3.75}$$

Thus

$$f(l) = \int_{-1/2}^{+1/2} e^{-i2\pi lu}du = \frac{e^{-i2\pi lu}}{-i2\pi l}\bigg|_{-1/2}^{+1/2} = \frac{e^{-i\pi l} - e^{i\pi l}}{-i2\pi l}. \tag{3.76}$$

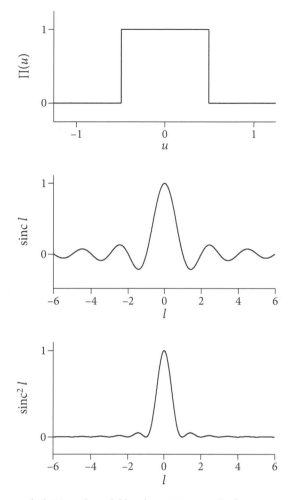

Figure 3.12. The symbol Π is shaped like the unit rectangle function it represents. The function $\mathrm{sinc}(l) \equiv \sin(\pi l)/(\pi l)$ is the Fourier transform of the unit rectangle function and is the electric-field pattern of a uniformly illuminated unit aperture. The power pattern of a uniformly illuminated unit aperture is shown in the bottom panel. For large $(D \gg \lambda)$ apertures, the zeros at $l = \pm 1, \pm 2, \ldots$ appear at the angles $\theta \approx \pm \lambda/D, \pm 2\lambda/D, \ldots$.

Next, difference the mathematical identities (Appendix B.3)

$$e^{i\pi l} = \cos(\pi l) + i \sin(\pi l),$$
$$e^{-i\pi l} = \cos(\pi l) - i \sin(\pi l)$$

to derive

$$e^{i\pi l} - e^{-i\pi l} = 2i \sin(\pi l).$$

Inserting this result into Equation 3.76 gives

$$f(l) = \frac{-2i \sin(\pi l)}{-2i\pi l} = \frac{\sin(\pi l)}{(\pi l)} \equiv \mathrm{sinc}(l). \tag{3.77}$$

The useful **sinc function** defined in Equation 3.77 is plotted in the middle panel of Figure 3.12.

The power pattern $p(l)$ is the square of the field pattern $f(l)$. The power pattern $p(l) = \text{sinc}^2(l)$ of a uniformly illuminated unit aperture is graphed in the bottom panel of Figure 3.12. The central peak of the power pattern between the first zeros at $l = \pm 1$ is called the **main beam**. The smaller peaks are called **sidelobes**. They are separated by zeros or **nulls** in the power pattern at $l = \pm 1, \pm 2, \ldots$.

Next apply the powerful **similarity theorem** for Fourier transforms: if $f(l)$ is the Fourier transform of $g(u)$, then

$$\frac{1}{|a|} f\left(\frac{l}{a}\right)$$

is the Fourier transform of $g(au)$, where $a \neq 0$ is a dimensionless scaling factor. According to the similarity theorem, making a function g wider ($0 < a < 1$) or narrower ($a > 1$) makes its Fourier transform f narrower and taller, or wider and shorter, respectively, always conserving the area under the transform. Consequently the **beamwidth** of an aperture antenna is inversely proportional to the aperture size in wavelengths and the on-axis field strength is directly proportional to the aperture size in wavelengths.

The scale factor for a uniformly illuminated one-dimensional aperture of width D operating at wavelength λ is $a = \lambda/D$, so the electric field pattern becomes

$$f(l) = \left(\frac{D}{\lambda}\right) \frac{\sin(\pi l D / \lambda)}{(\pi l D / \lambda)} \propto \frac{D}{\lambda} \text{sinc}\left(\frac{l D}{\lambda}\right).$$

If the aperture is large ($D/\lambda \gg 1$), the relevant angles θ are so small ($\theta \ll 1$ radian) that $l = \sin\theta \approx \theta$ and

$$f(\theta) = \frac{D}{\lambda} \text{sinc}\left(\frac{\theta D}{\lambda}\right). \tag{3.78}$$

The power pattern is proportional to the square of the electric field pattern, so

$$P(l) \propto \left(\frac{D}{\lambda}\right)^2 \text{sinc}^2\left(\frac{l D}{\lambda}\right).$$

If $\theta \ll 1$ radian, then

$$P(\theta) = \left(\frac{D}{\lambda}\right)^2 \text{sinc}^2\left(\frac{\theta D}{\lambda}\right). \tag{3.79}$$

Radio astronomers use the angle between the half-power points to specify the angular width of the main beam, calling it the **half-power beamwidth** (HPBW) or the **full width between half-maximum points** (FWHM). The narrow beamwidth $\theta_{\text{HPBW}} \ll 1$ rad of a large ($D \gg \lambda$) one-dimensional uniformly illuminated aperture satisfies

$$P(\theta_{\text{HPBW}}/2) = \frac{1}{2} = \text{sinc}^2\left(\frac{\theta_{\text{HPBW}} D}{2\lambda}\right), \tag{3.80}$$

$$0.443 \approx \frac{\theta_{\text{HPBW}} D}{2\lambda}, \tag{3.81}$$

$$\boxed{\theta_{\text{HPBW}} \approx 0.89\frac{\lambda}{D}.} \tag{3.82}$$

The similarity theorem implies the general scaling relation

$$\theta_{\text{HPBW}} \propto \frac{\lambda}{D}. \tag{3.83}$$

The constant of proportionality varies slightly with the **illumination taper**. Even an ideal aperture antenna of finite size has a finite resolving power that is limited by **diffraction**, the spreading of rays passing through a finite aperture, and Equation 3.82 specifies the **diffraction-limited** resolution of a uniformly illuminated aperture antenna.

The weak reciprocity theorem (Section 3.1.5) says that the preceding analysis of the transmitting power pattern of an aperture antenna also yields its receiving power pattern, or the variation of A_e with orientation. In receiving terms, the analog of the power pattern is called the **point-source response**. For a uniformly illuminated aperture, scanning a radio telescope beam in angle θ across a point source will cause the antenna temperature to vary as $\text{sinc}^2(\theta)$, and the width of the half-power response will equal the transmitting HPBW. The receiving HPBW is sometimes called the **resolving power** of a telescope because two equal point sources separated by the HPBW are just resolved by the **Rayleigh criterion** that the total response has a slight minimum midway between the point sources.

3.2.5 The Electric-Field and Power Patterns with Tapered Illumination

Practical feeds such as small waveguide horns or half-wave dipoles backed by small subreflectors cannot illuminate a large aperture uniformly. A better approximation to their illumination is the cosine-tapered field pattern (cosine-squared tapered power pattern)

$$g(u) = \frac{\pi}{2}\cos(\pi u), \quad -1/2 < u < +1/2, \tag{3.84}$$

and $g(u) = 0$ otherwise (Figure 3.13). The $(\pi/2)$ normalization factor in Equation 3.84 ensures that

$$\int_{-1/2}^{+1/2} g(u)du = 1. \tag{3.85}$$

The corresponding field pattern of a one-dimensional unit aperture is given by

$$f(l) = \int_{-1/2}^{+1/2} \frac{\pi}{2}\cos(\pi u)e^{-i2\pi lu}du. \tag{3.86}$$

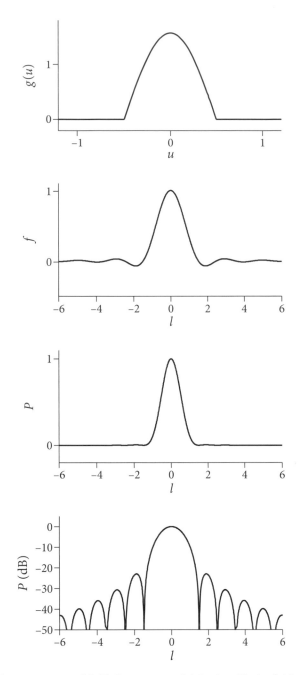

Figure 3.13. The cosine-tapered field illumination $g(u)$ (top) yields the field pattern $f(l)$ and power pattern $P(l)$ on a unit aperture. The low sidelobes of $P(l)$ are clearly visible only on a plot of $P(\text{dB})$ (bottom).

This Fourier transform can be evaluated as follows:

$$f(l) = \frac{\pi}{4} \int_{-1/2}^{+1/2} (e^{i\pi u} + e^{-i\pi u}) e^{-i2\pi l u} du \tag{3.87}$$

$$= \frac{\pi}{4} \left[\frac{e^{i\pi(1-2l)u}}{i\pi(1-2l)} \Big|_{-1/2}^{+1/2} + \frac{e^{-i\pi(1+2l)u}}{-i\pi(1+2l)} \Big|_{-1/2}^{+1/2} \right] \tag{3.88}$$

$$= \frac{\pi}{4} \left[\frac{e^{i\pi(1/2-l)} - e^{-i\pi(1/2-l)}}{i2\pi(1/2-l)} + \frac{e^{i\pi(1/2+l)} - e^{-i\pi(1/2+l)}}{i2\pi(1/2+l)} \right] \tag{3.89}$$

$$= \frac{\pi}{4} \left[\frac{2i\sin[\pi(1/2-l)]}{i2\pi(1/2-l)} + \frac{2i\sin[\pi(1/2+l)]}{i2\pi(1/2+l)} \right] \tag{3.90}$$

$$= \frac{\pi}{4} \left[\frac{\cos(\pi l)}{\pi(1/2-l)} + \frac{\cos(\pi l)}{\pi(1/2+l)} \right] = \frac{\pi}{4} \cos(\pi l) \left(\frac{\pi}{\pi^2/4 - \pi^2 l^2} \right) \tag{3.91}$$

to yield the field pattern

$$f(l) = \frac{\cos(\pi l)}{1 - 4l^2} \tag{3.92}$$

of a one-dimensional unit aperture with cosine-tapered illumination given by Equation 3.84. Both the field pattern and the power pattern

$$P(l) = [f(l)]^2 = \left[\frac{\cos(\pi l)}{1 - 4l^2} \right]^2 \tag{3.93}$$

are shown in Figure 3.13. The sidelobes are so weak that a plot of $P(\mathrm{dB}) = 10 \log_{10}(P)$ is needed to show them clearly (bottom panel of Figure 3.13).

Tapering increases the half-power beamwidth. If $D \gg \lambda$, the normalized power pattern is

$$P(\theta) = \left[\frac{\cos(\pi \theta D/\lambda)}{1 - 4(\theta D/\lambda)^2} \right]^2, \tag{3.94}$$

and

$$P\left(\frac{\theta_{\mathrm{HPBW}}}{2} \right) = \frac{1}{2} = \left[\frac{\cos[\pi \theta_{\mathrm{HPBW}} D/(2\lambda)]}{1 - 4[\theta_{\mathrm{HPBW}} D/(2\lambda)]^2} \right]^2 \tag{3.95}$$

can be solved numerically to yield

$$\boxed{\theta_{\mathrm{HPBW}} \approx 1.2 \frac{\lambda}{D}.} \tag{3.96}$$

This beamwidth is typical of most radio telescopes.

The perfectly sharp cutoff of illumination at the edge of the aperture shown in the top panel of Figure 3.13 cannot be achieved in practice. Any illumination extending beyond the reflector is called **spillover**. In the case of a receiving antenna, a prime-focus feed looking down at an aperture also sees spillover radiation from the surrounding ground. Most soils are good absorbers, which emit blackbody radiation at the ambient temperature $T \sim 300$ K, and ground radiation can add significantly to the system noise temperature of a radio telescope. The purpose of the 15-m high annular **ground screen** surrounding the Arecibo reflector (Color Plate 2) is to intercept most of the spillover radiation and redirect it to the cold sky in order minimize the system temperature.

3.3 TWO-DIMENSIONAL APERTURE ANTENNAS

3.3.1 The Field Pattern of a Two-Dimensional Aperture

The method used to show that the field pattern of a one-dimensional aperture is the one-dimensional Fourier transform of the aperture field illumination (Equation 3.73) can easily be generalized to the more realistic case of a two-dimensional aperture:

$$f(l, m) \propto \int_{-\infty}^{\infty} \int_{-\infty}^{\infty} g(u, v)e^{-i2\pi(lu+mv)}\, du\, dv, \tag{3.97}$$

where m is the y-axis analog of l on the x-axis, and

$$v \equiv \frac{y}{\lambda}. \tag{3.98}$$

In words, Equation 3.97 states that *the electric field pattern of a two-dimensional aperture is the two-dimensional Fourier transform of the aperture field illumination.*

3.3.2 The Uniformly Illuminated Rectangular Aperture

The two-dimensional counterpart of a uniformly illuminated one-dimensional aperture is a uniformly illuminated rectangular aperture with side lengths D_x and D_y. Dividing lengths in the aperture plane by the wavelength λ yields the normalized coordinates $u \equiv x/\lambda$ and $v \equiv y/\lambda$. The direction from the origin of the (u, v) plane to any distant point can be specified by $l \equiv \sin\theta_x$ and $m \equiv \sin\theta_y$, where θ_x is the angle from the (y, z) plane and θ_y is the angle from the (x, z) plane (Figure 3.14). If the illumination $g(x, y)$ is constant over the aperture, the integrals over u and v in the Fourier transform are separable and

$$f(l, m) \propto \operatorname{sinc}\left(\frac{lD_x}{\lambda}\right) \operatorname{sinc}\left(\frac{mD_y}{\lambda}\right). \tag{3.99}$$

Squaring the electric field pattern gives the relative (normalized to unity at the peak) power pattern

$$P_n(l, m) = \operatorname{sinc}^2\left(\frac{lD_x}{\lambda}\right) \operatorname{sinc}^2\left(\frac{mD_y}{\lambda}\right). \tag{3.100}$$

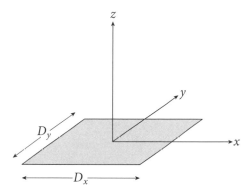

Figure 3.14. A two-dimensional rectangular aperture with side lengths D_x and D_y. Dividing lengths in the aperture plane by the wavelength λ yields the normalized coordinates $u \equiv x/\lambda$ and $v \equiv y/\lambda$. The direction from the origin to any distant point can be specified by $l \equiv \sin\theta_x$ and $m \equiv \sin\theta_y$, where θ_x is the angle from the (y, z) plane and θ_y is the angle from the (x, z) plane.

The absolute power gain G in any direction can be calculated from the relative power pattern by invoking energy conservation:

$$\int G \, d\Omega = 4\pi = G_0 \int_{-1}^{+1} \int_{-1}^{+1} P_n(l, m) dl \, dm, \qquad (3.101)$$

$$4\pi = G_0 \int_{-1}^{+1} \left[\frac{\sin(\pi l D_x/\lambda)}{\pi l D_x/\lambda} \right]^2 dl \int_{-1}^{+1} \left[\frac{\sin(\pi m D_y/\lambda)}{\pi m D_y/\lambda} \right]^2 dm. \qquad (3.102)$$

Defining the temporary variable a as

$$a \equiv \frac{\pi l D_x}{\lambda}, \quad \text{so } da = \frac{\pi D_x}{\lambda} dl, \qquad (3.103)$$

gives, for $D_x \gg \lambda$,

$$\int_{-1}^{+1} \left[\frac{\sin(\pi l D_x/\lambda)}{\pi l D_x/\lambda} \right]^2 dl \approx \left[\int_{-\infty}^{\infty} \frac{\sin^2 a}{a^2} da \right] \frac{\lambda}{\pi D_x} = \frac{\lambda}{D_x} \qquad (3.104)$$

because the value of the definite integral in square brackets is π. [To prove this, simply apply Rayleigh's theorem (Equation A.7) to the Fourier transform pair $\text{sinc}(l)$ (Equation 3.77) and $\Pi(u)$ (Equation 3.74).] Then

$$4\pi = G_0 \frac{\lambda^2}{D_x D_y}. \qquad (3.105)$$

Thus the peak power gain is

$$G_0 = \frac{4\pi D_x D_y}{\lambda^2}, \qquad (3.106)$$

and the power pattern of a uniformly illuminated rectangular aperture with side lengths D_x and D_y is

$$G = \frac{4\pi D_x D_y}{\lambda^2} \text{sinc}^2 \left(\frac{l D_x}{\lambda} \right) \text{sinc}^2 \left(\frac{m D_y}{\lambda} \right), \qquad (3.107)$$

and

$$G \approx \frac{4\pi \, D_x D_y}{\lambda^2} \mathrm{sinc}^2\left(\frac{\theta_x D_x}{\lambda}\right) \mathrm{sinc}^2\left(\frac{\theta_y D_y}{\lambda}\right) \tag{3.108}$$

when θ_x and θ_y are much smaller than 1 radian.

In general, the peak power gain of an aperture antenna is proportional to the geometric area A_{geom} ($A_{\mathrm{geom}} = D_x D_y$ in this case) of the aperture. The constant of proportionality is $4\pi/\lambda^2$ for a uniformly illuminated aperture and somewhat less for any other illumination pattern.

Using Equation 3.46

$$A_e = \frac{\lambda^2 G}{4\pi}, \tag{3.109}$$

we find that the on-axis effective collecting area is

$$A_0 = \frac{\lambda^2 G_0}{4\pi} = \frac{4\pi \lambda^2 D_x D_y}{4\pi \lambda^2} = D_x D_y = A_{\mathrm{geom}}. \tag{3.110}$$

The peak effective area of an ideal uniformly illuminated aperture equals its geometric area, independent of wavelength. With any other illumination taper, the effective area is smaller than but proportional to the geometric area. It is useful to define the **aperture efficiency** η_A as the ratio of the effective area to geometric area:

$$\eta_A \equiv \frac{A_0}{A_{\mathrm{geom}}}. \tag{3.111}$$

Thus $\eta_A = 1$ for an ideal uniformly illuminated aperture and $\eta_A < 1$ otherwise. The aperture efficiencies of most radio telescopes are $\eta_A \leq 70\%$, although phased-array feeds control the illumination well enough to let ASKAP (Color Plate 6) reach $\eta_A \approx 80\%$.

Large ($D \gg \lambda$) rectangular waveguide horns are nearly uniformly illuminated unblocked apertures, so their actual gains and effective collecting areas can be calculated accurately. This makes them useful for measuring the absolute flux densities of strong sources such as Cas A and Cyg A and defining the practical flux-density scales used by radio astronomers [6].

Most apertures associated with reflectors and lenses are circular. The power pattern of a uniformly illuminated circular aperture is known as the **Airy pattern**.[1]

3.3.3 Gaussian Beam Solid Angle and Beamwidth

For any realistic illumination taper, the beam solid angle (Equation 3.42)

$$\Omega_A \equiv \int_{4\pi} \frac{A_e(\theta, \phi)}{A_0} \, d\Omega$$

---

[1] See http://www.olympusfluoview.com/java/resolution3d/index.html for an interactive plot showing how the Airy pattern behaves as a function of wavelength and aperture size.

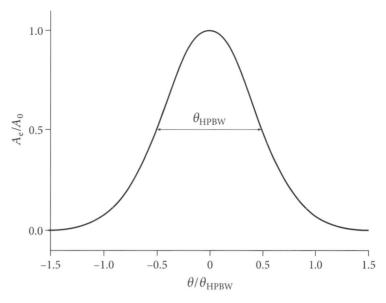

Figure 3.15. The beams of most radio telescopes are nearly Gaussian, and their beamwidths are usually specified by the angle θ_{HPBW} between the half-power points. Abscissa: offset θ from the beam center in units of the HPBW. Ordinate: Effective aperture A_{e} normalized by the peak effective aperture A_0.

of a radio telescope is about equal to the square of the half-power beamwidth θ_{HPBW}. In fact, the beams of most radio telescopes are nearly Gaussian and can be written as

$$\frac{A_{\mathrm{e}}}{A_0} = \exp(-x\theta^2), \tag{3.112}$$

where θ is the angle from the beam center and x is a scaling factor such that $A_{\mathrm{e}}/A_0 = 1/2$ when $\theta = \pm\theta_{\mathrm{HPBW}}/2$ (Figure 3.15):

$$\frac{1}{2} = \exp\left[-x\left(\frac{\theta_{\mathrm{HPBW}}}{2}\right)^2\right]. \tag{3.113}$$

Thus

$$x = \frac{4\ln 2}{\theta_{\mathrm{HPBW}}^2}, \tag{3.114}$$

$$\frac{A_{\mathrm{e}}}{A_0} = \exp\left[-4\ln 2\left(\frac{\theta}{\theta_{\mathrm{HPBW}}}\right)^2\right], \tag{3.115}$$

and

$$\Omega_{\mathrm{A}} = \int_{\theta=0}^{\infty} \int_{\phi=0}^{2\pi} \exp\left[-4\ln 2\left(\frac{\theta}{\theta_{\mathrm{HPBW}}}\right)^2\right] \theta\, d\phi\, d\theta. \tag{3.116}$$

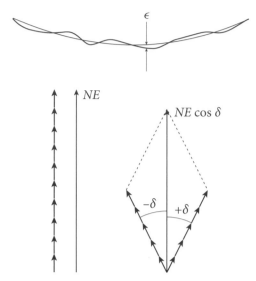

Figure 3.16. Deviations ϵ of the actual reflector surface (thick curve) from the best-fit paraboloid (thin curve) degrade short-wavelength performance (top). Vector sums of the electric fields E produced by N elements of perfect and imperfect apertures are shown bottom left. Bumps in the imperfect aperture produce phase shifts $\pm\delta \approx \pm 4\pi\epsilon/\lambda$ which lower the vector sum of electric fields from NE to $NE\cos\delta$ (bottom right).

Integrating over ϕ and substituting the dummy variable $y = 4\ln 2(\theta/\theta_{\text{HPBW}})^2$ yields

$$\Omega_{\text{A}} = 2\pi \left(\frac{\theta_{\text{HPBW}}^2}{8\ln 2} \right) \int_{y=0}^{\infty} \exp(-y)\, dy, \tag{3.117}$$

so the beam solid angle of a Gaussian beam is

$$\Omega_{\text{A}} = \left(\frac{\pi}{4\ln 2} \right) \theta_{\text{HPBW}}^2 \approx 1.133\, \theta_{\text{HPBW}}^2. \tag{3.118}$$

3.3.4 Reflector Accuracy Requirements

Real radio telescopes don't have perfectly smooth paraboloidal reflectors. Small deviations from the best-fit paraboloid may be caused by permanent manufacturing errors, changing gravitational deformations as the reflector is tilted, thermal distortions resulting from solar heating, and bending by strong winds. There will be some shortest wavelength λ_{min} below which these **surface errors** degrade the reflector performance so severely that the telescope becomes unusable. The **reflector surface efficiency** η_{s} is defined as the power gain of the actual reflector divided by the power gain of a perfect paraboloidal reflector with the same size and illumination. The following calculation of how η_{s} varies with the rms (root mean square) surface error in wavelengths (ϵ/λ) is based on the classic method of Ruze [96]. Where the actual reflector surface deviates from the best-fit paraboloid by a distance ϵ (Figure 3.16), the path length of the reflected wave will be in error by almost 2ϵ and the phase error

δ (radians) of the reflected wave will be

$$\delta \approx \frac{2\pi}{\lambda}(2\epsilon) = \frac{4\pi\epsilon}{\lambda}. \tag{3.119}$$

An oversimplified example would be a bumpy surface, half covered with small bumps of height $\epsilon \ll \lambda$ and half covered with small dips of the same depth ϵ. Then the contribution of each area element to the far (electric) field (Figure 3.16) is reduced by a factor $\cos \delta$. In the limit $\delta \ll 1$ rad, $\cos \delta \approx 1 - \delta^2/2 + \cdots$ and

$$\frac{E(\delta)}{E(0)} \approx 1 - \frac{\delta^2}{2} + \cdots, \tag{3.120}$$

so the relative power gain is

$$\frac{G(\delta)}{G(0)} \approx \left[\frac{E(\delta)}{E(0)}\right]^2 \approx 1 - \delta^2 \approx 1 - \left(\frac{4\pi\epsilon}{\lambda}\right)^2. \tag{3.121}$$

This rough estimate shows that the surface errors must be an order-of-magnitude smaller than the shortest usable wavelength, a severe requirement indeed.

A more realistic calculation makes use of the fact the most errors have roughly Gaussian amplitude distributions. Suppose that the surface errors have a Gaussian probability distribution $P(\epsilon)$ with rms σ:

$$P(\epsilon) = \frac{1}{\sqrt{2\pi}\sigma} \exp\left(-\frac{\epsilon^2}{2\sigma^2}\right). \tag{3.122}$$

Then the relative field strength is obtained as the weighted sum over all possible ϵ:

$$\langle E/E(0)\rangle \approx \int_{-\infty}^{\infty} \cos\left(\frac{4\pi\epsilon}{\lambda}\right) \cdot \frac{1}{\sqrt{2\pi}\sigma} \exp\left(-\frac{\epsilon^2}{2\sigma^2}\right) d\epsilon. \tag{3.123}$$

Substituting $e^{iz} = \cos z + i \sin z$ turns this integral into a more familiar one, the Fourier transform of a Gaussian:

$$\langle E/E(0)\rangle \approx \int_{-\infty}^{\infty} \exp\left(-i\frac{4\pi\epsilon}{\lambda}\right) \cdot \frac{1}{\sqrt{2\pi}\sigma} \exp\left(-\frac{\pi\epsilon^2}{2\pi\sigma^2}\right) d\epsilon. \tag{3.124}$$

Note that the $i \sin z$ part drops out immediately because it is antisymmetric in an otherwise symmetric integral. To make this look even more familiar, let $s \equiv 2/\lambda$, $x \equiv \epsilon$, and $a \equiv (\sqrt{2\pi}\sigma)^{-1}$. Then

$$\langle E/E(0)\rangle \approx \int_{-\infty}^{\infty} \exp(-i2\pi s x) \exp\left(-\pi(ax)^2\right) dx. \tag{3.125}$$

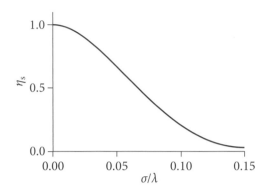

Figure 3.17. The surface efficiency η_s declines rapidly as the rms surface error in wavelengths σ/λ exceeds $1/16 \approx 0.06$.

Recall that the Fourier transform of $f(x) = \exp(-\pi x^2)$ is $F(s) = \exp(-\pi s^2)$ (Appendix B.4) and apply the similarity theorem (Equation A.11) to get

$$\langle E/E(0)\rangle = \frac{1}{|a|\sqrt{2\pi}\sigma} \exp\left[-\pi\left(\frac{s}{a}\right)^2\right] \tag{3.126}$$

$$= \exp[-2\pi^2\sigma^2 s^2] \tag{3.127}$$

$$= \exp\left(-\frac{8\pi^2\sigma^2}{\lambda^2}\right). \tag{3.128}$$

Power is proportional to E^2 so the reflector surface efficiency is simply

$$\boxed{\eta_s = \exp\left[-\left(\frac{4\pi\sigma}{\lambda}\right)^2\right].} \tag{3.129}$$

Equation 3.129 is often called the **Ruze equation**; it is plotted in Figure 3.17. The surface efficiency η_s is closely related to the **Strehl ratio** S used by optical astronomers to specify the peak intensity loss caused by optical aberrations or atmospheric turbulence. The Strehl ratio is normally expressed in terms of the rms wavefront error in wavelengths ω, which is about twice the rms surface error in wavelengths σ/λ, so Equation 3.129 implies

$$S = \exp[-(2\pi\omega)^2]. \tag{3.130}$$

A traditional rule-of-thumb for the shortest wavelength $\lambda_{\min}$ at which a radio telescope works reasonably well is

$$\sigma \approx \frac{\lambda_{\min}}{16} \tag{3.131}$$

because the surface efficiency at $\lambda = \lambda_{\min}$ is only

$$\eta_s \approx \exp\left[-\left(\frac{\pi}{4}\right)^2\right] \approx 0.54 \tag{3.132}$$

and falls exponentially at shorter wavelengths. For example, the 100-m diameter GBT is intended to operate at frequencies as high as $\nu \approx 100\,\text{GHz}$, or $\lambda_{\min} \approx 3\,\text{mm}$. To meet this specification, the rms deviation from a perfect paraboloid must not exceed $\sigma \approx 3\,\text{mm}/16 \approx 200\,\mu\text{m}$, the thickness of two sheets of paper. The power gain of a perfect paraboloidal reflector is proportional to ν^2. If the reflector surface has a Gaussian error distribution with rms σ, then its gain increases as ν^2 at low frequencies, reaches a maximum at

$$\lambda = 4\pi\sigma, \tag{3.133}$$

and decreases quickly at higher frequencies.

3.3.5 Pointing-Accuracy Requirements

Real radio telescopes don't have perfectly accurate pointing. Small errors in tracking a target source reduce the gain in the source direction and contribute to the uncertainty in flux-density measurements of compact sources. Tracking errors are just as important as surface errors in limiting the short-wavelength performance of large radio telescopes.

The power patterns of most radio telescopes are nearly Gaussian near the peak. In terms of the beamwidth between half-power points θ_{HPBW}, the relative gain at a point offset by angle ρ from the beam axis is

$$\boxed{\frac{G}{G_0} = \exp\left[-4\ln 2\left(\frac{\rho}{\theta_{\text{HPBW}}}\right)^2\right].} \tag{3.134}$$

If the one-dimensional tracking error in each coordinate (e.g., azimuth or elevation angle) has a Gaussian distribution with rms σ_1, the tracking error ρ in two dimensions has a **Rayleigh distribution**

$$P(\rho) = \frac{\rho}{\sigma_1^2} \exp\left(-\frac{\rho^2}{2\sigma_1^2}\right). \tag{3.135}$$

The mean squared tracking error is

$$\langle \rho^2 \rangle = \int_0^\infty \rho^2 P(\rho)\, d\rho = 2\sigma_1^2. \tag{3.136}$$

The rms value of the two-dimensional tracking error is $\sigma_2 = 2^{1/2}\sigma_1$, so small tracking errors reduce the average on-source gain by the factor

$$\langle G/G_0 \rangle = \left[1 + 4\ln 2\left(\frac{\sigma_2}{\theta_{\text{HPBW}}}\right)^2\right]^{-1}. \tag{3.137}$$

More importantly, the fluctuating on-source gain caused by tracking errors contributes a fractional uncertainty[2]

$$\frac{\sigma_S}{S} = \frac{z}{(1 + 2z)^{1/2}},$$ (3.138)

where

$$z \equiv 4 \ln 2 \left(\frac{\sigma_2}{\theta_{\text{HPBW}}} \right)^2,$$ (3.139)

to a measurement of source flux density S. Thus an rms tracking error of $0.2\,\theta_{\text{HPBW}}$ will contribute a 10% rms flux-density uncertainty. For 5% accuracy, $\sigma_2/\theta_{\text{HPBW}} \approx 0.14$ (or $\sigma_1/\theta_{\text{HPBW}} \approx 0.10$ in each coordinate) is needed.

For example, we can calculate the largest tracking error in arcsec compatible with making flux-density measurements with 5% rms errors using the GBT 100-m telescope at $\nu = 33$ GHz. From Equation 3.138, $\sigma_S/S = 0.05$ when $\sigma_2/\theta_{\text{HPBW}} \approx 0.14$. The half-power beamwidth of the GBT at $\nu = 33$ GHz ($\lambda \approx 9.1$ mm) is

$$\theta_{\text{HPBW}} \approx \frac{1.2D}{\lambda} = \frac{1.2 \cdot 0.0091\,\text{m}}{100\,\text{m}} \approx 1.09 \times 10^{-4}\,\text{rad} \approx 23\,\text{arcsec}.$$ (3.140)

Thus the total tracking error must be smaller than $\sigma_2 = 0.14 \times 23\,\text{arcsec} = 3.2\,\text{arcsec}$, or $\sigma_1 \approx 2^{-1/2}\sigma_2 \approx 2.2\,\text{arcsec} \approx 10^{-5}\,\text{rad}$ in azimuth and in elevation angle.

The thermal expansion coefficient of steel is about $10^{-5}\,\text{C}^{-1}$, so changing the temperature differential across the steel GBT support structure by only 1 centigrade degree could produce a $10^{-5}\,\text{rad} \approx 2\,\text{arcsec}$ pointing shift. For this reason, high-frequency observers must monitor pointing calibration sources and correct the GBT pointing every hour or so, particularly just after sunrise on sunny days. Wind gusts also degrade pointing accuracy, but they fluctuate on much shorter timescales.

3.4 WAVEGUIDES

Waveguides are low-loss shielded "pipes" used to transport electromagnetic waves between antennas and receivers or between sections of a receiver. The simplest waveguide is a hollow rectangular tube with conducting walls (Figure 3.18, top) separated by distance a in the horizontal (x) direction and $b \leq a/2$ in the vertical (y) direction. At the conducting walls, the parallel component of any electric field inside the waveguide must be zero. Three permitted distributions of the electric field strength E along the horizontal axis are shown as curves in the top panel of Figure 3.18, which is similar to Figure 2.16 for standing waves in a cavity. However, only the longest-wavelength **dominant mode** with $n_x = 1$ is normally used, and higher-order modes with $n_x = 2, 3, \dots$ are deliberately suppressed because they travel down the waveguide with different group velocities.

The bottom panel of Figure 3.18 presents the plan view of the dominant radiation mode traveling through the waveguide with a wave normal in the direction of the large arrow and a wave node ($|E| = 0$) indicated by the dashed line.

---

[2]http://wwwlocal.gb.nrao.edu/ptcs/ptcssn/ptcssn3.pdf.

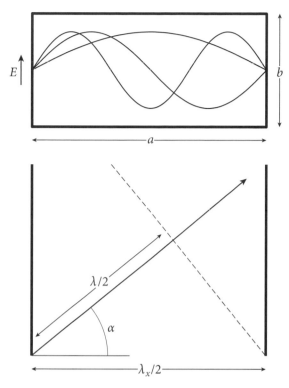

Figure 3.18. The upper drawing shows the cross section of a rectangular waveguide having interior width a in the x-direction and height $b \leq a/2$ in the y-direction. The component of any electric field $\vec{E}$ parallel to a conducting wall must go to zero at the wall, just as with radiation in a cavity (Figure 2.16). The curves indicate three allowed x distributions of electric field strength (called **modes**). The lower drawing is a plan view of the waveguide. It is analogous to Figure 2.17 showing waves in a conducting cavity. Radiation of wavelength λ must travel through the waveguide in the direction indicated by the large arrow to satisfy the boundary condition $n_x = 2a/\lambda_x = (2a/\lambda)\cos\alpha$, where $n_x = 1, 2, 3, \ldots$ (Equation 2.66). Only the TE10 mode shown, with $n_x = 1$ ($a = \lambda_x/2$) and no variation of $E = E_y$ with height ($n_y = 0$), is normally used.

It is analogous to Figure 2.17 showing waves in a conducting cavity. Radiation of wavelength λ traveling through the waveguide in the direction indicated by the arrow must satisfy the boundary condition $n_x = 2a/\lambda_x = (2a/\lambda)\cos\alpha$, where $n_x = 1, 2, 3, \ldots$ (Equation 2.66). When $n_x = 1$, $\lambda_x/2 = a = (\lambda/2)\cos\alpha$.

The maximum wavelength ($\cos\alpha = 1$) that can propagate ($\alpha \geq 0$) in the waveguide is the **cutoff wavelength**

$$\lambda_c = 2a, \tag{3.141}$$

and the corresponding minimum frequency

$$\nu_c = c/\lambda_c \tag{3.142}$$

is called the **cutoff frequency**. Waveguides are extremely effective high-pass filters.

The **group velocity** of propagation down the waveguide is

$$v_g = c \sin \alpha = c(1 - \cos^2 \alpha)^{1/2} = c \left[1 - \left(\frac{v_c}{v} \right)^2 \right]^{1/2}, \qquad (3.143)$$

which varies quite rapidly with frequency as v approaches v_c ($\sin \alpha$ approaches 0). The waveguide **phase velocity** $v_p = c^2/v_g \geq c$, so the **guide wavelength**

$$\lambda_w = \frac{c}{v} \left[1 - \left(\frac{v_c}{v} \right)^2 \right]^{-1/2} \qquad (3.144)$$

is somewhat greater than the free-space wavelength.

To minimize **dispersion** (the variation of v_g with frequency), waveguides are rarely used at frequencies below $v \approx 1.25 v_c$. Higher-order modes with $n_x = 2, 3, \ldots$ have frequencies $v > 2v_c, 3v_c, \ldots$ and propagate with different group velocities. To suppress them, waveguides are not used for frequencies $v > 2v_c$, the cutoff frequency of the $n_x = 2$ mode. Practical waveguides are usually limited to frequencies $v < 1.9v_c$. The requirement $b \leq a/2$ ensures that $\lambda/2 > b$ for all $v < 2v_c$ so $n_y = 0$, so only this TE10 (Transverse Electric field with $n_x = 1, n_y = 0$) mode can propagate. The TE10 mode electric field is vertically polarized and its strength is independent of y.

The combination of these upper and lower frequency limits restrict most waveguide applications to octave bandwidths, and waveguides of different sizes cover different octaves. Many of the **waveguide band names** in use today originated as deliberately confusing code names for World War II radar bands. They and their frequency ranges are listed in Appendix F.5. For example, the standard X-band waveguide has interior dimensions $a = 0.9$ inches ≈ 2.286 cm, $b = 0.4$ inches ≈ 1.016 cm. Its cutoff wavelength is $\lambda_c = 2a \approx 4.572$ cm and its cutoff frequency is $v_c = c/\lambda_c \approx 6.557$ GHz. Its **nominal frequency range** extends from $1.25 v_c \approx 8.2$ GHz to $1.9 v_c \approx 12.4$ GHz. Unfortunately, the waveguide band names are so deeply embedded in radio-astronomy jargon that radio observers cannot avoid them any more than optical astronomers can avoid "magnitudes."

Each feed and receiver on a radio telescope covers only one waveguide band, so several feeds and receivers are needed to span the much wider useful frequency range of the telescope itself. At the VLA, the frequency range from 1 to 50 GHz is covered by eight sets of feeds and receivers in eight waveguide bands: L (1–2 GHz), S (2–4 GHz), C (4–8 GHz), X (8–12 GHz), Ku (12–18 GHz), K (18–26.5 GHz), Ka (26.5–40 GHz), and Q (40–50 GHz).

3.5 RADIO TELESCOPES

The radio band is too wide (five decades in wavelength) to be covered effectively by a single telescope design. The surface brightnesses and angular sizes of radio sources span an even wider range, so a combination of single telescopes and aperture-synthesis interferometers are needed to detect and image them. It is not practical to build a single radio telescope that is even close to optimum for all of radio astronomy.

Figure 3.19. The horn antenna at Bell Labs, Holmdel, NJ used by Penzias and Wilson to discover the 3 K cosmic microwave background radiation in 1965. Reprinted with permission of Alcatel-Lucent USA Inc.

The ideal radio telescope should have a large collecting area to detect faint sources. The effective collecting area $A_e(\theta, \phi)$ of any antenna averaged over all directions (θ, ϕ) is (Equation 3.41)

$$\langle A_e \rangle = \frac{\lambda^2}{4\pi}, \tag{3.145}$$

so large peak collecting areas imply extremely directive antennas at short wavelengths. Only at long wavelengths ($\lambda > 1$ m) is it feasible to construct sensitive antennas from reasonable numbers of small, nearly isotropic elements such as dipoles. Jansky's $\lambda \approx 15$-m "wire" antenna (Figure 1.7) is an array of phased dipoles. It produces a wide fan beam near the horizon but has a large collecting area because λ^2 is so large. Directive aperture antennas are needed for adequate sensitivity at higher frequencies.

The simplest aperture antenna is a **waveguide horn**. Radiation incident on the opening is guided by a tapered waveguide. At the narrow end of the tapered horn is a waveguide with parallel walls, and inside this waveguide is a quarter-wave ground-plane vertical antenna that converts the electromagnetic wave into an electrical current that is sent to the receiver via a cable.

Horn antennas pick up very little ground radiation because, unlike most paraboloidal dishes, their apertures are not partially blocked by external feeds and feed-support structures, which scatter ground radiation into the receiver. This freedom from ground pickup allowed Penzias and Wilson [80] to show that the zenith antenna temperature of the Bell Labs horn (Figure 3.19) was 3.5 K higher at

$\nu \approx 4\,\mathrm{GHz}$ than expected—the first detection of the cosmic microwave background radiation. The aperture of a waveguide horn is not blocked by any feed-support structure, so it is also easier to calculate the gain of a horn antenna from first principles than to calculate the gain of a partially blocked reflecting antenna. Thus small horn antennas have been used by radio astronomers to measure the **absolute flux densities** of very strong sources such as Cas A. Radio astronomers observing with large dishes typically do not measure the absolute flux densities of sources, only their **relative flux densities** by comparison with secondary calibration sources whose flux densities relative to that of Cas A are known in advance. The painstaking process of measuring the absolute flux densities of Cas A and comparing them with the flux densities of weaker point sources suitable for calibrating observations made with large radio telescopes was described in detail by Baars et al. [6].

Most radio telescopes use circular paraboloidal reflectors to obtain large collecting areas and high angular resolution over a wide frequency range. Because the feed is on the reflector axis, the feed and legs supporting it partially block the path of radiation falling onto the reflector. This **aperture blockage** has a number of undesirable consequences:

1. The effective collecting area is reduced because some of the incoming radiation is blocked.
2. The beam pattern is degraded by increased sidelobe levels.
3. Radiation from the ground that is scattered off the feed and its support structure increases the system noise.
4. Radiation from the Sun and artificial sources of radio frequency interference (RFI) far from the main beam will be mixed with the desired signal.

Radio telescopes are so large that paraboloids with high f/D ratios are impractical; typically $f/D \approx 0.4$. Thus radio "dishes" are relatively deep, as shown in Figure 3.20. Another consequence of a low f/D ratio is a tiny field of view at the prime focus. The instantaneous imaging capability of a large single dish is severely limited by the small number of feeds that can fit into the tiny focal circle.

Nearly all radio telescopes have **alt-az mounts** consisting of a horizontal azimuth track on which the telescope turns in **azimuth** (the angle measured clockwise from north in the horizontal plane) and a horizontal elevation axle about which the telescope tips in **altitude** or **elevation angle** (two names for the angle above the horizon). The 140-foot telescope in Green Bank is unique among large radio telescopes in having an **equatorial mount** (Figure 3.20). The advantage of a equatorial mount is tracking simplicity—the **declination axis** is fixed and the **hour-angle axis** turns at a constant rate while tracking a distant celestial source. (The **hour angle** is the angle past the meridian, measured in hours. The **meridian** is the great circle passing through the north pole, south pole, and zenith.) In contrast, both the altitude and the azimuth of a celestial source change nonlinearly with time. When the 140-foot telescope was being designed, the ability of computers to perform the real-time calculations needed for an alt-az telescope to track a source accurately was in doubt. The disadvantage of a equatorial mount is mechanical—the sloped hour-angle yoke and polar axle with its huge tail bearing are very difficult to build and support.

Figure 3.20 clearly shows the Cassegrain optical system of the 140-foot telescope. Radiation reflected from the main dish is reflected a second time from the convex **Cassegrain subreflector** located just below the focal point down to feed horns

Figure 3.20. The 140-foot (43-m) telescope in Green Bank, WV is the largest telescope with an equatorial mount. Image credit: NRAO/AUI/NSF.

and receivers near the vertex of the paraboloid. A subreflector system has some advantages over a prime-focus system:

1. The magnifying subreflector can multiply the effective f/D ratio; values of $f/D \sim 2$ are typical. This greatly increases the size of the focal ellipsoid. Multiple feeds can be located within the focal ellipsoid to produce multiple simultaneous beams for faster imaging.
2. The subreflector is many wavelengths in diameter so it can be used to tailor the illumination taper to optimize the trade-off between high aperture efficiency and low sidelobes.
3. Receivers can be located near the vertex, not the focal point, where they are easier to access.
4. Feed spillover radiation is directed toward the cold sky instead of the warm ground, lowering overall system temperatures.
5. The subreflector can be **nutated** (rocked back and forth) rapidly to switch the beam between two adjacent positions on the sky. Such differential observations in

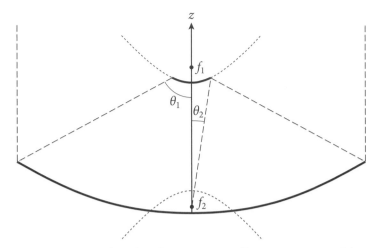

Figure 3.21. Cross section of a radio telescope rotationally symmetric around the z-axis and having a Cassegrain subreflector. Parallel rays from a distant radio source are reflected by a circular paraboloid whose prime focus is at the point marked f_1. The convex Cassegrain subreflector is a circular hyperboloid located below the prime focus. It reflects these rays to the feed located at the secondary focus f_2 just above the vertex of the paraboloid. The angle $2\theta_1$ subtended by the main reflector viewed from the prime focus is much larger than the angle $2\theta_2$ subtended by the subreflector viewed from the secondary focus, so Cassegrain feeds have to be much larger than primary feeds.

time and space can be used to remove receiver baseline drift in time and large-scale spatial fluctuations of atmospheric noise.

6. The subreflector can be tilted to select one of several feeds at the secondary focus, so that the observing frequency band can be changed rapidly.

A subreflector system has some disadvantages:

1. Relatively large feeds are required to produce the narrow beams needed to illuminate the subreflector, which typically subtends only a small angle as viewed from the vertex.

2. **Standing waves** in the leaky cavity formed by the reflector and subreflector cause sinusoidal ripples with frequency period $\Delta v \approx c/(2f)$ in the observed spectra of strong continuum radio sources. These ripples can be minimized by alternately defocusing the subreflector radially by $\pm\lambda/8$ and averaging the data from both subreflector positions.

3. A Cassegrain subreflector blocks the prime-focus position, so prime-focus feeds cannot be used when the Cassegrain subreflector is in position.

The geometry of a symmetrical radio telescope with a Cassegrain subreflector is shown in Figure 3.21. The paraboloidal shape of the primary reflector was determined by the requirement that all incoming rays parallel to the z-axis travel the same distance to reach the prime focus at f_1. Likewise, the secondary reflector shape is determined by the requirement that these rays travel the same distance to reach the **secondary focus** at f_2. For a subreflector located below the prime focus, the required shape is a hyperboloid whose major axis coincides with the major axis

Figure 3.22. The Parkes 64-m telescope. Photo © Shaun Amy.

of the paraboloid. The equation

$$\frac{z^2}{a^2} - \frac{r^2}{b^2} = 1 \qquad (3.146)$$

with $a > b$ defines such a hyperboloid. From any point on the hyperboloid, the difference between the distance to f_2 and the distance to f_1 is $2a$. The distance between the foci is $2(a^2 + b^2)^{1/2}$. The two free parameters a and b can be adjusted to set both the diameter of the subreflector as needed to intercept rays from the edge of the primary and the height of the secondary focus on the z-axis. The **magnification** provided by the subreflector is

$$M = \frac{\tan(\theta_1/2)}{\tan(\theta_2/2)}, \qquad (3.147)$$

where θ_1 is the half angle subtended by the primary viewed from f_1 and θ_2 is the half angle subtended by the secondary viewed from f_2. A small subreflector is light, easy to tilt, and reduces standing waves, but it subtends a small angle $2\theta_2$ at f_2 so a feed horn several wavelengths in diameter is required to illuminate it properly.

The Parkes 210-foot (since renamed to 64-m) telescope (Figure 3.22) in Australia was built about the same time as the 140-foot telescope, but its alt-az mount and centrally concentrated reflector backup structure pointed the way to the design of modern radio telescopes.

Elevation-dependent gravitational deformations degrade the short-wavelength performance of tilting reflectors. The deformations can be controlled by designing the backup structure so that the deformed surface remains paraboloidal. The deformations cause the focal point to shift slightly in elevation, but this shift can be accommodated by moving the feed slightly to track the focus. The

Figure 3.23. The 100-m telescope near Effelsberg, Germany. The first deliberately homologous telescope, it works to $\lambda \sim 7$ mm. Note the large Gregorian subreflector above the prime focus. Photo by Matthias Kadler.

first large **homologous telescope** deliberately designed to deform this way is the 100-m telescope (Figure 3.23) of the Max Planck Institut für Radioastronomie (MPIfR) near Effelsberg, Germany. Despite its huge size, its passive surface remains accurate enough to work at wavelengths as short as $\lambda = 7$ mm over a range of elevations. The 100-m telescope has a concave **Gregorian subreflector** above the prime focus. The geometry of a symmetric Gregorian system is shown in Figure 3.24. As with the Cassegrain subreflector, the Gregorian reflector shape is determined by the requirement that all parallel axial rays travel the same distance to reach the **secondary focus** at f_2. For a subreflector located above the prime focus, the required shape is an ellipsoid whose major axis coincides with the major axis of the paraboloid. The equation

$$\frac{z^2}{a^2} + \frac{r^2}{b^2} = 1 \tag{3.148}$$

with $a > b$ defines such an ellipsoid. From any point on the ellipsoid, the sum of the distance to f_2 and the distance to f_1 is $2a$. The distance between the foci is $2(a^2 - b^2)^{1/2}$.

The Arecibo radio telescope (Color Plate 2 and Figure 3.25) was originally designed as a radar facility to study the ionosphere via Thomson scattering of 430 MHz ($\lambda = 70$ cm) radio waves by free electrons. Thermal motions of truly free electrons would greatly Doppler broaden the bandwidth of the radar echo and lower

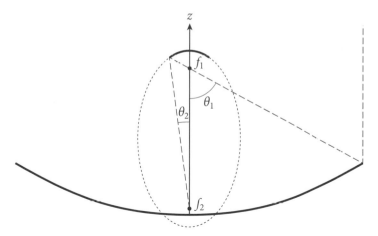

Figure 3.24. Cross section of a radio telescope rotationally symmetric around the z-axis and having a Gregorian subreflector. Parallel rays from a distant radio source are reflected by the circular paraboloid whose prime focus is at the point marked f_1. The Gregorian subreflector is a circular ellipsoid located above the prime focus. It reflects these rays to the feed located at the secondary focus f_2 just above the vertex of the paraboloid.

Figure 3.25. The Arecibo feed-support platform can steer the beam anywhere up to 20 degrees from the zenith even though the spherical reflector is fixed. The curved azimuth arm rotates about the vertical under a circular ring at the base of the fixed triangular structure. The carriage house under the left side of the azimuth arm carries a waveguide line feed that corrects for spherical aberration. The dome under the carriage house on the right side contains the Gregorian secondary mirror and tertiary correcting mirror, illuminated by waveguide horn feeds. The carriage houses can move along tracks at the bottom of the azimuth arm to change the zenith angle of the beam.

the received signal-to-noise ratio, so a very large antenna was built for sensitivity. However, ionospheric electrons are coupled to the much heavier ions on scales larger than the ionospheric Debye length, which is only a few mm. This is much smaller than the 70 cm wavelength, so the actual bandwidth is determined by thermal motions of the much heavier ions and is lower by two orders of magnitude. Thus a far smaller dish would have sufficed! Astronomers have benefited from this oversight and use Arecibo's huge collecting area at frequencies up to about 10 GHz for Solar-System radar (planets, moons, asteroids), pulsar studies, HI 21-cm line observations of galaxies, and other observations that need high sensitivity. The spherical reflector can be very large because it is does not move. A sphere is symmetric about any axis passing through its center, so the Arecibo beam can be steered by moving the feed instead of the reflector. The curved feed-support arm visible in Figure 3.25 is 300 feet long and rotates in azimuth below the fixed triangular structure. The feeds are mounted under two carriage houses that move along tracks on the bottom of the feed arm and permit tracking at zenith angles up to 20 degrees. The feed illumination spills over the edge of the fixed reflector at high zenith angles, so a large **ground screen** surrounds the spherical reflector to reflect the spillover onto the cold sky and keep it away from the warm and noisy ground.

A spherical reflector focuses a distant point source onto a radial line segment, so a radial line feed (see Figure 3.25) up to 96 feet long is needed to illuminate the entire aperture efficiently from the prime focus. The line feed is a slotted waveguide tapered to control the group velocity (Equation 3.143) and phase up radiation arriving from all over the reflector. However, long slotted-waveguide line feeds are inherently narrowband, and ohmic losses in the long slotted waveguide increase the system temperature significantly at short wavelengths. The "golf ball" under the feed arm at Arecibo (Figure 3.25) houses an enormous Gregorian subreflector and a tertiary reflector that allow low-noise wideband point feeds to illuminate an ellipse about 200 m by 225 m in size on the main reflector.

The 100-m Robert C. Byrd Green Bank Telescope (GBT) (Color Plate 1) is the successor to the collapsed 300-foot telescope in Green Bank, and it incorporates a number of new design features to optimize its sensitivity and short-wavelength performance.

The actual reflector is a $110 \text{ m} \times 100 \text{ m}$ off-axis section of an imaginary symmetric paraboloid 208 m in diameter. Projected onto a plane normal to the beam, it is a 100-m diameter circle. Because the projected edge of the actual reflector is 4 m away from the axis of the 208-m paraboloid, the focal point does not block the aperture. The GBT enjoys the same clear-aperture benefits of waveguide horns— a very clean beam and low spillover noise—but is much larger than any practical horn antenna. The clean beam is especially valuable for suppressing radio-frequency interference (RFI) and stray radiation from very extended sources, such as HI emission from the Galaxy.

The vertical cross section of the GBT plotted in Figure 3.26 shows how the offset Gregorian subreflector does not block any radiation falling onto the primary reflector. The Gregorian subreflector is above the prime focus at f_1, so prime-focus operation is possible by raising a swinging boom carrying the prime-focus feeds into position below the subreflector, although this temporarily blocks the Gregorian subreflector. The huge feed-support arm is over 60 m long, the focal length of the 208-m paraboloid. The feed-support arm has a much larger cross section than

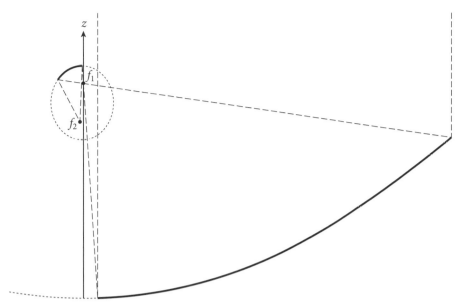

Figure 3.26. Vertical cross section showing the symmetry plane of the GBT. The actual dish shown by the continuous curve is an asymmetric section of the symmetric parent paraboloid (dotted curve) whose diameter is 208 m. The inner edge of the GBT reflector is 4 m to the right of the z-axis of symmetry so the foci and feed-support structure to the left of the z-axis never block the incoming radiation. The primary focal length is $f_1 = 60$ m, and the distance from f_1 to the secondary focus f_2 is 11 m. The secondary focus is offset by 1.068 m from the symmetry axis to minimize **instrumental polarization**. The diameter of the Gregorian subreflector is 8 m. The secondary focus is far above the vertex of the parent paraboloid, but the off-axis feed support arm of the GBT is strong enough to support a large feed/receiver cabin (Figure 3.27) at this height.

the feed-support structures of symmetrical telescopes, which must be kept as thin as possible to minimize blockage. This GBT arm is very strong and can support heavy subreflectors, feeds, equipment rooms, and an elevator. At the top of the arm and above the prime focus is the concave Gregorian subreflector. This subreflector illuminates feeds emerging through the roof of a large receiver cabin attached to the feed arm a short distance below (Figure 3.27). Because these feeds are relatively close to the subreflector, even a moderately small subreflector subtends a large angle as viewed from the feeds, which can then be moderately small themselves. Most of the receivers and feeds needed to cover the frequency range $1 < \nu\,(\text{GHz}) < 100$ can fit into the receiver cabin simultaneously and are available for use on short notice.

The main reflector is supported by a backup structure that deforms homologously to ensure good efficiency at wavelengths as short as $\lambda = 2$ cm. The active reflecting surface consists of approximately two thousand panels, each about 2 m on a side. The corners of individual panels are mounted on computer-controlled actuators that can move the panels up or down as needed to continuously correct the overall shape of the surface. Photogrammetry was used to measure the surface at the rigging elevation (the elevation at which the surface was originally set). The gravitational deformations at other elevation angles predicted by the finite-element computer

Figure 3.27. The concave Gregorian subreflector just above the prime focus of the GBT images sources onto conical horn feeds extending through the top of the rectangular receiver cabin. The prime-focus feed arm is shown stowed out of the way of the subreflector. None of these offset structures block radiation reflected from the main aperture. Image credit: NRAO/AUI/NSF.

model of the GBT are continuously removed by the actuators as the telescope moves. As a result, the rms surface error is only $\sigma \approx 0.2$ mm and the GBT has a high surface efficiency at wavelengths as short as $\lambda \approx 3$ mm.

The 30-m IRAM (Institut de Radioastronomie Millimétrique) telescope (Figure 3.28) is the largest telescope operating at 3, 2, 1, and 0.8 mm. Its rms surface error is only 55 μm, and its pointing accuracy is about 1 arcsec.

3.6 RADIOMETERS

Natural radio emission from the cosmic microwave background, discrete astronomical sources, the Earth's atmosphere, and the ground is random broadband noise that

Figure 3.28. The 30-m IRAM telescope on Pico Veleta in Spain. Image credit: IRAM.

is nearly indistinguishable from the noise generated by a warm resistor (Section 2.5) or by receiver electronics. A radio receiver used to measure the average power of the noise coming from a radio telescope in a well-defined frequency range is called a **radiometer**. The noise voltage has a Gaussian amplitude distribution with zero mean, and it fluctuates on the very short timescales (nanoseconds) comparable with the inverse of the radiometer bandwidth $\Delta \nu$. A **square-law detector** in the radiometer squares the input noise voltage to produce an output voltage proportional to the input noise power. Noise power is always greater than zero, and the noise from most astronomical sources is **stationary**, meaning that its mean power is steady when averaged over much longer timescales τ (seconds to hours). The Nyquist–Shannon **sampling theorem** (Appendix A.3) states that any function having finite bandwidth $\Delta \nu$ and duration τ can be represented by $2\Delta\nu\tau$ independent samples spaced in time by $(2\,\Delta\nu)^{-1}$. By averaging a large number $N = (2\Delta\nu\,\tau)$ of independent noise samples, an ideal radiometer can determine the average noise power with a fractional uncertainty as small as $(N/2)^{-1/2} = (\Delta\nu\,\tau)^{-1/2} \ll 1$ and detect faint sources that increase the antenna temperature by only a tiny fraction of the total noise power. The ideal radiometer equation expresses this result in terms of the radiometer bandwidth and the averaging time. Gain variations in practical radiometers, fluctuations in atmospheric emission, and confusion by unresolved radio sources may significantly degrade the actual sensitivity compared with that predicted by the ideal radiometer equation.

3.6.1 Band-Limited Noise

The voltage at the output of a radio telescope is the sum of noise voltages from many independent random contributions. The **central limit theorem** [15] states that the amplitude distribution of such noise is nearly Gaussian. Figure 3.29 (lower panel)

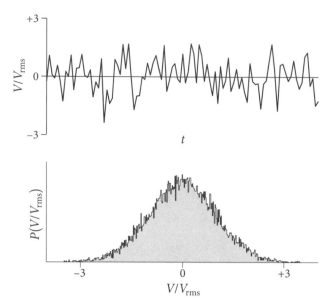

Figure 3.29. The output voltage V of a radio telescope varies rapidly on short timescales, as indicated by the upper plot showing 100 independent samples of band-limited noise drawn from a Gaussian probability distribution $P(V/V_{\text{rms}})$ (lower plot) having zero mean and fixed rms V_{rms}. See Appendix B.5 for a mathematical description of the Gaussian distribution.

shows the histogram of about 20,000 independent voltage samples randomly drawn from a Gaussian parent distribution having rms V_{rms} and mean $\langle V \rangle = 0$. Figure 3.29 (upper panel) shows $N = 100$ successive samples drawn from the Gaussian noise distribution. This sequence of voltages is representative of band-limited noise in the frequency range from 0 to $\Delta \nu$ during a time interval τ such that $(\Delta \nu\, \tau) = N/2 = 50$, e.g., noise with all frequencies up to $\Delta \nu = 1\,\text{MHz}$ sampled every $(2\Delta \nu)^{-1} = 0.5\,\mu\text{s}$ for $\tau = 50\,\mu\text{s}$. This is what the band-limited noise output voltage of a radio telescope looks like.

It is convenient to describe noise power in units of temperature. The noise power per unit bandwidth generated by a resistor of temperature T is $P_\nu = kT$ in the low-frequency limit, so we can define the **noise temperature** of *any* noiselike source in terms of its power per unit bandwidth P_ν:

$$T_{\text{N}} \equiv \frac{P_\nu}{k}, \tag{3.149}$$

where $k \approx 1.38 \times 10^{-23}$ joule K^{-1} is Boltzmann's constant.

The temperature equivalent to the *total* noise power from all sources referenced to the input of a radiometer connected to the output of a radio telescope is called the **system noise temperature** T_{s}. It is the sum of many contributors to the antenna temperature plus the radiometer noise temperature T_{r}:

$$T_{\text{s}} = T_{\text{cmb}} + T_{\text{rsb}} + \Delta T_{\text{source}} + [1 - \exp(-\tau_{\text{A}})]T_{\text{atm}} + T_{\text{spill}} + T_{\text{r}} + \cdots . \tag{3.150}$$

There are seven antenna-temperature contributions listed explicitly in Equation 3.150:

1. $T_{cmb} \approx 2.73$ K is from the nearly isotropic cosmic microwave background.
2. T_{rsb} is the average sky brightness temperature contributed by all "background" radio sources. Extragalactic sources add [29]

$$\left(\frac{T_{rsb}}{0.1\,\text{K}}\right) \approx \left(\frac{\nu}{1.4\,\text{GHz}}\right)^{-2.7} \tag{3.151}$$

   in all directions, and the Galactic plane is a bright diffuse source at low ($\nu < 0.5$ GHz) frequencies [44].
3. ΔT_{source} is from the astronomical source being observed, written with a Δ to emphasize that it is usually much smaller than the total system noise: $\Delta T_{source} \ll T_s$. For example, in the $\nu_{RF} \approx 4.85$ GHz sky survey made with the 300-foot telescope, the system noise was $T_s \approx 60$ K, but the faintest detected sources added only $\Delta T_{source} \approx 0.01$ K.
4. $[1 - \exp(-\tau_A)]T_{atm}$ is the brightness of atmospheric emission in the telescope beam (Section 2.2.3).
5. T_{spill} accounts for spillover radiation that the feed picks up in directions beyond the edge of the reflector, primarily from the ground.
6. T_r is the **radiometer noise temperature** attributable to noise generated by the radiometer itself, referenced to the radiometer input. All radiometers generate noise, and any radiometer can be represented by an equivalent circuit consisting of a noiseless radiometer whose input is connected to a resistor of temperature T_r. Radiometer noise is usually minimized by cooling the radiometer to cryogenic temperatures. However, radiometers are not just matched resistors, so T_r may be either lower or higher than the physical temperature of the radiometer itself.
7. "$\cdots$" represents any other noise sources that might be important. An example is emission resulting from ohmic losses in the long slotted waveguide feed at Arecibo (Figure 3.25).

3.6.2 Radiometers

The purpose of the simplest **total-power radiometer** is to measure the time-averaged power of the input noise in some well-defined radio frequency (RF) range

$$\nu_{RF} - \frac{\Delta\nu}{2} \quad \text{to} \quad \nu_{RF} + \frac{\Delta\nu}{2}, \tag{3.152}$$

where $\Delta\nu$ is the receiver **bandwidth**. For example, the receivers used on the 300-foot telescope to make the $\lambda \approx 6$ cm continuum survey of the northern sky had a center radio frequency $\nu_{RF} \approx 4.85 \times 10^9$ Hz and a bandwidth $\Delta\nu \approx 6 \times 10^8$ Hz.

The simplest radiometer (Figure 3.30) consists of four stages in series: (1) a low-loss **bandpass filter** that passes input noise only in the desired frequency range; (2) a **square-law detector** whose output voltage V_o is proportional to the square of its input voltage; that is, V_o is proportional to its input power; (3) a signal averager or **integrator** that smooths the rapidly fluctuating detector output; and (4) a voltmeter or other device to measure and record the smoothed voltage. After passing through an input filter of width $\Delta\nu < \nu_{RF}$, the noise voltage is no longer

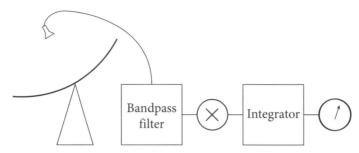

Figure 3.30. The simplest radiometer filters the broadband noise coming from the telescope, multiplies the filtered voltage by itself (square-law detection), smooths the detected voltage, and measures the smoothed voltage. The function of the detector is to convert the noise voltage, which has zero mean, to noise power, which is proportional to the square of voltage.

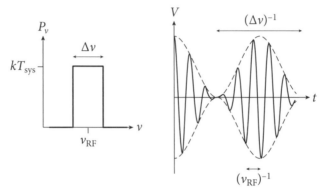

Figure 3.31. The voltage output $V(t)$ of the filter with center frequency ν_{RF} and bandwidth $\Delta\nu < \nu_{RF}$ is a sinusoid with frequency ν_{RF} whose envelope (dashed curves) fluctuates on timescales $(\Delta\nu)^{-1} > (\nu_{RF})^{-1}$.

completely random; it looks more like a sine wave of frequency $\approx \nu_{RF}$ whose amplitude envelope (dashed curve in Figure 3.31) varies randomly on timescales $\Delta t \approx (\Delta\nu)^{-1} > \nu_{RF}^{-1}$. The positive and negative envelopes are similar so long as $\Delta\nu \ll \nu_{RF}$.

The filtered output is sent to a square-law detector whose output voltage V_o is proportional to its input power. For a narrowband (quasi-sinusoidal) input voltage $V_i \approx \cos(2\pi\nu_{RF}t)$ at frequency ν_{RF}, the detector output voltage would be $V_o \propto \cos^2(2\pi\nu_{RF}t)$. This can be rewritten as $[1 + \cos(4\pi\nu_{RF}t)]/2$, a function whose mean value is proportional to the average power of the input signal. In addition to the DC (zero-frequency) component there is an oscillating component at twice the input frequency ν_{RF}. The detector output spectrum for a finite bandwidth $\Delta\nu$ and a typical waveform is shown in Figure 3.32. The oscillations under the envelope approach zero every $\Delta t \approx (2\nu_{RF})^{-1}$. Thus the oscillating component of the detector output is centered near the frequency $2\nu_{RF}$. The detector output also has frequency components near zero (DC) because the mean output voltage is greater than zero.

Both the rapidly varying component at frequencies near $2\nu_{RF}$ and its envelope vary on timescales that are normally much shorter than the timescales on which the average signal power ΔT varies. The unwanted rapid variations can be

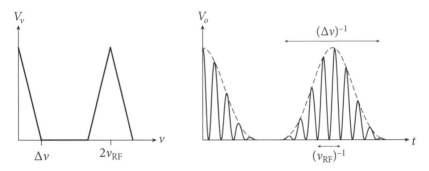

Figure 3.32. The output voltage V_o of a square-law detector (Figure 3.33) is proportional to the square of the input voltage. It is always positive, so its mean (DC, or zero-frequency component) is positive and proportional to the input power. The high frequency ($\nu \approx 2\nu_{RF}$) fluctuations add no information about the source and are filtered out in the next stage.

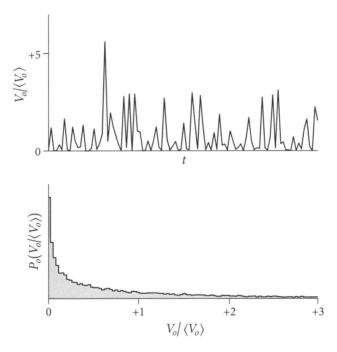

Figure 3.33. The upper plot shows the output voltage V_o of a square-law detector whose input is the Gaussian noise shown by the upper plot in Figure 3.29. The output voltage histogram (lower plot) is peaked sharply near zero and has a long positive tail. The mean detected voltage $\langle V_o \rangle$ equals the mean squared input voltage, and the rms of the detected voltage distribution is $2^{1/2} \langle V_o \rangle$. For a full derivation of the detector output distribution and its rms, see Appendix B.6.

suppressed by taking the arithmetic mean of the detected envelope over some timescale $\tau \gg (\Delta\nu)^{-1}$ by integrating or averaging the detector output. This integration might be done electronically by smoothing with an RC (resistance plus capacitance) filter or numerically by sampling and digitizing the detector output voltage and then computing its running mean.

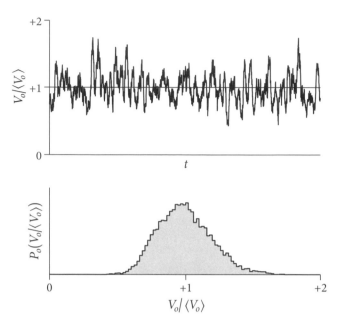

Figure 3.34. The smoothed output voltage from the integrator varies on timescale τ with small amplitude σ_T given by the ideal radiometer equation. The top part of this figure shows the detected voltage smoothed by an $N = 50$ sample running mean, and the bottom part shows the amplitude distribution of the smoothed voltage. This amplitude distribution has mean $\langle V_o \rangle$ and rms $(2/N)^{1/2} \langle V_o \rangle = \langle V_o \rangle / 5$. As N grows, the smoothed amplitude distribution approaches a Gaussian. The sampling theorem (Appendix A.3) states that $N = (2\Delta\nu\,\tau)$ so $(\Delta\nu\,\tau) = 25$ for this example.

Integration greatly reduces the receiver output fluctuations. In the time interval τ there are $N = (2\Delta\nu\,\tau)$ independent samples of the total noise power T_s, each of which has an rms error $\sigma_T \approx 2^{1/2} T_s$. The rms error in the average of $N \gg 1$ independent samples is reduced by the factor $\sqrt{N}$, so the rms receiver output fluctuation σ_T (see Appendix B.6 for a formal derivation of this result) is only

$$\sigma_T = \frac{2^{1/2} T_s}{N^{1/2}}. \tag{3.153}$$

In terms of bandwidth $\Delta\nu$ and **integration time** τ,

$$\boxed{\sigma_T \approx \frac{T_s}{\sqrt{\Delta\nu\,\tau}}} \tag{3.154}$$

after smoothing. The central limit theorem of statistics implies that heavily smoothed ($\Delta\nu\,\tau \gg 1$) output voltages also have a nearly Gaussian amplitude distribution. This important equation is called the **ideal radiometer equation** for a total-power receiver. The weakest detectable signals ΔT only have to be several (typically five) times the output rms σ_T given by the radiometer equation, not several times the total system noise T_s. The product $(\Delta\nu\,\tau)$ may be quite large in practice (10^8 is not unusual), so signals as faint as $\Delta T \sim 5 \times 10^{-4} T_s$ would be detectable. Figures 3.34

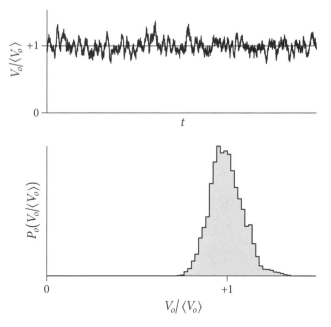

Figure 3.35. When the same detector output is smoothed over $N = 200$ samples instead of $N = 50$ samples (Figure 3.34), the mean remains the same but the rms falls by a factor of $4^{1/2} = 2$ to $\langle V_o \rangle / 10$. In this example $(\Delta \nu \, \tau) = 100$.

and 3.35 illustrate the effects of smoothing the detector output by taking running means of lengths $N = 50$ and $N = 200$ samples.

3.6.3 Some Caveats

The ideal radiometer equation suggests that the sensitivity of a radio observation improves as $\tau^{1/2}$ forever. In practice, systematic errors set a floor to the noise level that can be reached. Receiver gain changes, erratic fluctuations in atmospheric emission, or "confusion" by the unresolved background of continuum radio sources usually limit the sensitivity of single-dish continuum observations.

3.6.3.1 Receiver Gain and Atmospheric Fluctuations

Radiometers contain a series of amplifiers that multiply the weak input powers $P_{\mathrm{in}} = k T_{\mathrm{s}} \Delta \nu \sim 10^{-14}\,\mathrm{W}$ to milliwatt levels. The output voltage of a total-power receiver is directly proportional to the overall power gain G of the receiver. If G isn't perfectly constant, the change in output voltage caused by a **gain fluctuation** ΔG in a practical radiometer produces a false signal whose apparent temperature

$$\sigma_G = T_{\mathrm{s}} \left(\frac{\Delta G}{G} \right) \tag{3.155}$$

is indistinguishable from a comparable change σ_T caused by noise in an ideal radiometer. Receiver gain fluctuations and noise fluctuations are independent random processes, so their **variances** (the variance is the square of the rms) add, and the total receiver output fluctuation becomes

$$\sigma_T^2 = \sigma_{\text{noise}}^2 + \sigma_G^2 \tag{3.156}$$

$$= T_s^2 \left[\frac{1}{\Delta \nu \, \tau} + \left(\frac{\Delta G}{G} \right)^2 \right]. \tag{3.157}$$

The **practical total-power radiometer equation** is thus

$$\sigma_T \approx T_s \left[\frac{1}{\Delta \nu \, \tau} + \left(\frac{\Delta G}{G} \right)^2 \right]^{1/2}. \tag{3.158}$$

Clearly, radiometer gain fluctuations will degrade the sensitivity of an observation unless

$$\left(\frac{\Delta G}{G} \right) \ll \frac{1}{\sqrt{\Delta \nu \, \tau}}. \tag{3.159}$$

For example, the 5 GHz receiver used to make the sky survey with the 300-foot telescope had $\Delta \nu \approx 6 \times 10^8$ Hz and $\tau \approx 0.1$ s, so the fractional gain fluctuations on timescales up to a few seconds (the time to scan one baseline length) had to satisfy

$$\frac{\Delta G}{G} \ll \frac{1}{\sqrt{6 \times 10^8 \, \text{Hz} \cdot 0.1 \, \text{s}}} = 1.3 \times 10^{-4}. \tag{3.160}$$

This is difficult to achieve in practice. Gain fluctuations typically have "$1/f$" power spectra, where f is the postdetection frequency, so they are larger on longer timescales and increasing τ eventually results in a *higher* output noise level. The gain stability of a receiver is often specified by the "$1/f$ knee" frequency f_k, the postdetection frequency at which $\sigma_{\text{noise}} = \sigma_G$. Integrations longer than $\tau \approx 1/(2\pi f_k)$ will likely increase the receiver output fluctuations. Depending on the stability and bandwidth of the radiometer, ~ 1 Hz $< f_k <\sim 1$ kHz.

Fluctuations in atmospheric emission also add to the noise in the output of a simple total-power receiver. Water vapor is the main culprit because it is not well mixed in the atmosphere, and noise from water-vapor fluctuations can be a significant problem at frequencies of ~ 5 GHz and up.

One way to minimize the effects of fluctuations in both receiver gain and atmospheric emission is to make a *differential* measurement by comparing signals from two adjacent feeds. The method of switching rapidly between beams or loads is called **Dicke switching** after Robert Dicke, its inventor. Figure 3.36 shows the block diagram of a beam-switching Dicke radiometer. If the system temperatures are T_1 and T_2 in the two positions of the switch, then the receiver output is proportional to $T_1 - T_2 \ll T_1$ and the effect of gain fluctuations is only

$$\sigma_G \approx (T_1 - T_2) \frac{\Delta G}{G} \ll T_1 \frac{\Delta G}{G}. \tag{3.161}$$

Likewise, the atmospheric emission in two nearly overlapping beams through the troposphere is nearly the same, so most of the tropospheric fluctuations cancel out. The main drawback with Dicke switching is that the receiver output fluctuations, relative to the source signal in a single beam, are doubled because the source signal

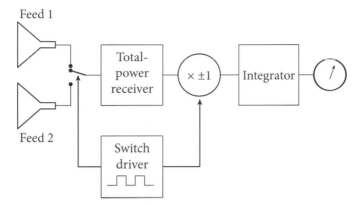

Figure 3.36. Block diagram of a beam-switching **differential radiometer**. The total-power receiver is switched between two feeds, one pointing at the source and one displaced by a few beamwidths to avoid the source but measure emission from nearly the same sample of atmosphere. The output of the total-power receiver is multiplied by +1 when the receiver is connected to the on-source feed and by −1 when it is connected to the reference feed. Fluctuations in atmospheric emission and in receiver gain are effectively suppressed for frequencies below the switching rate, which is typically in the range 10 to 1000 Hz.

is being received only half the time while the noise power is present all the time. The ideal radiometer equation for a Dicke switching receiver is

$$\sigma_T = \frac{2T_s}{\sqrt{\Delta\nu\,\tau}}. \tag{3.162}$$

3.6.3.2 Confusion

Single-dish radio telescopes have large collecting areas but relatively broad beams at long wavelengths. Nearly all discrete continuum sources are extragalactic and extremely distant, so they are distributed randomly and isotropically on the sky. The sky-brightness fluctuations caused by numerous faint sources in every telescope beam are called **confusion**, and confusion usually limits the sensitivity of single-dish continuum observations at frequencies below $\nu \sim 10\,\mathrm{GHz}$. Figure 3.37 is a profile plot of confusion fluctuations in a low-resolution image. Figure 3.38 shows contours from a portion of that low-resolution image superimposed on an overlapping high-resolution gray-scale image.

Although the amplitude distribution of confusion is distinctly non-Gaussian, the "rms" confusion σ_c calculated by ignoring the long positive tail is a widely quoted. At cm wavelengths, the rms confusion in a Gaussian telescope beam with FHWM θ is

$$\left(\frac{\sigma_c}{\mathrm{mJy\ beam}^{-1}}\right) \approx \begin{cases} 0.2\left(\dfrac{\nu}{\mathrm{GHz}}\right)^{-0.7}\left(\dfrac{\theta}{\mathrm{arcmin}}\right)^2 & (\theta > 0.17\ \mathrm{arcmin}), \\[4mm] 2.2\left(\dfrac{\nu}{\mathrm{GHz}}\right)^{-0.7}\left(\dfrac{\theta}{\mathrm{arcmin}}\right)^{10/3} & (\theta < 0.17\ \mathrm{arcmin}). \end{cases} \tag{3.163}$$

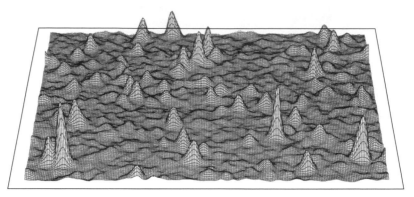

Figure 3.37. A profile plot covering 45 deg^2 of sky imaged by the 300-foot telescope at 1.4 GHz with $\theta = 12$ arcmin resolution [25]. The ubiquitous fluctuations with rms $\sigma \approx 20$ mJy beam^{-1} are caused by the superposition of numerous faint sources, not receiver noise.

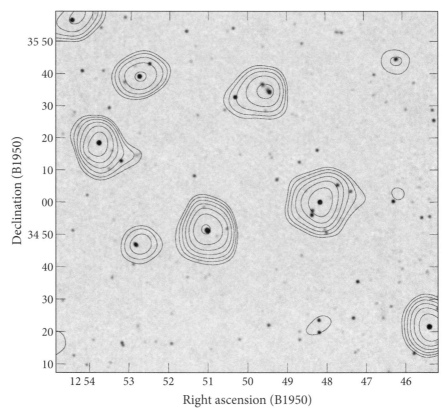

Figure 3.38. The contour image [25] is a 4 deg^2 subset of the area shown in Figure 3.37. The contours start at 45 mJy beam^{-1} $\approx 2\sigma_c$ and are spaced by factors of $2^{1/2}$, so sources with fewer than four contours are below the $5\sigma_c$ confusion limit. The gray-scale plot is a 1.4 GHz VLA image made with $\theta = 45$ arcsec resolution. Some of the faint "sources" seen by the 300-foot telescope are blends of two or more fainter sources resolved by the VLA.

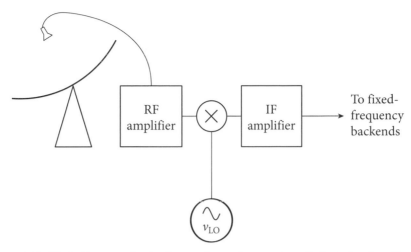

Figure 3.39. Block diagram of a simple superheterodyne receiver. Only the local oscillator is tuned to change the observing frequency range.

Individual sources fainter than the **confusion limit** $\approx 5\sigma_c$ cannot be detected reliably, no matter how low the receiver noise. Most continuum observations of faint sources at frequencies below $\nu \sim 10\,\mathrm{GHz}$ are made with interferometers instead of single dishes because interferometers can synthesize much smaller beamwidths θ and hence have significantly lower confusion limits.

Confusion by steady continuum sources has a much smaller effect on observations of spectral lines or rapidly varying sources such as pulsars.

3.6.4 Superheterodyne Receivers

Few actual radiometers are as simple as those described above. Nearly all practical radiometers are **superheterodyne** receivers (Figure 3.39), in which the RF amplifier is followed by a **mixer** that multiplies the RF signal by a sine wave of frequency ν_{LO} generated by a **local oscillator** (LO). The product of two sine waves contains the sum and difference frequency components,

$$2\sin(2\pi \nu_{\mathrm{LO}} t)\sin(2\pi \nu_{\mathrm{RF}} t) = \cos[2\pi(\nu_{\mathrm{LO}} - \nu_{\mathrm{RF}})t] - \cos[2\pi(\nu_{\mathrm{LO}} + \nu_{\mathrm{RF}})t], \quad (3.164)$$

so the mixer acts as a frequency shifter. For example, if $\nu_{\mathrm{LO}} = 12\,\mathrm{GHz}$ and $\nu_{\mathrm{RF}} = 9\,\mathrm{GHz}$, the mixer output frequency, called the **intermediate frequency** (IF), will be $\nu_{\mathrm{LO}} - \nu_{\mathrm{RF}} = 3\,\mathrm{GHz}$.

The advantages of superheterodyne receivers include

1. shifting the signals to lower frequencies $\nu_{\mathrm{IF}} < \nu_{\mathrm{RF}}$ where they are easier to amplify, transmit over long distances, filter, and digitize;
2. tunability over a wide range of ν_{RF};
3. tuning by adjusting *only* the local oscillator frequency so that
4. the IF amplifier and **back-end** devices such as multichannel filter banks or digital spectrometers can all operate over fixed frequency ranges.

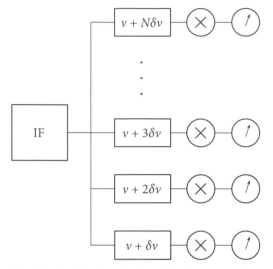

Figure 3.40. An analog filter bank splits the broadband output of an IF amplifier into N contiguous frequency channels of width $\delta \nu$ each. In effect, each channel is a narrowband IF amplifier whose output voltage is detected (multiplied by itself), smoothed, and recorded.

3.6.5 Spectrometers

The simplest superheterodyne radiometer measures the total power in its normally broad IF passband of width $\Delta \nu$. A **spectrometer** is a backend that divides that passband into N adjacent narrow frequency ranges of width $\delta \nu \leq \Delta \nu / N$ and simultaneously measures the power in all N channels to quickly locate and resolve spectral features such as atomic and molecular lines (Section 7.1).

The most straightforward spectrometer is a **filter bank** of narrowband analog filters connected in parallel and with center frequencies uniformly spaced by $\delta \nu$ (Figure 3.40). Each **channel** acts as a separate IF and has its own detector. However, the channel gains, bandpasses, and detector responses of an analog filter bank must be very closely matched and stable to yield smooth spectral baselines, so analog filter banks with more than $N \sim 10^2$ channels are difficult to build and tune. Analog filter banks are also inflexible because their channel bandwidths $\delta \nu$ and numbers N cannot be changed easily. Flexible spectrometers with $N \sim 10^3$ or even $N \sim 10^4$ frequency channels require digital signal processing (**DSP**) techniques.

For many years, most digital spectrometers were **autocorrelation spectrometers** using the Wiener–Khinchin theorem (Equation A.18) to compute power spectra from digitally sampled time series (see Appendix A.3) of the band-limited IF output. A sampled copy of a portion of the input radio signal is delayed by a series of progressively longer time delays, the delayed signals are multiplied with the original signal, and their products are integrated. This series of operations is an **autocorrelation** (Appendix A.7). If the digital samples contain only one or two bits (two or three levels) each, autocorrelation can be performed in hardware and often in a single chip, with relatively simple digital logic. These autocorrelation functions (ACFs) can be integrated to build up signal-to-noise and then finally converted into a power spectrum via a discrete Fourier transform (usually an FFT; see Appendix A.2) of the ACF via the Wiener–Khinchin theorem. Autocorrelation spectrometers allow the integration of very deep (i.e., long-duration) spectra using relatively simple digital hardware and without computing many "costly" FFTs directly on incoming Nyquist-sampled data; only one FFT is computed at the very end of the integration. Similar techniques, but using cross-correlation of the signals from different antennas, are often used to calculate spectra from radio interferometers.

With the continuing improvements in the speeds and capabilities of DSP systems, spectra are increasingly being computed directly via FFTs of a Nyquist-sampled band. The Fourier amplitudes are squared to make power spectra, and the power spectra are accumulated for deep spectral integrations. Such systems are known as **Fourier transform spectrometers**, and the FFTs can be computed in a variety of ways. Many recent spectrometers use Field Programmable Gate Arrays (**FPGAs**) to compute the FFTs, integrate, and compute polarization products, all on a single chip. Other hybrid designs use FPGAs to divide the band into coarse channels and pass those effectively Nyquist-sampled subbands off to CPUs, other FPGAs, or Graphical Processing Units (GPUs) for further processing, such as coherent dedispersion and folding of pulsar data in a pulsar back-end, or much finer frequency resolution and perhaps even active interference removal for high-sensitivity spectroscopy applications. The new VErsatile GBT Astronomical Spectrometer (VEGAS) is a hybrid Fourier transform spectrometer. The capabilities of such systems, especially given the fidelity provided by sampling with eight or more bits precision, is making them the new standard back-end technology for radio astronomy.

3.6.6 Measuring Radiometer Noise

The radiometer itself usually contributes significantly to the total system noise temperature T_{sys}. Any radiometer can be modeled by an equivalent circuit consisting of an ideal noiseless radiometer plus an input matched load resistor at temperature T_r, where T_r is called the **radiometer input noise temperature**.

The simplest way to measure T_r is to connect a matched "hot" load resistor whose physical temperature is T_h to the radiometer input and record the detector output voltage V_h, and then replace it with a "cold" load whose physical temperature is T_c and record the output voltage V_c. Often the hot load is just a resistor at room temperature $T_h \approx 290$ K and the cold load is a resistor immersed in liquid nitrogen at its boiling temperature $T_c \approx 77$ K.

For each measurement, the square-law detector output voltage is proportional to the total input noise power generated by the actual load plus the imaginary resistor whose temperature is T_r. In the low-frequency Nyquist approximation $P_\nu = kT$, so

$$V_h = P_\nu \, \Delta\nu \, G = k(T_h + T_r) \, \Delta\nu \, G, \tag{3.165}$$

$$V_c = P_\nu \, \Delta\nu \, G = k(T_c + T_r) \, \Delta\nu \, G, \tag{3.166}$$

where $\Delta\nu$ is the bandwidth and G is the overall gain of the radiometer. Both G and $\Delta\nu$ cancel out in the Y **factor** defined by

$$Y \equiv \frac{V_h}{V_c} = \frac{T_h + T_r}{T_c + T_r}, \tag{3.167}$$

so they do not have to be measured. Equation 3.167 can be solved for the radiometer noise temperature

$$T_r = \frac{T_h - YT_c}{Y - 1}.$$
(3.168)

This technique for measuring T_r is called the Y-**factor method**.

Communications engineers often specify the radiometer **noise factor** F_n defined by

$$F_n \equiv \frac{T_r + T_0}{T_0},$$
(3.169)

where the standard temperature defined as $T_0 \equiv 290\,\text{K}$ is close to room temperature. The numerator in Equation 3.169 is proportional to the detected output voltage of the radiometer connected to an ambient-temperature load and the denominator is the output of a noiseless radiometer connected to an ambient-temperature load. In terms of F_n, the radiometer noise temperature is

$$T_r = (F_n - 1)T_0.$$
(3.170)

The related radiometer **noise figure** NF used by many commercial manufacturers of amplifiers and radiometers is just the noise factor F_n expressed in dB:

$$\text{NF} \equiv 10\log_{10}(F_n).$$
(3.171)

3.7 INTERFEROMETERS

Every practical single-dish radio telescope (Section 3.5) has relatively low angular resolution and pointing accuracy, small field-of-view, and limited sensitivity. The largest fully steerable dish has diameter $D \approx 100\,\text{m}$ and its angular resolution is diffraction limited to $\theta \approx \lambda/D$ radians, so impossibly large diameters would be needed to achieve sub-arcsecond resolution at radio wavelengths. Pointing and source-tracking accuracy is also a problem for a large single dish. The telescope beam should be able to follow a radio source on the sky within $\sigma \approx \theta/10$ for reasonably accurate photometry or imaging. The accuracy with which the actual beam direction during an observation can be recovered by later data analysis determines the accuracy with which the sky position of a radio source can be measured. Gravitational sagging, telescope deformations caused by differential solar heating, and torques caused by wind gusts combine to limit the mechanical tracking and pointing accuracies of the best radio telescopes to $\sigma \sim 1$ arcsec. Most optical telescopes can make high-resolution images covering large areas of sky rapidly because their large fields-of-view $\Omega_{\text{FoV}} \gg \theta^2$ cover millions or billions of pixels. In contrast, most single-dish radio telescopes have only one or several beams. The geometric area of a single dish is just $\pi D^2/4$, while the geometric area $N\pi D^2/4$ of an interferometer with N dishes can be arbitrarily large. The *continuum* sensitivity of a single dish is strongly limited by confusion at frequencies below about 10 GHz.

Aperture-synthesis interferometers comprising $N \geq 2$ moderately small dishes have mitigated these and many other practical problems associated with single dishes, such as vulnerability to fluctuations in atmospheric emission and receiver gain, radio-frequency interference, and pointing shifts caused by atmospheric refraction. For example, the Westerbork Synthesis Radio Telescope (Color Plate 3) consists of $N = 14$, $D = 25$ m telescopes on east–west baselines up to $b \approx 3$ km in length. Its total collecting area is that of a single dish with diameter $D_{\mathrm{tot}} \approx N^{1/2} D \approx 92$ m. It has the angular resolution of a diffraction-limited telescope 3 km in diameter. It has the large instantaneous field-of-view of a 25-m telescope, so it can image $\sim (b/D)^2 \sim 10^6$ pixels at once with only one receiver on each telescope. It can measure positions of radio sources with subarcsecond accuracy despite the much larger source-tracking errors of the individual telescopes.

Historically, the total bandwidths and numbers of simultaneous frequency channels of aperture-synthesis interferometers with many dishes were lower than those of single dishes. Recent advances in correlator electronics and computing have largely overcome these practical limitations, so new or updated interferometers such as ALMA (Color Plate 5) and the JVLA (Color Plate 4) are playing an increasingly dominant role in observational radio astronomy. The primary uses of single dishes today are

(a) observing pulsars, which are time variable so they are easy to separate from confusion by time-independent continuum sources;
(b) spectroscopic observations of extended low-brightness sources, again largely immune to confusion;
(c) complementing interferometers by providing "zero-spacing" data on very extended sources or by serving as elements of very long baseline arrays.

3.7.1 The Two-Element Quasi-Monochromatic Interferometer

The simplest radio **interferometer** is a pair of radio telescopes whose voltage outputs are **correlated** (multiplied and averaged), and even the most elaborate interferometers with $N \gg 2$ antennas, often called **elements**, can be treated as $N(N-1)/2$ independent two-element interferometers. Figure 3.41 shows two identical dishes separated by the **baseline vector** $\vec{b}$ of length b that points from antenna 1 to antenna 2. Both dishes point in the same direction specified by the unit vector $\hat{s}$, and θ is the angle between $\vec{b}$ and $\hat{s}$. Plane waves from a distant point source in this direction must travel an extra distance $\vec{b} \cdot \hat{s} = b \cos \theta$ to reach antenna 1, so the output of antenna 1 is the same as that of antenna 2, but it lags in time by the **geometric delay**

$$\tau_{\mathrm{g}} = \frac{\vec{b} \cdot \hat{s}}{c}. \tag{3.172}$$

For simplicity, we first consider a **quasi-monochromatic** interferometer, one that responds only to radiation in a very narrow band $\Delta \nu \ll 2\pi/\tau_{\mathrm{g}}$ centered on frequency $\nu = \omega/(2\pi)$. Then the output voltages of antennas 1 and 2 at time t can be written as

$$V_1 = V \cos[\omega(t - \tau_{\mathrm{g}})] \quad \text{and} \quad V_2 = V \cos(\omega t). \tag{3.173}$$

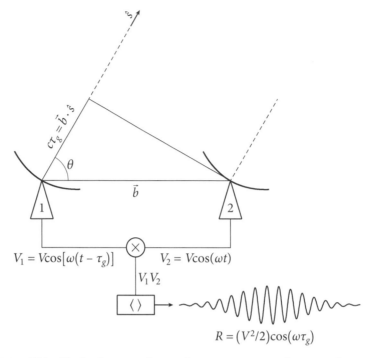

Figure 3.41. This block diagram shows the components of a two-element quasi-monochromatic **multiplying interferometer** observing in a very narrow radio frequency range centered on $\nu = \omega/(2\pi)$. $\hat{s}$ is the unit vector in the direction of a distant point source and $\vec{b}$ is the **baseline vector** pointing from antenna 1 to antenna 2. The output voltage V_1 of antenna 1 is the same as the output voltage V_2 of antenna 2, but it is retarded by the geometric delay $\tau_g = \vec{b} \cdot \hat{s}/c$ representing the additional light-travel delay to antenna 1 for a plane wavefront from a source at angle θ from the baseline vector. These voltages are amplified, multiplied ($\times$), and time averaged ($\langle\,\rangle$) by the **correlator** to yield an output response whose amplitude R is proportional to the flux density of the point source and whose **phase** ($\omega\tau_g$) depends on the delay and the frequency. The quasi-sinusoidal output **fringe** shown occurs if the source direction in the interferometer frame is changing at a constant rate $d\theta/dt$. The broad Gaussian envelope of the fringe shows the primary-beam attenuation as the source passes through the beam of the dishes.

These output voltages are amplified versions of the antenna input voltages; they have not passed through square-law detectors. Instead, a **correlator** multiplies these two voltages to yield the product

$$V_1 V_2 = V^2 \cos[\omega(t - \tau_g)] \cos(\omega t) = \left(\frac{V^2}{2}\right)[\cos(2\omega t - \omega\tau_g) + \cos(\omega\tau_g)] \quad (3.174)$$

that follows directly from the trigonometric identity $\cos x \cos y = [\cos(x + y) + \cos(x - y)]/2$. The correlator also takes a time average long enough ($\Delta t \gg (2\omega)^{-1}$) to remove the high-frequency term $\cos(2\omega t - \omega\tau_g)$ from the **correlator response** (output voltage) R and keep only the slowly varying term

$$R = \langle V_1 V_2 \rangle = \left(\frac{V^2}{2} \right) \cos(\omega \tau_g). \tag{3.175}$$

The voltages V_1 and V_2 are proportional to the electric field produced by the source multiplied by the voltage gains of the two antennas and receivers. Thus the correlator output amplitude $V^2/2$ is proportional to the flux density S of the point source multiplied by $(A_1 A_2)^{1/2}$, where A_1 and A_2 are the effective collecting areas of the two antennas.

Notice that the time-averaged response R of a **multiplying interferometer** is zero. There is no DC output, so fluctuations in receiver gain do not act on the whole system temperature T_s as for a total-power observation with a single dish (Equation 3.155). Uncorrelated noise power from very extended radio sources such as the cosmic microwave background and the atmosphere over the telescopes, also averages to zero in the correlator response. Short interference pulses with duration $t \ll |b|/c$ are also suppressed because each pulse does not reach both telescopes simultaneously. Likewise, a multiplying radio interferometer differs from a classical **adding interferometer**, such as the optical Michelson interferometer, that *adds* the uncorrelated noise power contributions.

The correlator output voltage $R = (V^2/2) \cos(\omega \tau_g)$ varies sinusoidally as the Earth's rotation changes the source direction relative to the baseline vector. These sinusoids are called **fringes**, and the **fringe phase**

$$\phi = \omega \tau_g = \frac{\omega}{c} b \cos \theta \tag{3.176}$$

depends on θ as follows:

$$\frac{d\phi}{d\theta} = \frac{\omega}{c} b \sin \theta \tag{3.177}$$

$$= 2\pi \left(\frac{b \sin \theta}{\lambda} \right). \tag{3.178}$$

The **fringe period** $\Delta \phi = 2\pi$ corresponds to an angular shift $\Delta \theta = \lambda/(b \sin \theta)$. The fringe phase is an exquisitely sensitive measure of source position if the **projected baseline** $b \sin \theta$ is many wavelengths long. Note that fringe phase and hence measured source position is not affected by small tracking errors of the individual telescopes. It depends on time, and times can be measured by clocks with much higher accuracy than angles (ratios of lengths of moving telescope parts) can be measured by rulers. Also, an interferometer whose baseline is horizontal is not affected by the plane-parallel component of atmospheric refraction, which delays the signals reaching both telescopes equally. Consequently, interferometers can determine the positions of compact radio sources with unmatched accuracy, as shown in Figure 1.6. Absolute positions with errors as small as $\sigma_\theta \approx 10^{-3}$ arcsec and differential positions with errors down to $\sigma_\theta \approx 10^{-5}$ arcsec $< 10^{-10}$ rad have frequently been measured.

If the individual antennas comprising an interferometer were isotropic, the interferometer point-source response would be a sinusoid spanning the sky. Such an interferometer is sensitive to only one Fourier component of the sky brightness distribution: the component with angular period $\lambda/(b \sin \theta)$. The response R of a two-element interferometer with directive antennas is that sinusoid multiplied by the product of the voltage patterns of the individual antennas. Normally the two antennas are identical, so this product is the power pattern of the individual antennas and is called the **primary beam** of the interferometer. The primary beam is usually a Gaussian much wider than a fringe period, as indicated in Figure 3.41. The convolution theorem (Equation A.15) states that the Fourier transform of the product of two functions is the convolution of their Fourier transforms, so the interferometer with directive antennas responds to a finite range of angular frequencies centered on $(b \sin \theta/\lambda)$. Because the antenna diameters D must be smaller than the baseline b (else the antennas would overlap), the angular frequency response cannot extend to zero and the interferometer cannot detect an isotropic source—the bulk of the 3 K cosmic microwave background for example. The missing short spacings $(b < D)$ can be provided by a single-dish telescope with diameter $D > b$. Thus the $D = 100$ m GBT can fill in the missing baselines $b < 25$ m that the $D = 25$ m VLA dishes cannot obtain.

Improving the instantaneous point-source response pattern of an interferometer requires more Fourier components; that is, more baselines. An interferometer with N antennas contains $N(N-1)/2$ pairs of antennas, each of which is a two-element interferometer, so the instantaneous **synthesized beam** (the point-source response obtained by averaging the outputs of all of the two-element interferometers) rapidly approaches a Gaussian as N increases. The instantaneous point-source responses of a two-element interferometer with projected baseline length b, a three-element interferometer with three baselines (projected lengths $b/3$, $2b/3$, and b), and a four-element interferometer with six baselines (projected lengths $b/6$, $2b/6$, $3b/6$, $4b/6$, $5b/6$, and b) are shown in Figure 3.42.

Most radio sources are **stationary**; that is, their brightness distributions do not change significantly on the timescales of astronomical observations. For stationary sources, a two-element interferometer with movable antennas could make $N(N-1)/2$ observations to duplicate one observation with an N-element interferometer.

3.7.2 Slightly Extended Sources and the Complex Correlator

The response $R_c = (V^2/2) \cos(\omega \tau_g)$ of the quasi-monochromatic two-element interferometer with a "cosine" correlator (Figure 3.41 and Equation 3.175) to a spatially incoherent slightly extended (much smaller than the primary beamwidth) source with sky brightness distribution $I_\nu(\hat{s})$ near frequency $\nu = \omega/(2\pi)$ is obtained by treating the extended source as the sum of independent point sources:

$$R_c = \int I(\hat{s}) \cos(2\pi \nu \vec{b} \cdot \hat{s}/c) d\Omega = \int I(\hat{s}) \cos(2\pi \vec{b} \cdot \hat{s}/\lambda) d\Omega. \qquad (3.179)$$

Notice that the even cosine function in this response is sensitive only to the even (inversion-symmetric) part I_E of an arbitrary source brightness distribution, which can be written as the sum of even and odd (antisymmetric) parts: $I = I_E + I_O$. To detect the odd part I_O we need a "sine" correlator whose output is odd, $R_s = (V^2/2) \sin(\omega \tau_g)$. This can be implemented by a second correlator that follows a $\pi/2$ rad $= 90°$ phase delay inserted into the output of one antenna because

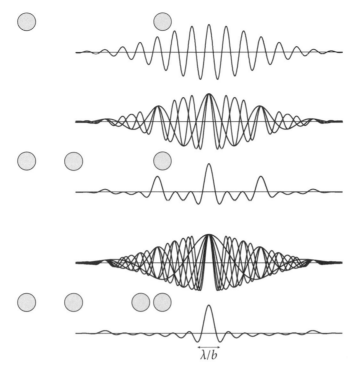

Figure 3.42. The instantaneous point-source responses of interferometers with overall projected length b and two, three, or four antennas distributed as shown are indicated by the thick curves. The synthesized main beam of the four-element interferometer is nearly Gaussian with angular resolution $\theta \approx \lambda/b$, but the sidelobes are still significant and there is a broad negative "bowl" caused by the lack of spacings shorter than the diameter of an individual antenna. Thus the **synthesized beam** is sometimes called the **dirty beam**. The instantaneous dirty beam of the multielement interferometer is the arithmetic mean of the individual responses of its component two-element interferometers. The individual responses of the three two-element interferometers comprising the three-element interferometer and of the six two-element interferometers comprising the four-element interferometer are plotted as thin curves.

$\sin(\omega\tau_g) = \cos(\omega\tau_g - \pi/2)$. Then

$$R_s = \int I(\hat{s}) \sin(2\pi \vec{b} \cdot \hat{s}/\lambda) d\Omega. \tag{3.180}$$

The combination of cosine and sine correlators is called a **complex correlator** because it is mathematically convenient to treat the cosines and sines as complex exponentials using Euler's formula (Appendix B.3)

$$e^{i\phi} = \cos\phi + i\sin\phi. \tag{3.181}$$

The **complex visibility** is defined by

$$\mathcal{V} \equiv R_c - iR_s \tag{3.182}$$

which can be written in the form

$$\mathcal{V} = Ae^{-i\phi}, \tag{3.183}$$

where

$$A = (R_c^2 + R_s^2)^{1/2} \tag{3.184}$$

is the **visibility amplitude** and

$$\phi = \tan^{-1}(R_s/R_c) \tag{3.185}$$

is the **visibility phase**. The response to an extended source with brightness distribution $I(\hat{s})$ of the two-element quasi-monochromatic interferometer with a complex correlator is the complex visibility

$$\boxed{\mathcal{V} = \int I(\hat{s}) \exp(-i2\pi \vec{b} \cdot \hat{s}/\lambda) \, d\Omega.} \tag{3.186}$$

3.7.3 Effects of Finite Bandwidths and Averaging Times

Equation 3.186 for quasi-monochromatic interferometers may be generalized to interferometers with finite bandwidths and integration times, which are necessary for high sensitivity. In the small but finite frequency range $\Delta \nu$ centered on frequency ν_c, Equation 3.186 becomes

$$\mathcal{V} = \int \left[\int_{\nu_c - \Delta\nu/2}^{\nu_c + \Delta\nu/2} I_\nu(\hat{s}) \exp(-i2\pi \vec{b} \cdot \hat{s}/\lambda) \, d\nu \right] d\Omega \tag{3.187}$$

$$= \int \left[\int_{\nu_c - \Delta\nu/2}^{\nu_c + \Delta\nu/2} I_\nu(\hat{s}) \exp(-i2\pi \nu \tau_g) \, d\nu \right] d\Omega. \tag{3.188}$$

If the source brightness and the response of the interferometer are nearly constant over $\Delta\nu$, the integral over frequency is just the Fourier transform of a rectangle function, so

$$\mathcal{V} \approx \int I_\nu(\hat{s}) \mathrm{sinc}(\Delta\nu \, \tau_g) \exp(-i2\pi \nu_c \tau_g) d\Omega. \tag{3.189}$$

For a finite bandwidth $\Delta\nu$ and delay τ_g, the fringe amplitude is attenuated by the factor $\mathrm{sinc}(\Delta\nu \, \tau_g)$. This attenuation can be eliminated in any one direction $\hat{s}_0$ called the **delay center** or the **phase reference position** by introducing a compensating delay $\tau_0 \approx \tau_g$ in the signal path of the "leading" antenna, as shown in Figure 3.43. As the Earth turns, τ_0 must be continuously adjusted to track τ_g within a tolerance $|\tau_0 - \tau_g| \ll (\Delta\nu)^{-1}$. This is usually done with digital electronics.

The geometric delay varies with direction, so delay compensation can be exact in only one direction. The angular radius $\Delta\theta$ of the usable field-of-view is determined by the variation of τ_g with offset $\Delta\theta$ from the direction $\hat{s}_0$. Because $c\tau_g = \vec{b} \cdot \vec{s} = b\cos\theta$, $|c\Delta\tau_g| = b\sin\theta\,\Delta\theta$. Requiring

$$\Delta\nu \, \Delta\tau_g \ll 1 \tag{3.190}$$

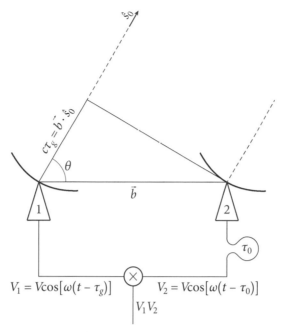

Figure 3.43. The compensating delay τ_0, shown here as an extra loop of cable between antenna 2 and the correlator, must track the geometric delay τ_g in the direction $\hat{s}_0$ of the delay center accurately enough to keep $|\tau_0 - \tau_g| \ll (\Delta\nu)^{-1}$ in order to minimize attenuation.

implies

$$\Delta\nu \, (b \sin\theta) \, \Delta\theta / c \ll 1. \tag{3.191}$$

Substituting $\lambda\nu = c$ and using $\theta_s \approx \lambda/(b \sin\theta)$ for the synthesized beamwidth, we get the requirement

$$\boxed{\frac{\Delta\theta}{\theta_s} \ll \frac{\nu}{\Delta\nu}.} \tag{3.192}$$

At larger angular offsets $\Delta\theta$ from the phase reference position, **bandwidth smearing** will radially broaden the synthesized beam by convolving it with a rectangle of angular width $\Delta\theta \, \Delta\nu/\nu$.

Satisfactory wide-field images can be made with a larger total bandwidth only by dividing that bandwidth into a number of narrower frequency channels each satisfying Equation 3.192. For example, the synthesized beamwidth of the VLA "B" configuration (maximum baseline length $b \approx 10\,\text{km}$) at $\lambda = 20\,\text{cm}$ ($\nu = 1.5\,\text{GHz}$) is $\theta_s \approx [(0.2\,\text{m})/(10^4\,\text{m})]\,\text{rad} \approx 4\,\text{arcsec}$. To image out to an angular radius $\Delta\theta = 15\,\text{arcmin} = 900\,\text{arcsec}$ equal to the half-power radius of the VLA primary beam requires channel bandwidths

$$\Delta\nu \ll \frac{\nu\theta_s}{\Delta\theta} = \frac{1.5 \times 10^9\,\text{Hz} \cdot 4\,\text{arcsec}}{900\,\text{arcsec}} \approx 7\,\text{MHz}. \tag{3.193}$$

Likewise, the correlator averaging time Δt must be kept short enough that the Earth's rotation will not move the source position in the frame of the interferometer by as much as the synthesized beamwidth $\theta_s \approx \lambda/b$. For example, if the delay is set to track the north celestial pole, a source $\Delta\theta$ away from the north pole will appear to move at an angular rate $2\pi\,\Delta\theta/P$, where $P \approx 23^h56^m04^s \approx 86164\,\text{s}$ is the Earth's sidereal rotation period. Excessive correlator averaging times will cause **time smearing** that tangentially broadens the synthesized beam. To minimize time smearing in an image of angular radius $\Delta\theta$, we require

$$\boxed{\frac{2\pi\,\Delta t}{P} \approx \frac{\Delta t}{1.37 \times 10^4\,\text{s}} \ll \frac{\theta_s}{\Delta\theta}.} \tag{3.194}$$

Continuing with the previous example, to image out to an angular radius $\Delta\theta = 900\,\text{arcsec}$ when $\theta_s = 4\,\text{arcsec}$ requires averaging times Δt short enough that

$$\Delta t \ll \frac{\theta_s}{\Delta\theta} \cdot 1.37 \times 10^4\,\text{s} = \frac{4\,\text{arcsec}}{900\,\text{arcsec}} \cdot 1.37 \times 10^4\,\text{s} \approx 60\,\text{s}. \tag{3.195}$$

3.7.4 Earth-Rotation Aperture Synthesis

The Earth's rotation varies the projected baseline coverage of an interferometer whose elements are fixed on the ground. In particular, all baselines of an interferometer whose baselines are confined to an east–west line will remain in a single plane perpendicular to the Earth's north–south rotation axis as the Earth turns daily. Confining all baselines to two dimensions has the computational advantage that the brightness distribution of a source is simply the two-dimensional Fourier transform of the measured visibilities.

Figure 3.44 illustrates **Earth-rotation aperture synthesis** by an east–west two-element interferometer at latitude $+40°$ as viewed from a source at declination $\delta = +30°$. Let u be the east–west component of the projected baseline in wavelengths and v be the north–south component of the projected baseline in wavelengths.

During the 12-hour period centered on source transit, the interferometer traces out a complete ellipse on the (u, v) **plane**. The maximum value of u equals the actual antenna separation in wavelengths, and the maximum value of v is smaller by the projection factor $\sin\delta$, where δ is the source declination. If the interferometer has more than two elements, or if the spacing of the two elements is changed daily, the (u, v) coverage will become a number of concentric ellipses having the same shape. Thus the synthesized beam obtained by east–west Earth-rotation aperture synthesis can approach an elliptical Gaussian. The synthesized beamwidth is $\approx u^{-1}$ radians east–west and $\approx u^{-1}\csc\delta$ radians in the north–south direction. The synthesized beam is circular for a source near the celestial pole, but the north–south beamwidth is very large for a source near the celestial equator.

3.7.5 Interferometers in Three Dimensions

The VLA (Very Large Array) shown in Color Plate 4 is Y-shaped and is instantaneously a nearly coplanar two-dimensional array of 27 25-m telescopes on the high Plains of San Augustin in New Mexico. It baselines are not confined to an east–west line but it is nearly coplanar, so "snapshot" observations much shorter than a sidereal day can be treated as two dimensional. On longer timescales, Earth rotation causes

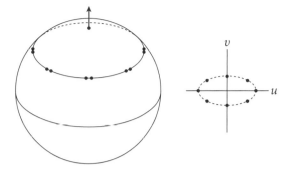

Figure 3.44. Viewed from a distant radio source, at declination $\delta = +30°$ for this drawing, the Earth rotates counterclockwise with a period of one sidereal day about the north–south axis indicated by the arrow emerging from the north pole. The antennas of a two-element east–west interferometer at latitude $+40°$ are shown, from left to right, as they would appear at hour angles -6^h, -3^h, 0^h, $+3^h$, and $+6^h$. Projected onto the plane of the page, which is normal to the line of sight, the interferometer baseline rotates continuously from purely north–south at -6^h through east–west at 0^h and back to north–south at $+6^h$. The projected antenna separation also changes. During this 12-hour period, the projected baseline traces an ellipse in the (u, v) plane as shown by the dashed curve, with points on the (u, v) ellipse highlighting the instantaneous coverage at -6^h, -3^h, 0^h, $+3^h$, and $+6^h$. The v-axis of the ellipse is smaller by a factor $\sin \delta$ than the u-axis.

the VLA baselines to fill a three-dimensional volume. The north–south baselines allow imaging with a nearly circular synthesized beam even near the celestial equator. Color Plate 4 shows the "D" configuration spanning about 1 km. The telescopes can be moved along railroad tracks to form the "C", "B", and "A" configurations spanning 3.4, 11, and 36 km, respectively for higher angular resolution. The VLA recently underwent a major upgrade to become the JVLA (the "J" stands for "Jansky"), with new wideband receivers completely covering the frequency range 1 to 50 GHz and a far more powerful and versatile correlator. It is up to an order of magnitude more sensitive than the original narrow-band VLA.

The (u, v, w) coordinate system used to describe any baseline vector $\vec{b}$ in three dimensions is shown in Figure 3.45. The w-axis is in the reference direction $\hat{s}_0$ usually chosen to contain the target radio source. The u- and v-axes point east and north in the (u, v) plane normal to the w-axis. u, v, and w are the components of $\vec{b}/\lambda$, the baseline vector in wavelength units. An arbitrary unit vector $\hat{s}$ has components (l, m, n) as drawn, where $n = \cos \theta = (1 - l^2 - m^2)^{1/2}$. The components (l, m, n) are called **direction cosines**.

Because

$$d\Omega = \frac{dl \, dm}{(1 - l^2 - m^2)^{1/2}},$$ (3.196)

the three-dimensional generalization of Equation 3.186 is

$$\mathcal{V}(u, v, w) = \int \int \frac{I_\nu(l, m)}{(1 - l^2 - m^2)^{1/2}} \exp[-i2\pi(ul + vm + wn)] dl \, dm.$$ (3.197)

This is *not* a three-dimensional Fourier transform.

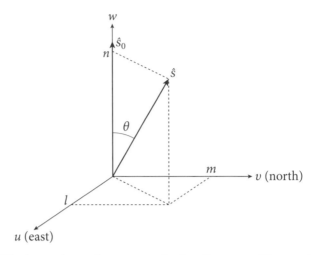

Figure 3.45. The (u, v, w) coordinate system for interferometers. The w-axis points in the reference direction $\hat{s}_0$ usually containing the source to be imaged. Projected onto the plane normal to the w-axis, u is the east–west baseline in wavelengths and v is the north–south baseline in wavelengths. l, m, and n are projections of the unit vector $\hat{s}$ onto the u-, v-, and w-axes, respectively.

However, if $w = 0$, Equation 3.197 becomes a two-dimensional Fourier transform, which can be inverted to give the source brightness distribution in terms of the measured visibilities:

$$\frac{I_\nu(l, m)}{(1 - l^2 - m^2)^{1/2}} = \int \int \mathcal{V}(u, v, 0) \exp[+i2\pi(ul + vm)] du\, dv. \qquad (3.198)$$

That is the case for an Earth-rotation aperture synthesis by an east–west interferometer if we choose $\hat{s}_0$ to coincide with the Earth's rotation axis, in which case $(1 - l^2 - m^2)^{1/2} = \cos\theta = \sin\delta$, where δ is the declination of the reference position.

For any interferometer, if we consider only directions close to $\hat{s}_0$, then $n = \cos\theta \approx 1 - \theta^2/2$ and

$$\mathcal{V}(u, v, w) \approx \exp(-i2\pi w) \int \int \frac{I_\nu(l, m)}{(1 - l^2 - m^2)^{1/2}}$$

$$\times \exp[-i2\pi(ul + vm - w\theta^2/2)]\, dl\, dm. \qquad (3.199)$$

The factor $\exp(-i2\pi w\theta^2/2)$ can be kept close to unity by keeping $w\theta^2 \ll 1$; that is, by imaging only a small field of view whose radius is $\theta \ll w^{-1/2} \approx (\lambda/b)^{1/2}$. For example, $\theta \ll 0.01$ radians is sufficiently small for an interferometer baseline 10^4 wavelengths long. Then

$$\mathcal{V} \exp(i2\pi w) = \int \int \frac{I_\nu(l, m)}{(1 - l^2 - m^2)^{1/2}} \exp[-i2\pi(ul + vm)] dl\, dm. \qquad (3.200)$$

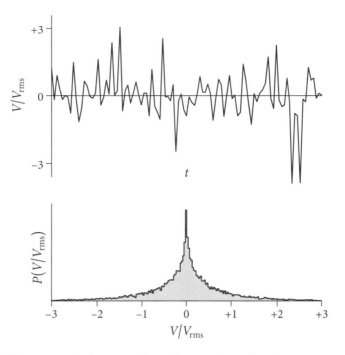

Figure 3.46. The unsmoothed output voltage of a correlator whose inputs are uncorrelated Gaussian noise has a symmetric distribution with zero mean, and the rms fluctuation is a factor $2^{1/2}$ times smaller than that of a square-law detector (Figure 3.33).

A field wider than $\theta \ll w^{-1/2}$ can be imaged with two-dimensional Fourier transforms by breaking it up into smaller facets, much like a fly's eye, and merging the facets to make the final image.

3.7.6 Sensitivity

The **point-source sensitivity** of a two-element interferometer can be derived from the radiometer equation for a total-power receiver on a single antenna because a square-law detector is equivalent to a correlator multiplying two identical input voltages supplied by one antenna. Consider an interferometer with two identical elements, each of which also has a square-law detector, observing a point source. The correlator multiplies the voltages from the two antennas, while each square-law detector multiplies the voltage from one antenna by itself, so the correlated/detected output voltages of the interferometer and each single dish are equal in strength. Thus the effective collecting area A_e of the two-element interferometer equals the effective collecting area of each element. However, the noise voltages from the two interferometer elements are almost completely uncorrelated (only the point source contributes correlated noise), while the noise voltages going into the square-law detectors are completely correlated (identical). The correlator output voltage distribution before smoothing is shown in Figure 3.46, and Figure 3.47 shows the correlator output voltage distribution after smoothing over $N = 50$ samples. In the limit where the antenna temperature ΔT contributed by the point source is much smaller than the system noise T_s, the correlator output noise is $2^{1/2}$ lower than the

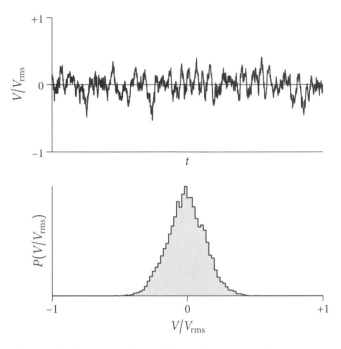

Figure 3.47. The smoothed output voltage of a correlator approaches a Gaussian with zero mean, and the rms noise is reduced by the square root of the number of independent samples averaged together. This figure shows noise from an $N = 50$ sample running mean. The rms fluctuation is a factor $2^{1/2}$ times smaller than that of a square-law detector (Figure 3.34).

square-law detector noise from each antenna. For an unpolarized point source of flux-density S, then $k\Delta T = S A_e/2$, so for a single antenna,

$$\sigma_S = \frac{2kT_s}{A_e(\Delta\nu\,\tau)^{1/2}} \tag{3.201}$$

and for a two-element interferometer,

$$\sigma_S = \frac{2^{1/2}kT_s}{A_e(\Delta\nu\,\tau)^{1/2}}. \tag{3.202}$$

The point-source sensitivity of a two-element interferometer is therefore $2^{1/2}$ times better than the sensitivity of each antenna, but $2^{1/2}$ times worse than that of a single dish whose area is that of two antennas. The reason the two-element interferometer is less sensitive than a single dish having the same total collecting area is that the information contained in the two independent square-law detector outputs has been discarded. Together they have $2^{1/2}$ times the sensitivity of a single dish. Combined with the independent correlator output, the total sensitivity is $(2 + 2)^{1/2} =$ twice the sensitivity of a single dish, or exactly the sensitivity of a single dish whose area equals the total area of the two-element interferometer.

An interferometer with N dishes contains $N(N - 1)/2$ independent two-element interferometers. So long as the signal from each dish can be amplified *coherently* before it is split up to be multiplied by the signals from the $N - 1$ other antennas, its point-source rms noise is

$$\sigma_S = \frac{2kT_s}{A_e[N(N-1)\Delta\nu\,\tau]^{1/2}}.$$

(3.203)

In the limit of large N, $[N(N-1)]^{1/2} \to N$ and the point-source sensitivity of an interferometer approaches that of a single antenna whose area equals the total effective area NA_e of the N interferometer antennas. For example, the VLA with $N = 27$ dishes each $d = 25$ m in diameter has the point-source sensitivity of a single dish whose diameter is $D = [N(N-1)]^{1/4}d = [27(26)]^{1/4} \cdot 25\,\text{m} = 129\,\text{m}$. Had the square-law detector outputs been used as well, the point-source sensitivity of the N-element interferometer would be exactly the same as the sensitivity of a single dish having the same total collecting area.

Practical interferometers are slightly less sensitive than this because their correlators use digital multipliers that sample and quantize the input voltage, not perfect analog multipliers. For example, a digital multiplier that samples at twice the Nyquist rate with three quantization levels ($-1, 0, +1$) is only 0.89 times as sensitive as a perfect analog multiplier. The chapter "Digital Signal Processing" in Thompson et al. [106] covers this and other consequences of quantization in detail.

Although the point-source sensitivity of an interferometer is comparable with the point-source sensitivity of a single dish having the same total area, beware that the **brightness sensitivity** of an interferometer is much worse because the synthesized beam solid angle of an interferometer is much smaller than the beam solid angle of a single dish of the same total effective area. The angular resolution of an interferometer with maximum baseline b is $\approx \lambda/b$ and the angular resolution of the single dish with diameter D is $\approx \lambda/D$, so the beam solid angle of the interferometer is smaller by a factor $\approx (D/b)^2$. This is roughly the **area filling factor** of the interferometer, defined as the ratio of the area covered by all of the antennas to the area spanned by the interferometer array. For example, the VLA in its $b \approx 11$ km "B" configuration has a filling factor $\approx (129\,\text{m}/1.1 \times 10^4\,\text{m})^2 \approx 1.2 \times 10^{-4}$. A high-resolution interferometer cannot detect a source of low surface brightness, no matter how high its total flux density.

The intensity axis of *any* astronomical image has dimensions of spectral brightness or specific intensity (e.g., units of Jy per beam solid angle or MJy sr^{-1} or K), not flux density (e.g., Jy). The point-source rms σ_S in Equation 3.203 corresponds to image flux density per beam solid angle, e.g., Jy beam^{-1}. Published radio images usually have intensity axes in units of Jy beam^{-1} because the flux density of a point source equals its brightness in those units and because σ_S is independent of beam solid angle. However, a proper spectral brightness depends only on the source. The "spectral brightness" specified in Jy beam^{-1} has the dimensions of spectral brightness, but *beware that this is not a proper spectral brightness* because it depends on the synthesized beam solid angle and not just on the radio source. Infrared astronomers frequently specify image intensity in MJy sr^{-1}, which is a proper brightness. The brightness temperature T is a convenient proper brightness for radio images. The rms brightness-temperature sensitivity σ_T of an image made with beam solid angle Ω_A follows directly from σ_S and the Rayleigh–Jeans approximation:

$$\sigma_T = \left(\frac{\sigma_S}{\Omega_A}\right)\frac{\lambda^2}{2k}.$$

(3.204)

Most interferometer images are restored with Gaussian beams. The beam solid angle (Equation 3.34) of a Gaussian beam with HPBW θ_{HPBW} is (Equation 3.118)

$$\Omega_A = \frac{\pi\theta_{\mathrm{HPBW}}^2}{4\ln 2},$$

so

$$\sigma_T = \left(\frac{2\ln 2\, c^2}{\pi k\nu^2}\right)\frac{\sigma_S}{\theta_{\mathrm{HPBW}}^2}.$$

(3.205)

For example, all of the 1.4 GHz NRAO VLA Sky Survey (NVSS) images have rms noise $\sigma_S \approx 0.45\,\mathrm{mJy\,beam^{-1}}$ and were restored with a circular Gaussian beam whose half-power beamwidth is $\theta_{\mathrm{HPBW}} = 45\,\mathrm{arcsec} \approx 2.18 \times 10^{-4}\,\mathrm{rad}$. Consequently, NVSS rms brightness temperature noise is

$$\sigma_T \approx \left[\frac{2\ln 2\,(3\times 10^8\,\mathrm{m\,s^{-1}})^2}{\pi\cdot 1.38\times 10^{-23}\,\mathrm{J\,K^{-1}}\cdot(1.4\times 10^9\,\mathrm{Hz})^2}\right]\frac{0.45\times 10^{-29}\,\mathrm{W\,Hz^{-1}}}{(2.18\times 10^{-4}\,\mathrm{rad})^2} \approx 0.14\,\mathrm{K}.$$

This is good enough to detect ($5\sigma_T \approx 0.7\,\mathrm{K}$) normal spiral galaxies with median $\langle T_\mathrm{b}\rangle \sim 1\,\mathrm{K}$ at 1.4 GHz. Beware that a high-resolution (low Ω_A) image with a good point-source sensitivity (low σ_S) may still have a poor brightness-temperature sensitivity (high σ_T).

4 Free-Free Radiation

4.1 THERMAL AND NONTHERMAL EMISSION

Larmor's formula (Equation 2.143)

$$P = \frac{2q^2\dot{v}^2}{3c^3} \tag{4.1}$$

states that electromagnetic radiation with power P is produced by accelerating (or decelerating; hence the German name **bremsstrahlung** meaning "braking radiation") an electrical charge q. Free charged particles can be accelerated by electrostatic or magnetic forces, gravitational acceleration being negligible by comparison. Electrostatic bremsstrahlung is the subject of this chapter, and its magnetic counterpart **magnetobremsstrahlung** or "magnetic braking radiation" (e.g., synchrotron radiation) is covered in Chapter 5.

Thermal emission is produced by a source whose emitting particles are in local thermodynamic equilibrium (LTE) (Section 2.2.2); otherwise **nonthermal emission** is produced. Most astronomical sources of electrostatic bremsstrahlung are thermal because the radiating electrons have the **Maxwellian velocity distribution** (Appendix B.8) of particles in LTE. The relativistic electrons in most astronomical sources of magnetobremsstrahlung have power-law energy distributions and hence are not in LTE, so synchrotron sources are often called nonthermal sources. However, electrostatic and magnetic bremsstrahlung are not synonymous with thermal and nonthermal radiation, respectively. For example, electrons with a relativistic Maxwellian energy distribution are in LTE and can emit thermal synchrotron radiation. Remember also that a thermal source does not have a blackbody spectrum if the source opacity is small and the emission coefficient depends on frequency.

4.2 HII REGIONS

The electrostatic force is so much stronger than gravity that free charges in interstellar gas quickly rearrange themselves so that the negative charges of free electrons in an ionized cloud neutralize the positive charges of ions on all scales larger than the Debye length $\lambda_D \leq 1$ m (Equation 4.46). As an electron (charge $-e \approx -4.8 \times 10^{-10}$

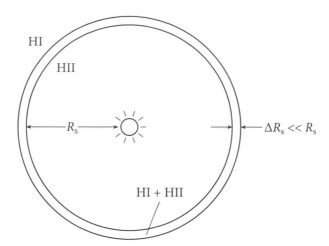

Figure 4.1. A Strömgren sphere of ionized hydrogen (HII) with **Strömgren radius** R_S surrounded by a thin shell of partially ionized hydrogen (HI + HII) surrounded by neutral hydrogen (HI).

statcoulomb) passes by an ion (charge $+Ze$ for an atom with Z electrons removed), the Coulomb force (Equation 2.134) causes an acceleration of magnitude

$$|\dot{v}| = \frac{F}{m_e} = \frac{Ze^2}{m_e r^2},\tag{4.2}$$

where $m_e \approx 9.1 \times 10^{-28}$ g is the electron mass and r is the distance between the electron and the ion. The resulting emission is called **free–free radiation** because the electron is free both before and after the interaction; it is not captured by the ion. If the ionized interstellar cloud is reasonably dense, the electrons and ions interact often enough that they quickly come into LTE at some common temperature, so free–free radiation is usually thermal emission.

Interstellar gas is primarily hydrogen and helium, plus trace amounts of heavier elements such as carbon, nitrogen, oxygen, neon, silicon, and iron. Astronomers call all of these heavier elements **metals**, meaning elements that readily form positive ions, even though most are not metallic in the usual sense of being solid, malleable, ductile, and electrically conducting solids at room temperature. Much of the interstellar hydrogen is in the form of neutral atoms (called HI in astronomical terminology) or diatomic molecules (H_2), but some is ionized. The singly ionized hydrogen atom H^+ is referred to as HII by astronomers, doubly ionized oxygen O^{++} is called OIII, triply ionized carbon C^{+++} is called CIV, etc.

In 1939 the astronomer Bengt Strömgren realized that the interstellar medium can be divided into distinct regions in which hydrogen is either (1) mostly atomic or molecular, with nearly all of the hydrogen atoms in the ground electronic state or (2) almost completely ionized. Furthermore, the boundaries separating these HI and HII regions are very thin. Sometimes the **HII regions** surrounding stars are called **Strömgren spheres** (Figure 4.1) after his early theoretical models. What is the microscopic physical basis for this picture?

A hydrogen atom in the ground state has the smallest and most tightly bound electronic orbit around the nuclear proton that is consistent with a stationary electronic wave function. (See Section 7.2.1 and Figure 7.1 to review the Bohr model of hydrogen atoms.) The permitted electronic energy levels are characterized by their **principal quantum numbers** $n = 1, 2, 3, \ldots$, where $n = 1$ corresponds to the ground state. Although quantum mechanics forbids an electron in the ground state from radiating according to the classical Larmor formula, it permits radiative decay from higher levels $n = 2, 3, \ldots$; and Larmor's equation fairly accurately predicts the radiative lifetimes of excited hydrogen atoms. The orbital radius a_n of an electron in the nth energy level is $a_n = n^2 a_0$, where $a_0 \approx 5.29 \times 10^{-9}$ cm is called the **Bohr radius**. Applying Larmor's equation shows that the radiative lifetime τ is proportional to a_n^3 and hence to n^6. Thus the (incorrect) classical result $\tau \approx 5.5 \times 10^{-11}$ s for radiation from the $n = 1$ ground state can be scaled to estimate the radiative lifetimes of excited states. For example, the approximate radiative lifetime of the $n = 2$ state would be $\tau \approx 2^6 \cdot 5.5 \times 10^{-11}$ s $\approx 3 \times 10^{-9}$ s, in reasonable agreement with the accurate quantum-mechanical result $\tau \approx 2 \times 10^{-9}$ s. Excited hydrogen atoms spontaneously decay very quickly to the ground state by emitting radiation so, at any instant, almost all neutral atoms are in the ground state.

Hydrogen atoms in the ground state can be ionized by photons with energy $E \geq 13.6$ eV (1 electron Volt $\approx 1.60 \times 10^{-12}$ erg). Such energetic photons have frequencies higher than the **Rydberg frequency** $R_\infty c = E/h \approx 3.29 \times 10^{15}$ Hz (Equation 7.11) and wavelengths shorter than $\lambda = 912$ Å $= 912 \times 10^{-10}$ m, a far-ultraviolet (UV) wavelength. These **Lyman continuum photons** are produced in significant numbers by the Wien tail of blackbody radiation from stars hotter than $T \sim 3 \times 10^4$ K. The rate Q_H at which a star with spectral luminosity L_ν produces photons that can ionize hydrogen atoms in the ground state is

$$Q_H = \int_{R_\infty c}^{\infty} \left(\frac{L_\nu}{h\nu} \right) d\nu. \tag{4.3}$$

If a star emits Q_H Lyman continuum photons per second, it will photoionize hydrogen atoms with number density n_H throughout some volume V surrounding the star. The helium mixed with the hydrogen can often be ignored because its ionization potential is so high, $E \approx 24.5$ eV, that only exceptionally hot stars can ionize significant amounts of helium.

The absorption cross section of a neutral hydrogen atom to photons with energies just above 13.6 eV is large enough, $\sigma \approx 10^{-17}$ cm^2, that each ionizing photon is absorbed and produces a new ion shortly after it passes from the ionized Strömgren sphere into the surrounding HI region. The thickness ΔR_S of the partially ionized shell surrounding a Strömgren sphere (Figure 4.1) is

$$\Delta R_S \approx (n_H \sigma)^{-1}. \tag{4.4}$$

For example, if the neutral hydrogen density is $n_H = 10^3$ atoms cm^{-3}, then

$$\Delta R_S \approx (10^3 \text{ cm}^{-3} \times 10^{-17} \text{ cm}^2)^{-1} \approx 10^{14} \text{ cm} \ll 1 \text{ pc}. \tag{4.5}$$

Light travels $\approx 10^{14}$ cm per hour, so an ionizing photon typically survives only about an hour in such an HI cloud before being absorbed.

Once ionized from HI into free protons (H$^+$ ions) and electrons, the HII region has a much lower opacity to ionizing photons. Thus a new ionizing star turning on in a uniform-density HI cloud will fully ionize a sphere whose Strömgren radius (Figure 4.1) grows with time until equilibrium between ionization and recombination is reached. This is sometimes called an **ionization bounded** HII region. If the surrounding HI cloud is small enough that the star can ionize it completely, the HII region is said to be **matter bounded** or **density bounded**.

Inside the HII region, electrons and protons occasionally collide and recombine at a volumetric **recombination rate** $\dot{n}_H$ that can be written as

$$\dot{n}_H \approx \alpha_H n_e n_p, \tag{4.6}$$

where $\dot{n}_H$ is the number of recombinations per unit time per unit volume (e.g., $cm^{-3} s^{-1}$),

$$\alpha_H \approx 3 \times 10^{-13} \, cm^3 \, s^{-1} \tag{4.7}$$

is the **recombination coefficient** for hydrogen, and the collision rate per unit volume is proportional to the product $n_e n_p$ of the electron and proton densities. For example, if $n_e = n_p = 10^3 \, cm^{-3}$,

$$\dot{n}_H \approx 3 \times 10^{-13} \, cm^3 \, s^{-1} \times 10^3 \, cm^{-3} \times 10^3 \, cm^{-3} \approx 3 \times 10^{-7} \, cm^{-3} \, s^{-1}. \tag{4.8}$$

This example shows that the **recombination time**

$$\tau \equiv \frac{n_e}{\dot{n}_H} \approx 3.3 \times 10^9 \, s \approx 10^2 \, yr \tag{4.9}$$

is usually much shorter than the $> 10^6$ year lifetime of an ionizing star. (A useful relation to remember is 1 year $\approx 10^{7.5}$ s.) The volume V of an ionization-bounded HII region grows until the total ionization and recombination rates in the Strömgren sphere are equal. In equilibrium,

$$Q_H = \dot{n}_H V = \alpha_H n_e n_p \frac{4}{3} \pi R_S^3 \tag{4.10}$$

yields a Strömgren radius

$$R_S \approx \left(\frac{3 Q_H}{4\pi \alpha_H n_e^2} \right)^{1/3}. \tag{4.11}$$

For example, an O5 star (very hot and luminous) emits $Q_H \approx 6 \times 10^{49}$ ionizing photons per second. If $n_e \approx 10^3 \, cm^{-3}$,

$$R_S \approx \left[\frac{3 \cdot 6 \times 10^{49} \, s^{-1}}{4\pi \cdot 3 \times 10^{-13} \, cm^3 \, s^{-1} (10^3 \, cm^{-3})^2} \right]^{1/3} \approx 3.6 \times 10^{18} \, cm \approx 1.2 \, pc.$$

This example illustrates that $R_S \gg \Delta R_S$; that is, the radius of the fully ionized Strömgren sphere is much larger than the thickness of its partially ionized skin.

Two distinct kinds of stars produce most of the HII regions in our Galaxy:

1. The most massive ($M \geq 15 M_\odot$) short-lived (lifetimes $\leq 10^7$ yr) main-sequence stars are big enough ($R \sim 10 R_\odot$) and hot enough ($T \geq 3 \times 10^4$ K) to be very luminous sources of ionizing UV. Such stars were recently formed by gravitational collapse and fragmentation of interstellar clouds containing neutral gas and dust grains.

2. Old lower-mass ($1 < M/M_\odot < 8$) stars whose main-sequence lifetimes are less than the age of our Galaxy ($\approx 10^{10}$ yr) eventually become red giants and finally white dwarfs. Young white dwarfs are small ($R \sim 10^{-2} R_\odot$) but hot enough to ionize the stellar envelope material that was ejected during the red giant stage, and these ionized regions are called **planetary nebulae** because many looked like planets to early astronomers using small telescopes.

Most ionizing stars are approximately blackbody emitters and their ionizing photons from the high-frequency Wien tail have energies only somewhat greater than the $E = 13.6$ eV minimum needed to ionize a hydrogen atom from its ground state. Momentum conservation during ionization ensures that nearly all of the photon energy in excess of 13.6 eV is converted into kinetic energy of the ejected electron. Collisions between these hot photoelectrons, and between electrons and ions, thermalize the ionized gas and gradually bring it into local thermodynamic equilibrium (LTE). Consequently, the thermalized electrons have a Maxwellian energy distribution. Eventually this heating is balanced by radiative cooling. Collisions of electrons with "metal" ions can excite low-lying (a few eV) energy states that decay slowly via forbidden transitions and emit visible photons that may escape from the nebula. Examples of visible cooling lines include the green lines of O III at $\lambda = 4959$ Å and 5007 Å, first discovered in nebulae and named *nebulium* lines because these forbidden lines hadn't been observed in the laboratory and were thought to be from a new element found only in nebulae (just as *helium* lines in the solar spectrum were once ascribed to a new element found in the Sun). The Balmer hydrogen recombination lines Hα at $\lambda = 6563$ Å and Hβ at $\lambda = 4861$ Å also contribute to the characteristic colors of H II regions.

Thermal equilibrium between heating and cooling of H II regions is usually reached at a temperature close to $T \approx 10^4$ K [76] that is much higher than the initial temperature $T < 100$ K of the neutral interstellar gas. The heated gas expands, reversing any infall onto the ionizing star and sending shocks into the surrounding cold gas, thereby both inhibiting and stimulating the subsequent production of stars in the region. Typical H II regions have sizes ~ 1 pc, electron number densities $\sim 10^3$ cm^{-3}, and masses up to $10^4 M_\odot$.

The free–free radio emission from an H II region is a tracer of the electron temperature, electron density, and ionized volume. It constrains the production rate Q_{H} of ionizing photons and, with an *assumption* about the **initial mass function (IMF)** (the mass distribution of new stars), the total star-formation rate in an H II region. The radio data are important for reliable quantitative estimates of star formation because they do not suffer from extinction by interstellar dust.

Ultra-compact (UC) H II regions [23] are small (diameter $2R_S \leq 0.03$ pc) but dense ($n_e > 10^4$ cm^{-3}) H II regions ionized by O and B stars so young that they are still optically obscured by the dusty molecular clouds from which they formed. The dust makes them strong far-infrared sources, and free–free radio emission penetrates the dust so they can be imaged and studied at radio wavelengths. UC H II regions are valuable tracers of the formation and early evolution of massive stars, and of their interactions with their environment. On a much larger scale, extragalactic **ultra-dense (UD) H II regions** [58] are ionized by young super star clusters (SSCs) so dense and massive that they may be the progenitors of long-lived globular clusters, and so luminous that they may disrupt the ISM in dwarf galaxies.

Planetary nebulae are HII regions surrounding the hot (up to $T \sim 10^5$ K) white-dwarf cores of low-mass ($1 < M_\odot < 8$) stars that have ejected their outer envelopes as stellar winds. White dwarfs are small, about the size of the Earth, so they are much less luminous than massive O stars and the ionized nebular masses are only $0.1 < M/M_\odot < 1$. Planetary nebulae are nonetheless fairly luminous indicators of the last stages in the lives of low-mass stars. They are potentially useful as a record of low-mass star formation throughout the history of our Galaxy. However, the optical selection of planetary nebulae is affected by dust extinction. Far-infrared and radio selection may avoid this limitation. Planetary nebulae are not particularly luminous radio sources, but they are the most numerous *compact* radio continuum sources in our Galaxy.

4.3 FREE–FREE RADIO EMISSION FROM HII REGIONS

Thermal bremsstrahlung from ionized hydrogen is often called **free–free emission** because it is produced by free electrons scattering off ions without being captured— the electrons are free before the interaction and remain free afterward. What are the basic properties of free–free radio emission from an astrophysical HII region? Despite all of the simplifications introduced in Section 4.2, this problem can't be solved without several additional approximations. A certain amount of astrophysical "intuition" is required to distinguish important from negligible effects. For example, the energy lost by an electron when it interacts with an ion is much smaller than the initial electron energy. Radiation from electron–electron collisions and from ions can be ignored. Formally divergent integrals over impact parameters can be avoided with physical limits to the range of integration and still yield reasonably accurate, but not exact, results. Most astrophysical conditions are so far removed from personal experience (How hot does 10^4 K feel? How big is a parsec compared with the height of a tree? How much is $M_\odot \approx 2 \times 10^{33}$ g compared with the mass of a person?) that astrophysical intuition depends on being familiar with numerical values for the relevant parameters, so it is possible to decide quickly what is important and what can be neglected. As Linus Pauling said, "The way to get good ideas is to get lots of ideas and throw the bad ones away." The following analysis of free–free emission from an HII region illustrates approaches and techniques used to solve "messy" astrophysical problems. For similar but alternative analyses, see the chapters on bremsstrahlung in Rybicki and Lightman [97] and in Wilson et al. [116].

Why should an HII region emit radio radiation at all? The answer is, because charged particles are being accelerated electrostatically, and nonrelativistic free accelerated charges radiate power according to Larmor's formula (Equation 2.143). Electrostatic interactions among many kinds of charged particles take place in an HII region, but most do not emit significant amounts of radiation. The magnitude of the acceleration $\dot{v}$ is inversely proportional to the particle mass m. The lightest ion is the hydrogen ion. Its mass is the **proton mass** $m_p \approx 1.66 \times 10^{-24}$ g, which is about 2×10^3 times the **electron mass** $m_e \approx 9.11 \times 10^{-28}$ g. In any electron–ion collision the electron will therefore radiate at least $(m_p/m_e)^2 \approx 4 \times 10^6$ as much power as the ion, so all ionic radiation can be neglected. Interactions between identical particles also do not radiate significantly because the accelerations of the two particles are equal in magnitude but opposite in direction: $\dot{v}_1 = -\dot{v}_2$. Their radiated

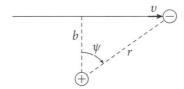

Figure 4.2. A light, fast electron passing by a slow, heavy ion. Low-energy radio photons are produced by weak scattering in which the velocity vector $\vec{v}$ of the electron changes little. The distance of closest approach b is called the **impact parameter** and the interval $\tau = b/v$ is called the **collision time**.

electric fields are equal in magnitude but opposite in sign, so the net radiated electric field approaches zero at distances much larger than the collision impact parameter (Figure 4.2) and the radiation from electron–electron collisions can be ignored. The bottom line is that *only the electron–ion collisions are important, and only the electrons radiate significantly.*

4.3.1 Radio Radiation From a Single Electron–Ion Interaction

Figure 4.2 shows an electron passing by a far more massive ion of charge Ze, where $Z = 1$ for a singly ionized atom such as hydrogen. Each electron–ion interaction will generate a single pulse of radiation. The total energy emitted and the approximate frequency spectrum of the pulse will be derived in this section. The exact spectrum of an individual pulse is not needed because the broad distribution of electron energies and impact parameters smears out spectral details in the total spectrum of an H$\textsc{ii}$ region.

Radio photons are produced by weak interactions because the energy $E = h\nu$ of a radio photon is much smaller than the average kinetic energy $\langle E_e \rangle$ of an electron in an H$\textsc{ii}$ region. The numerical comparison below is an example of how astrophysical "intuition" can simplify the electron–ion scattering problem.

The mean electron energy in a plasma of temperature T is

$$\langle E_e \rangle = \frac{3kT}{2}. \tag{4.12}$$

For an H$\textsc{ii}$ region with $T \sim 10^4$ K,

$$\langle E_e \rangle \approx \frac{3 \cdot 1.38 \times 10^{-16} \, \text{erg K}^{-1} \cdot 10^4 \, \text{K}}{2} \approx 2 \times 10^{-12} \, \text{erg} \approx 1 \, \text{eV}. \tag{4.13}$$

(This is another useful conversion factor to remember: 1 eV is the typical particle kinetic energy associated with the temperature $T \approx 10^4$ K.) By comparison, the energy of a radio photon of frequency $\nu = 10$ GHz is only

$$E = h\nu \approx 6.63 \times 10^{-27} \, \text{erg s} \cdot 10^{10} \, \text{Hz} \approx 6.63 \times 10^{-17} \, \text{erg} \approx 4 \times 10^{-5} \, \text{eV}. \tag{4.14}$$

The weak interactions that produce radio photons cause the trajectory of the electron to deflect by only a small angle ($\ll 1$ radian). As shown in Figure 4.2, the electron's path can be approximated by a straight line.

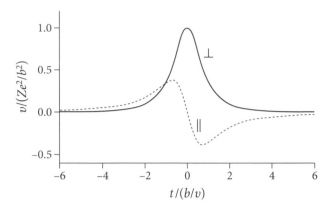

Figure 4.3. The acceleration of an electron by an ion may be resolved into components perpendicular ($\perp$) to and parallel ($\parallel$) to the electron's velocity. The perpendicular acceleration (Equation 4.16) yields a roughly Gaussian pulse whose power spectrum extends to low (radio) frequencies. The parallel acceleration (Equation 4.15) gives a roughly sinusoidal pulse with no "DC" component, so the resulting radiation is strongest at higher (infrared) frequencies and very weak at radio frequencies.

During the interaction, the electron will be accelerated electrostatically both parallel to and perpendicular to its nearly straight path:

$$F_{\parallel} = m_e \dot{v}_{\parallel} = \frac{-Ze^2}{r^2} \sin \psi = \frac{-Ze^2 \sin \psi \cos^2 \psi}{b^2}, \qquad (4.15)$$

$$F_{\perp} = m_e \dot{v}_{\perp} = \frac{Ze^2}{r^2} \cos \psi = \frac{Ze^2 \cos^3 \psi}{b^2}, \qquad (4.16)$$

where $\cos \psi = b/r$ and b is the **impact parameter** of the interaction, the minimum value of the distance r between the electron and the ion.

For any impact parameter b, these two equations can be solved to show that the maximum of $\dot{v}_{\parallel}$ is a nonnegligible 38% of the maximum of $\dot{v}_{\perp}$. Even so, the *radio* radiation arising from $\dot{v}_{\parallel}$ is completely negligible. Plotting the variation with time of $\dot{v}_{\parallel}$ and $\dot{v}_{\perp}$ during the interaction shows pulses with quite different shapes (Figure 4.3). The pulse duration is comparable with the collision time $\tau \approx b/v$. The $\dot{v}_{\parallel}$ pulse is roughly a sine wave of angular frequency $\omega \sim \tau^{-1} = v/b$, which is *much* higher than radio frequencies for all relevant impact parameters b (Equation 4.43). The parallel acceleration produces some infrared radiation but very little radio radiation. The $\dot{v}_{\perp}$ pulse is a single peak whose frequency spectrum extends from zero up to $\sim v/b$ because the Fourier transform of a Gaussian is also a Gaussian (Appendix B.4), so it is stronger at radio frequencies.

Inserting $\dot{v}_{\perp}$ from Equation 4.16 into Larmor's formula (Equation 2.143) gives the instantaneous power emitted by the acceleration perpendicular to the electron velocity:

$$P = \frac{2}{3} \frac{e^2 \dot{v}_{\perp}^2}{c^3} = \frac{2e^2}{3c^3} \frac{Z^2 e^4}{m_e^2} \left(\frac{\cos^3 \psi}{b^2} \right)^2. \qquad (4.17)$$

The total energy W emitted by the pulse is

$$W = \int_{\infty}^{\infty} P \, dt. \tag{4.18}$$

Because $(\Delta E_e)/E_e = E_\gamma/E_e \ll 1$ for radio photons, *the electron velocity is nearly constant*. The interaction diagram in Figure 4.2 shows that

$$v = \frac{dx}{dt} \quad \text{and} \quad \tan \psi = \frac{x}{b}, \tag{4.19}$$

so

$$v = \frac{b \, d \tan \psi}{dt} = \frac{b \sec^2 \psi \, d\psi}{dt} = \frac{b \, d\psi}{\cos^2 \psi \, dt}, \tag{4.20}$$

and

$$dt = \frac{b}{v} \frac{d\psi}{\cos^2 \psi}. \tag{4.21}$$

Inserting Equations 4.17 and 4.21 into 4.18 gives

$$W = \frac{2}{3} \frac{Z^2 e^6}{c^3 m_e^2 b^4} \int_{-\pi/2}^{\pi/2} \frac{b}{v} \frac{\cos^6 \psi}{\cos^2 \psi} d\psi = \frac{4}{3} \frac{Z^2 e^6}{c^3 m_e^2 b^3 v} \int_0^{\pi/2} \cos^4 \psi \, d\psi. \tag{4.22}$$

The integral in Equation 4.22 is evaluated in Appendix B.7; it is

$$\int_0^{\pi/2} \cos^4 \psi \, d\psi = \frac{3\pi}{16}. \tag{4.23}$$

Thus the **pulse energy** W radiated by a single electron–ion interaction characterized by impact parameter b and velocity v is

$$W = \frac{\pi Z^2 e^6}{4 c^3 m_e^2} \left(\frac{1}{b^3 v} \right). \tag{4.24}$$

This energy is emitted in a single pulse of duration $\tau \approx b/v$, so the pulse **power spectrum** (Appendix A.4) is nearly flat over all frequencies $v < v_{max} \approx (2\pi\tau)^{-1} \approx v/(2\pi b)$ and falls rapidly at higher frequencies. It is possible to calculate the actual Fourier transform of the pulse shape (solid curve in Figure 4.4), but doing so would only add an unnecessary complication to an already complicated calculation. The ranges of velocities v and impact parameters b characterizing electron–ion interactions in an H II region are so wide that averaging over all collision parameters will wash out fine details in the spectrum associated with any particular v and b.

The typical electron speed in a $T \sim 10^4$ K H II region is $v \approx 7 \times 10^7$ cm s^{-1} and the minimum impact parameter is $b_{min} \sim 10^{-7}$ cm, so $v_{max} \approx 10^{14}$ Hz, much higher than radio frequencies. In the approximation that the power spectrum is flat out to $v = v_{max}$ and zero at higher frequencies (dashed line in Figure 4.4), the average energy *per unit frequency* emitted during a single interaction is approximately

$$W_v \approx \frac{W}{v_{max}} = \left(\frac{\pi Z^2 e^6}{4 c^3 m_e^2 b^3 v} \right) \left(\frac{2\pi b}{v} \right), \tag{4.25}$$

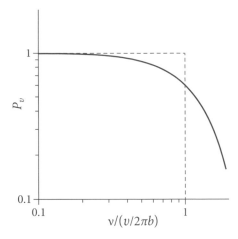

Figure 4.4. The actual power spectrum of the electromagnetic pulse generated by one electron–ion interaction is nearly flat up to frequency $\nu \approx v/(2\pi b)$, where v is the electron speed and b is the impact parameter, and declines at higher frequencies. The approximation $P_\nu = 1$ for all $\nu < v/(2\pi b)$ and $P_\nu = 0$ at higher frequencies (dashed line) is quite good at radio frequencies $\nu \ll v/(2\pi b)$.

which simplifies to

$$W_\nu \approx \frac{\pi^2}{2} \frac{Z^2 e^6}{c^3 m_e^2} \left(\frac{1}{b^2 v^2} \right), \quad \nu < \nu_{\max} \approx \frac{v}{2\pi b} \approx 10^{14} \text{ Hz}. \tag{4.26}$$

4.3.2 Radio Radiation From an H II Region

The strength and spectrum of radio emission from an H II region depends on the distributions of electron velocities v and collision impact parameters b (Figure 4.2). The distribution of v depends on the electron temperature T. The distribution of b depends on the electron number density n_e (cm^{-3}) and the ion number density n_i (cm^{-3}).

In LTE, the average kinetic energies of electrons and ions are equal. The electrons are much less massive, so their speeds are much higher and the ions can be considered nearly stationary during an interaction (Figure 4.5). The number of electrons passing any ion per unit time with impact parameter b to $b + db$ and speed range v to $v + dv$ is

$$n_e (2\pi b\, db)\, v\, f(v)\, dv, \tag{4.27}$$

where $f(v)$ is the normalized ($\int f(v) dv = 1$) speed distribution of the electrons. The number $\dot{n}_c(v, b)$ of such collisions per unit volume per unit time is

$$\dot{n}_c(v, b) = (2\pi b) v f(v)\, n_e n_i. \tag{4.28}$$

The spectral power at frequency ν emitted isotropically per unit volume is $4\pi j_\nu$, where j_ν is the emission coefficient defined by Equation 2.26. Thus

$$4\pi j_\nu = \int_{b=0}^{\infty} \int_{v=0}^{\infty} W_\nu(v, b) \dot{n}_c(v, b) dv\, db. \tag{4.29}$$

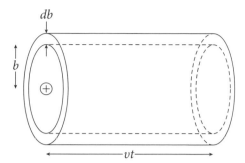

Figure 4.5. The number of electrons with speeds v to $v + dv$ passing by a stationary ion and having impact parameters in the range b to $b + db$ during the time interval t equals the number of electrons with speeds v to $v + dv$ in the cylindrical shell shown here.

Substituting the results for $W_\nu(v, b)$ (Equation 4.26) and $\dot{n}_c(v, b)$ (Equation 4.28) into Equation 4.29 gives

$$4\pi j_\nu = \int_{b=0}^{\infty} \int_{v=0}^{\infty} \left(\frac{\pi^2 Z^2 e^6}{2c^3 m_e^2 b^2 v^2} \right) 2\pi b \, db \, n_e n_i \, v f(v) dv \tag{4.30}$$

$$= \frac{\pi^3 Z^2 e^6 n_e n_i}{c^3 m_e^2} \int_{v=0}^{\infty} \frac{f(v)}{v} dv \int_{b=0}^{\infty} \frac{db}{b}. \tag{4.31}$$

Equation 4.31 exposes a problem: the integral

$$\int_{b=0}^{\infty} \frac{db}{b} \tag{4.32}$$

diverges logarithmically. There must be finite physical limits b_{min} and b_{max} (to be determined) on the range of the impact parameter b that prevent this divergence:

$$4\pi j_\nu = \frac{\pi^3 Z^2 e^6 n_e n_i}{c^3 m_e^2} \int_{v=0}^{\infty} \frac{f(v)}{v} dv \int_{b_{\text{min}}}^{b_{\text{max}}} \frac{db}{b}. \tag{4.33}$$

The distribution $f(v)$ of electron speeds in LTE is the **nonrelativistic Maxwellian distribution** (see Appendix B.8 for its derivation):

$$f(v) = \frac{4v^2}{\sqrt{\pi}} \left(\frac{m_e}{2kT} \right)^{3/2} \exp\left(-\frac{m_e v^2}{2kT} \right). \tag{4.34}$$

Equation 4.34 can be used to evaluate the integral over the electron speeds in Equation 4.33:

$$\int_{v=0}^{\infty} \frac{f(v)}{v} dv = \frac{4}{\sqrt{\pi}} \left(\frac{m_e}{2kT} \right)^{3/2} \int_{v=0}^{\infty} v \exp\left(-\frac{m_e v^2}{2kT} \right) dv. \tag{4.35}$$

Substituting $x \equiv m_e v^2 / (2kT)$ so $dx = m_e v \, dv / (kT)$ gives

$$\int_{v=0}^{\infty} \frac{f(v)}{v} \, dv = \frac{4}{\sqrt{\pi}} \left(\frac{m_e}{2kT} \right)^{3/2} \int_{x=0}^{\infty} \frac{kT}{m_e} e^{-x} dx \tag{4.36}$$

$$= \frac{2}{\sqrt{\pi}} \left(\frac{m_e}{2kT} \right)^{1/2} \int_{x=0}^{\infty} e^{-x} dx \tag{4.37}$$

$$= \left(\frac{2m_e}{\pi kT} \right)^{1/2}. \tag{4.38}$$

In conclusion, the **free–free emission coefficient** can be written as

$$\boxed{ j_v = \frac{\pi^2 Z^2 e^6 n_e n_i}{4c^3 m_e^2} \left(\frac{2m_e}{\pi kT} \right)^{1/2} \ln \left(\frac{b_{max}}{b_{min}} \right). } \tag{4.39}$$

The remaining problem is to estimate the minimum and maximum impact parameters b_{min} and b_{max}. These estimates don't have to be very precise because only their logarithms appear in Equation 4.39.

To estimate the minimum impact parameter b_{min}, notice that the net impulse (change in momentum) during a single electron–ion interaction

$$m_e \Delta v = \int_{-\infty}^{\infty} F \, dt \tag{4.40}$$

comes entirely from the perpendicular component $F_{\perp}$ of the electric force because the contribution from $E_{\parallel}$ is antisymmetric about $t = 0$ (Figure 4.3). Inserting $F_{\perp}$ from Equation 4.16 into Equation 4.40 gives

$$m_e \Delta v = \int_{-\infty}^{\infty} \left(\frac{Ze^2 \cos \psi}{r^2} \right) dt = Ze^2 \int_{-\infty}^{\infty} \frac{\cos^3 \psi}{b^2} dt. \tag{4.41}$$

Using Equation 4.21 to change the variable of integration from t to ψ gives

$$m_e \Delta v = \frac{Ze^2}{bv} \int_{-\pi/2}^{\pi/2} \cos \psi \, d\psi = \frac{2Ze^2}{bv}. \tag{4.42}$$

The maximum possible momentum transfer $m_e \Delta v$ during the free–free interaction is twice the initial momentum $m_e v$ of the electron, so the impact parameter of a free–free interaction cannot be smaller than

$$\boxed{ b_{min} \approx \frac{Ze^2}{m_e v^2}. } \tag{4.43}$$

This result is based on a purely classical treatment of the interaction (see also Jackson [55, section 13.1 and problem 13.1] for a more detailed discussion). The uncertainty principle ($\Delta x \Delta p \simeq \hbar$) implies an independent quantum-mechanical limit

$$b_{\min} = \frac{\hbar}{m_e v}, \tag{4.44}$$

but this lower limit is generally smaller than the classical limit in HII regions and hence may be ignored. This claim can be tested by computing the ratio of the classical to quantum limits for the rms electron velocity $v = (3kT/m_e)^{1/2}$:

$$\left(\frac{Ze^2}{m_e v^2}\right)\left(\frac{\hbar}{m_e v}\right)^{-1} = \frac{Ze^2}{\hbar v} = \frac{Ze^2}{\hbar}\left(\frac{m_e}{3kT}\right)^{1/2}. \tag{4.45}$$

In an HII region with $T \approx 10^4$ K and $Z = 1$, this ratio ≈ 3, so the classical limit is stronger. Only in much cooler ($T < 10^3$ K) plasmas is the quantum-mechanical limit important.

There are two effects that might determine the upper limit $b_{\max}$ to the impact parameter. Because electrostatic forces always dominate gravity on small scales, electrons in the vicinity of a nearly stationary ion are free to rearrange themselves to neutralize, or shield, the ionic charge. The characteristic scale length of this shielding is called the **Debye length**. From Jackson [55], the Debye length is

$$\lambda_D \approx \left(\frac{kT}{4\pi n_e e^2}\right)^{1/2}. \tag{4.46}$$

The Debye length is quite large in the low-density plasma of a typical HII region. For example, if $T \approx 10^4$ K and $n_e \approx 10^3$ cm^{-3},

$$\lambda_D \approx \left[\frac{1.38 \times 10^{-16} \text{ erg K}^{-1} \cdot 10^4 \text{ K}}{4\pi \cdot 10^3 \text{ cm}^{-3} \cdot (4.8 \times 10^{-10} \text{ statcoulomb})^2}\right]^{1/2} \approx 22 \text{ cm.} \tag{4.47}$$

An independent upper limit to the impact parameter is the largest value of b that can generate a significant amount of power at some relevant radio frequency ν. Recall that the pulse power per unit bandwidth is small above angular frequency $\omega \approx v/b$ so

$$\boxed{b_{\max} \approx \frac{v}{\omega} = \frac{v}{2\pi\nu},} \tag{4.48}$$

is the maximum impact parameter capable of emitting significant power at frequency ν.

The controlling upper limit $b_{\max}$ in any particular situation is the *smaller* of these two upper limits. Numerical values for the lower and upper limits in a typical HII region observed at $\nu = 1$ GHz are derived in the boxed example, and Equation 4.48 gives the smaller $b_{\max} \sim 10^{-2}$ cm.

Example. Estimate b_{min} and b_{max} for a pure HII region ($Z = 1$) with $T \approx 10^4$ K observed at a fairly low frequency $\nu = 1$ GHz $= 10^9$ Hz:

$$b_{min} \approx \frac{Ze^2}{m_e v^2} \approx \frac{e^2}{3kT}$$

$$\approx \frac{(4.8 \times 10^{-10} \text{ statcoulomb})^2}{3 \cdot 1.38 \times 10^{-16} \text{ erg K}^{-1} \cdot 10^4 \text{ K}} \approx 5.6 \times 10^{-8} \text{ cm},$$

$$b_{max} \approx \frac{v}{2\pi \nu} \approx \left(\frac{3kT}{m_e}\right)^{1/2} (2\pi \nu)^{-1}$$

$$\approx \left(\frac{3 \cdot 1.38 \times 10^{-16} \text{ erg K}^{-1} \cdot 10^4 \text{ K}}{9.1 \times 10^{-28} \text{ g}}\right)^{1/2} (2\pi \times 10^9 \text{ s}^{-1})^{-1}$$

$$\approx 1.1 \times 10^{-2} \text{ cm}.$$

The maximum impact parameter capable of generating power at this frequency is much smaller than the $\lambda_D \approx 22$ cm Debye length in an HII region with electron density $n_e \approx 10^3$ cm^{-3} (Equation 4.47), so the Debye length is irrelevant. A $T = 10^4$ K electron takes so long to move 22 cm that it would emit at unobservably low frequencies $\nu < 1$ MHz. The Debye length becomes relevant only in much denser plasmas such as the solar chromosphere ($n_e \approx 10^{12}$ cm^{-3}).

Our simple estimate of the ratio

$$\frac{b_{max}}{b_{min}} \approx \left(\frac{3kT}{m_e}\right)^{1/2} (2\pi \nu)^{-1} \left(\frac{3kT}{Ze^2}\right) \approx \left(\frac{3kT}{m_e}\right)^{3/2} \frac{m_e}{2\pi Ze^2 \nu} \tag{4.49}$$

is very close to the result of the very detailed derivation in Oster [75]. The ratio (b_{max}/b_{min}) is of order 10^5, which is much greater than the fractional velocity range $\sigma_v/v \approx 1$ in the Maxwellian velocity distribution (Figure 4.6). Also note that

$$\ln\left(\frac{b_{max}}{b_{min}}\right) \sim 12 \tag{4.50}$$

varies slowly with changes in either b_{max} or b_{min}, so small uncertainties in these limits have very little effect on the calculated emission coefficient of an HII region.

Because the HII region is in local thermodynamic equilibrium (LTE) at some temperature T, Kirchhoff's law (Equation 2.30) immediately yields the absorption coefficient κ (Equation 2.18) in terms of the emission coefficient and the blackbody brightness $B_\nu(T)$:

$$\kappa = \frac{j_\nu}{B_\nu(T)} \approx \frac{j_\nu c^2}{2kT\nu^2} \tag{4.51}$$

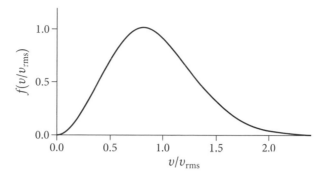

Figure 4.6. The nonrelativistic Maxwellian distribution of particle speeds in LTE (Equation 4.34), where $v_{\rm rms} = (3kT/m)^{1/2}$ is the rms speed of particles with mass m at temperature T.

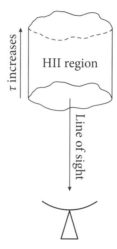

Figure 4.7. Astronomers often approximate H_{II} regions by uniform cylinders whose axis is the line of sight because this gross oversimplification finesses the radiative-transfer problem. It is for good reason that astronomers often feature in jokes beginning "Consider a spherical cow...."

in the Rayleigh–Jeans limit. Thus

$$\kappa = \frac{1}{\nu^2 T^{3/2}} \left[\frac{Z^2 e^6}{c} n_e n_i \frac{1}{\sqrt{2\pi (m_e k)^3}} \right] \frac{\pi^2}{4} \ln\left(\frac{b_{\rm max}}{b_{\rm min}} \right). \tag{4.52}$$

The limit $b_{\rm max}$ (Equation 4.48) is inversely proportional to frequency so the absorption coefficient is not exactly proportional to ν^{-2}. A good numerical approximation is $\kappa(\nu) \propto \nu^{-2.1}$.

The total opacity τ of an H_{II} region is the integral of $-\kappa$ along the line of sight, as illustrated in Figure 4.7:

$$\tau = -\int_{\rm los} \kappa \, ds \propto \int \frac{n_e n_i}{\nu^{2.1} T^{3/2}} ds \approx \int \frac{n_e^2}{\nu^{2.1} T^{3/2}} ds. \tag{4.53}$$

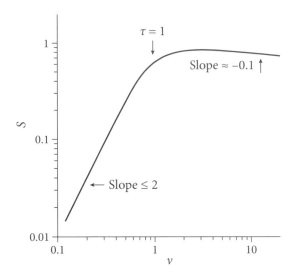

Figure 4.8. The radio spectrum of an H II region. It is a blackbody at low frequencies, with slope 2 if a uniform cylinder as shown in Figure 4.7 and < 2 otherwise. At some frequency ν the optical depth $\tau = 1$, and at much higher frequencies the spectral slope becomes ≈ -0.1 because the opacity coefficient $\kappa(\nu) \propto \nu^{-2.1}$. The source brightness at low frequencies equals the electron temperature. The brightness at high frequencies is proportional to the emission measure (Equation 4.57) of the H II region.

At frequencies low enough that $\tau \gg 1$, the H II region becomes opaque, its spectrum approaches that of a blackbody with brightness temperature approaching the electron temperature ($T_b \approx T \sim 10^4$ K), and its flux density obeys the Rayleigh–Jeans approximation $S \propto \nu^2$. At very high frequencies, $\tau \ll 1$, the H II region is nearly transparent, and

$$S \propto \frac{2kT\nu^2}{c^2}\tau(\nu) \propto \nu^{-0.1}. \tag{4.54}$$

On a log-log plot, the overall spectrum of a uniform H II region looks like Figure 4.8, with the spectral break corresponding to the frequency at which $\tau \approx 1$.

The spectral slope on a log-log plot is often called the **spectral index** and denoted by α, whose sign is defined ambiguously:

$$\alpha \equiv \pm \frac{d \log S}{d \log \nu}. \tag{4.55}$$

Beware the $\pm$ sign! Unfortunately both sign conventions are found in the literature, and you have to look carefully at each paper to find out which one is being used. With the $+$ sign convention, the low-frequency spectral index of a uniform H II region would be $\alpha = +2$. The $-$ sign convention was introduced in the early days of radio astronomy because most sources discovered at low frequencies are stronger at low frequencies than at high frequencies. Thus $\alpha = +0.7$ might mean

$$\frac{d \log S}{d \log \nu} = -0.7. \tag{4.56}$$

The $(+)$ spectral index of any inhomogeneous HII region will be $\alpha \approx -0.1$ well above the break frequency, but the break will be more gradual and the low-frequency slope will be somewhat less than $+2$ just below the break. For example, ionized winds from stars are quite inhomogeneous. Mass conservation in a constant-velocity, isothermal spherical wind implies that the electron density is inversely proportional to the square of the distance from the star: $n_e \propto r^{-2}$. The low-frequency spectral index of free–free emission by such a wind is closer to $+0.6$ than to $+2$.

The **emission measure** (EM) of an HII region is defined by the integral of n_e^2 along the line of sight expressed in astronomically convenient units:

$$\frac{\text{EM}}{\text{pc cm}^{-6}} \equiv \int_{\text{los}} \left(\frac{n_e}{\text{cm}^{-3}} \right)^2 d\left(\frac{s}{\text{pc}} \right). \tag{4.57}$$

Because κ is proportional to $n_e n_i \approx n_e^2$, the optical depth τ is proportional to the emission measure. The emission measure is commonly used to parameterize τ in astronomically convenient units:

$$\tau \approx 3.014 \times 10^{-2} \left(\frac{T}{\text{K}} \right)^{-3/2} \left(\frac{\nu}{\text{GHz}} \right)^{-2} \left(\frac{\text{EM}}{\text{pc cm}^{-6}} \right) \langle g_{\text{ff}} \rangle, \tag{4.58}$$

where the **free–free Gaunt factor** [18] $\langle g_{\text{ff}} \rangle$ is a parameter that absorbs the weak frequency dependence associated with the logarithmic term in κ:

$$\langle g_{\text{ff}} \rangle \approx \ln \left[4.955 \times 10^{-2} \left(\frac{\nu}{\text{GHz}} \right)^{-1} \right] + 1.5 \ln \left(\frac{T}{\text{K}} \right). \tag{4.59}$$

Mezger and Henderson [72] found a very good approximation for the **free–free opacity** τ that is easy to evaluate numerically:

$$\tau \approx 3.28 \times 10^{-7} \left(\frac{T}{10^4 \text{ K}} \right)^{-1.35} \left(\frac{\nu}{\text{GHz}} \right)^{-2.1} \left(\frac{\text{EM}}{\text{pc cm}^{-6}} \right). \tag{4.60}$$

Mezger and Henderson [72, Table 6] lists the errors introduced by this approximation over wide ranges of temperature and frequency.

Example. The interstellar medium of our Galaxy contains a diffuse ionized component, some of which is "warm" ($T \approx 10^4$ K) and some is "hot" ($T \approx 10^6$ K). These two *phases* are roughly in pressure equilibrium so the hot medium is less dense by a factor of $\sim 10^2$. The combination of high T_e and low n_e of the hot phase means that only the warm component contributes significantly to the free–free opacity of the ISM. The warm ionized gas is largely confined to the disk of our Galaxy, where we reside. There must be some frequency ν below which this disk becomes opaque and we cannot see out of our Galaxy, even in the direction perpendicular to the disk.

(Continued)

From the observed brightness spectrum in the direction perpendicular to the disk, Cane [21] found that $\tau \approx 1$ at $\nu \approx 2$ MHz. This result can be inserted into Equation 4.60 to estimate the rms electron density in the warm ISM:

$$1 \approx 3.28 \times 10^{-7} \cdot (1)^{-1.35} \cdot 0.002^{-2.1} \cdot \langle n_e^2 \rangle \cdot 1000 \text{ pc},$$

$$\langle n_e^2 \rangle^{1/2} \approx \left(\frac{0.002^{2.1}}{3.28 \times 10^{-4}} \right)^{1/2} \approx 0.08 \text{ cm}^{-3}.$$

From the optical depth τ and the electron temperature T it is possible to calculate the brightness temperature

$$T_b = T(1 - e^{-\tau}) \tag{4.61}$$

of free–free emission. The line-of-sight structure of an HII region is not normally known, so it is common to approximate the geometry of an HII region by a circular cylinder whose axis lies along the line of sight, and whose axis length equals its diameter. Suppose further that the temperature and density are constant throughout this volume. Then it is very easy to estimate physical parameters of the HII region (e.g., electron density, temperature, emission measure, production rate Q_H of ionizing photons) from the observed radio spectrum, once the distance to the HII region is known.

A useful approximation relating the production rate of ionizing photons to the **free–free spectral luminosity** L_ν at the high frequencies where $\tau \ll 1$ of an HII region in ionization equilibrium is [94]

$$\left(\frac{Q_H}{s^{-1}} \right) \approx 6.3 \times 10^{52} \left(\frac{T}{10^4 \text{ K}} \right)^{-0.45} \left(\frac{\nu}{\text{GHz}} \right)^{0.1} \left(\frac{L_\nu}{10^{20} \text{ W Hz}^{-1}} \right). \tag{4.62}$$

Example. Suppose an idealized HII region with temperature $T = 10^4$ K at distance $d = 10$ kpc produces the radio spectrum shown in Figure 4.8 where S is in Jy and ν is in GHz. What constraints on the HII region are provided by this spectrum? At low frequencies $\nu \ll 1$ GHz, the HII region is optically thick, so it is a blackbody radiator and obeys the Rayleigh–Jeans approximation from which the source solid angle can be derived. From Figure 4.8, $S \approx 0.1$ Jy at $\nu = 0.3$ GHz ($\lambda = 1$ m) so

$$\Omega = \frac{\lambda^2 S}{2kT} \approx \frac{(1 \text{ m})^2 \cdot 0.1 \times 10^{-26} \text{ W m}^{-2} \text{ Hz}^{-1}}{2 \cdot 1.38 \times 10^{-23} \text{ J K}^{-1} \cdot 10^4 \text{ K}} \approx 3.6 \times 10^{-9} \text{sr}.$$

(Continued)

Thus the angular diameter of the H II region is

$$\theta \approx \left(\frac{4\Omega}{\pi}\right)^{1/2} \approx 7 \times 10^{-5} \, \text{rad}$$

and its linear diameter is about 0.7 pc.

At high frequencies $\nu \gg 1 \, \text{GHz}$, the opacity $\tau \ll 1$ so the ratio of observed flux density to blackbody flux density is $\approx \tau$. At $\nu = 10 \, \text{GHz}$, the observed flux density is $S \approx 0.8 \, \text{Jy}$ while the Rayleigh–Jeans extrapolated flux density is

$$S \approx 0.1 \, \text{Jy} \left(\frac{\nu}{0.3 \, \text{GHz}}\right)^2 \approx 110 \, \text{Jy},$$

so the optical depth at 10 GHz is $\tau \approx 0.8/110 \approx 0.007$.

Free–free emission accounts for about 10% of the 1 GHz continuum luminosity in most spiral galaxies. It is the strongest component in the frequency range from $\nu \approx 30 \, \text{GHz}$ to $\nu \approx 200 \, \text{GHz}$, above which thermal emission from cool dust grains dominates (Figure 2.24). Free–free absorption flattens the low-frequency spectra of spiral galaxies, and the frequency at which $\tau \approx 1$ is higher in galaxies with high star-formation rates, especially if the star formation is confined to a compact region near the nucleus. If the free–free and synchrotron emission from a starburst galaxy are roughly cospatial, its radio brightness temperature at frequencies $\nu < 100 \, \text{GHz}$ is [24]

$$T_{\text{b}} \sim T[1 - \exp(-\tau)] \left[1 + 10\left(\frac{\nu}{\text{GHz}}\right)^{0.1+\alpha}\right], \tag{4.63}$$

where $T \approx 10^4 \, \text{K}$ and $\alpha \approx -0.8$ is the spectral index of the synchrotron radiation. Free–free absorption of the synchrotron radiation limits the maximum brightness temperature to $T_{\text{b}} \leq 10^5 \, \text{K}$ at frequencies $\nu \geq 1 \, \text{GHz}$. This limit can be used to identify the energy source powering a compact radio source at the center of a galaxy: if its brightness temperature is significantly higher than 10^5 K, it is powered by an AGN, not a compact starburst.

5 Synchrotron Radiation

5.1 MAGNETOBREMSSTRAHLUNG

Any accelerated charged particle emits electromagnetic radiation with power specified by Larmor's formula (Equation 2.143). In astrophysical situations, electromagnetic forces produce the strongest accelerations of charged particles. Acceleration by an electric field accounts for free–free radiation. Acceleration by a magnetic field produces **magnetobremsstrahlung**, the German word for "magnetic braking radiation." The lightest charged particles (electrons, and positrons if any are present) are accelerated more than relatively massive protons and heavier ions, so electrons (and possibly positrons) account for virtually all of the radiation observed. The character of magnetobremsstrahlung depends on the speeds of the electrons, so these somewhat different types of radiation are given specific names. **Gyro radiation** comes from electrons whose velocities are much smaller than the speed of light: $v \ll c$. Mildly relativistic electrons whose kinetic energies are comparable with their rest mass $m_e c^2$ emit **cyclotron radiation**, and ultrarelativistic electrons (kinetic energies $\gg m_e c^2$) produce **synchrotron radiation**.

Synchrotron radiation is ubiquitous in astronomy. It accounts for most of the radio emission from **active galactic nuclei** (AGNs) thought to be powered by supermassive black holes in galaxies and quasars, and it dominates the radio continuum emission from star-forming galaxies like our own at frequencies below $v \sim 30\,\mathrm{GHz}$. The magnetosphere of Jupiter is a synchrotron radio source. The optical emission from the Crab Nebula supernova remnant, the optical jet of the radio galaxy M87, and the optical through X-ray emission from many quasars is synchrotron radiation.

The relativistic electrons in nearly all synchrotron sources have power-law energy distributions, so they are not in local thermodynamic equilibrium (LTE). Consequently, synchrotron sources are often called "nonthermal" sources. However, a synchrotron source with a relativistic Maxwellian electron-energy distribution would be a thermal source, so "synchrotron" and "nonthermal" are not completely synonymous.

Even though synchrotron radiation is quite different from free–free emission, notice how many themes from the derivation of free–free source spectra are repeated for synchrotron sources—Larmor's equation is used to derive the total power and spectrum of radiation by a single electron, the spectrum of an optically thin source is obtained as the superposition of the spectra of individual electrons, the electron energy distribution is broad enough that the spectrum of an individual electron can be approximated by a delta function, Kirchhoff's law yields the absorption coefficient in terms of the emission coefficient (even though the synchrotron source is not in LTE!), and the simple "cylindrical cow" geometry is used to yield the spectrum of a source that is optically thick at low frequencies.

5.1.1 Gyro Radiation

Larmor's equation is valid only for gyro radiation from a particle with charge q moving with a small velocity $v \ll c$. The **magnetic force** $\vec{F}$ exerted on the particle by a magnetic field $\vec{B}$ is

$$\boxed{\vec{F} = \frac{q(\vec{v} \times \vec{B})}{c}.}$$

(5.1)

The magnetic force is perpendicular to the particle velocity, so $\vec{F} \cdot \vec{v} = 0$. Consequently the magnetic force does no work on the particle, does not change the particle's kinetic energy $mv^2/2$, and does not change the component of velocity $v_\parallel$ parallel to the magnetic field. Because both $|v|$ and $v_\parallel$ are constant, the magnitude of the velocity component $|v_\perp|$ perpendicular to the magnetic field must also be constant. In a uniform magnetic field, the particle moves along the magnetic field line on a helical path with constant linear and angular speeds. In the inertial frame moving with velocity $v_\parallel$, the particle orbits in a circle of radius r perpendicular to the magnetic field with the angular velocity ω needed to balance the centripetal and magnetic forces:

$$m|\dot{v}| = m\omega^2 r = \frac{q}{c}|\vec{v} \times \vec{B}| = \frac{q}{c}\omega r B;$$

(5.2)

the orbital angular frequency is

$$\omega = \frac{qB}{mc}.$$

(5.3)

Equation 5.3 implies that the orbital frequency is independent of the particle speed so long as $v \ll c$. The angular **gyro frequency** ω_G is *defined* by

$$\boxed{\omega_G \equiv \frac{qB}{mc}.}$$

(5.4)

This *definition* holds for *any* particle speed, so the gyro frequency equals the actual orbital frequency if and only if $v \ll c$.

The angular gyro frequency (rad s^{-1}) of an electron is

$$\omega_G = \frac{eB}{m_e c} = \frac{4.8 \times 10^{-10} \text{ statcoul} \cdot B}{9.1 \times 10^{-28} \text{ g} \cdot 3 \times 10^{10} \text{ cm s}^{-1}}$$

(5.5)

$$\approx 17.6 \times 10^6 \text{ rad s}^{-1} \cdot B \text{ (gauss)}.$$

(5.6)

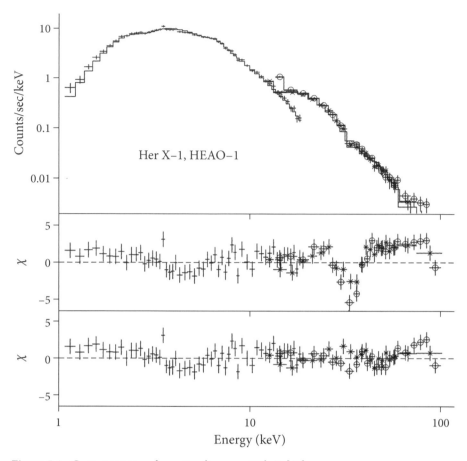

Figure 5.1. Gyro-resonance absorption line near 34 keV [41].

In terms of $\nu_G \equiv \omega_G/(2\pi)$ the electron gyro frequency in MHz is

$$\left(\frac{\nu_G}{\text{MHz}}\right) = 2.8\left(\frac{B}{\text{gauss}}\right). \tag{5.7}$$

The typical interstellar magnetic field strength in a normal spiral galaxy like ours is $B \approx 10\,\mu\text{G}$, so the electron gyro frequency is only $\nu_G = 2.8\,\text{MHz} \cdot 10 \times 10^{-6}$ gauss ~ 28 Hz. The associated gyro radiation cannot propagate through the ISM because its frequency $\nu \sim 28$ Hz is less than the plasma frequency (Equation 6.40).

Gyro radiation from nonrelativistic electrons is observable in very strong magnetic fields such as the $B \sim 10^{12}$ gauss magnetic field of a neutron star. For example, the binary X-ray source Hercules X-1 exhibits an X-ray absorption line at photon energy $E \approx 34$ keV (Figure 5.1). This spectral feature is thought to be a gyro-resonance absorption, in which case the frequency of this absorption line directly measures the magnetic field strength near the Her X-1 neutron star. The observed

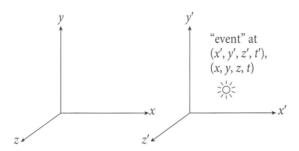

Figure 5.2. An "event" viewed by observers in two coordinate frames. The unprimed frame is the rest frame and the primed frame is moving to the right with velocity v.

photon energy corresponds to the frequency

$$\nu = \frac{E}{h} \approx \frac{34 \times 10^3 \text{ eV} \cdot 1.60 \times 10^{-12} \text{erg eV}^{-1}}{6.63 \times 10^{-27} \text{erg s}} \approx 8.2 \times 10^{18} \text{ Hz}. \quad (5.8)$$

Equating this frequency to the gyro frequency yields the magnetic field:

$$B = \frac{2\pi \nu_G \, m_e \, c}{e} \quad (5.9)$$

$$\approx \frac{2\pi \cdot 8.2 \times 10^{18} \text{ Hz} \cdot 9.1 \times 10^{-28} \text{ g} \cdot 3 \times 10^{10} \text{ cm s}^{-1}}{4.8 \times 10^{-10} \text{ statcoul}} \quad (5.10)$$

$$\approx 2.9 \times 10^{12} \text{ gauss}. \quad (5.11)$$

5.2 SYNCHROTRON POWER

Cosmic rays are celestial particles (e.g., electrons, protons, and heavier nuclei) with extremely high energies. Cosmic-ray electrons in the interstellar magnetic field emit the **synchrotron radiation** that accounts for most of the continuum emission from our Galaxy at frequencies below about 30 GHz. Larmor's formula can be used to calculate the synchrotron power and synchrotron spectrum of a single electron in the inertial frame in which the electron is instantaneously at rest, but the **Lorentz transform** of special relativity is needed to transform these results to the frame of an observer at rest in the Galaxy.

5.2.1 Lorentz Transforms

For any pointlike **event**, the Lorentz transform (see Appendix C for a derivation) relates the coordinates (x, y, z, t) in the unprimed inertial frame and the coordinates (x', y', z', t') in the primed frame moving with velocity v in the x-direction (Figure 5.2). They are

$$\boxed{x = \gamma(x' + vt'), \quad y = y', \quad z = z', \quad t = \gamma(t' + \beta x'/c),} \quad (5.12)$$

$$x' = \gamma(x - vt), \quad y' = y, \quad z' = z, \quad t' = \gamma(t - \beta x/c), \qquad (5.13)$$

where

$$\beta \equiv v/c \qquad (5.14)$$

and

$$\gamma \equiv (1 - \beta^2)^{-1/2} \qquad (5.15)$$

is called the **Lorentz factor**. The Lorentz transform is linear, so even for finite coordinate differences $(\Delta x, \Delta y, \Delta z, \Delta t)$ and $(\Delta x', \Delta y', \Delta z', \Delta t')$ between two events, the differential form of the Lorentz transform is

$$\Delta x = \gamma(\Delta x' + v\Delta t'), \quad \Delta y = \Delta y', \quad \Delta z = \Delta z', \quad \Delta t = \gamma(\Delta t' + \beta \Delta x'/c),$$

$$(5.16)$$

$$\Delta x' = \gamma(\Delta x - v\Delta t), \quad \Delta y' = \Delta y, \quad \Delta z' = \Delta z, \quad \Delta t' = \gamma(\Delta t - \beta \Delta x/c).$$

$$(5.17)$$

5.2.2 Relativistic Masses

The rest mass m_e of an electron can be converted to an energy by Einstein's famous **mass–energy equation**

$$E = mc^2. \qquad (5.18)$$

The energy corresponding to the electron's rest mass m_e is

$$E_0 = m_e c^2 = 9.1 \times 10^{-28} \text{ g} \cdot (3 \times 10^{10} \text{ cm s}^{-1})^2 = 8.2 \times 10^{-7} \text{ erg} \qquad (5.19)$$

$$= \frac{8.2 \times 10^{-7} \text{ erg}}{1.60 \times 10^{-12} \text{ erg (eV)}^{-1}} = 5.1 \times 10^5 \text{ eV} = 0.51 \text{ MeV}. \qquad (5.20)$$

Cosmic-ray electrons having masses $m = \gamma m_e \gg m_e$ (Equation C.28) and total energies $E \gg E_0 = 0.51$ MeV are called **ultrarelativistic**.

Ultrarelativistic electrons still move on spiral paths along magnetic field lines, but the angular frequencies ω_B of their orbits are lower because their inertial masses are multiplied by γ (Appendix C):

$$\omega_B = \frac{eB}{(\gamma m_e)c} = \frac{\omega_G}{\gamma}. \qquad (5.21)$$

The orbital frequency of a cosmic-ray electron with $\gamma = 10^5$ in a $B \approx 10 \,\mu$G interstellar magnetic field is only

$$\nu_B \equiv \frac{\omega_B}{2\pi} \approx 28 \times 10^{-5} \text{ Hz} \approx 1 \text{ cycle per hour}. \qquad (5.22)$$

Because $v \approx c$ whenever $\gamma \gg 1$, the orbital radius r of an ultrarelativistic electron is nearly independent of γ and can be quite large:

$$r \approx \frac{c}{\omega_B} \approx \frac{3 \times 10^{10} \text{ cm s}^{-1}}{2\pi \cdot 28 \times 10^{-5} \text{ Hz}} \approx 1.7 \times 10^{13} \text{ cm} \approx 1 \text{ AU}. \qquad (5.23)$$

Equation 5.21 is not promising for the production of observable synchrotron radiation: the high observed masses $m = \gamma m_e$ of relativistic electrons reduce their orbital frequencies and accelerations to extremely low values. However, two compensating relativistic effects can explain the strong synchrotron radiation observed at radio frequencies: (1) the total radiated power in the observer's frame is proportional to γ^2 and (2) relativistic beaming turns the low-frequency sinusoidal radiation in the electron frame into a series of extremely sharp pulses containing power at much higher frequencies $\sim \gamma^3 \nu_B = \gamma^2 \nu_G$ in the observer's frame. These relativistic corrections are derived in Sections 5.2.3 and 5.3.1.

5.2.3 Synchrotron Power Radiated by a Single Electron

Let primed coordinates describe the inertial frame in which the electron is (temporarily) nearly at rest. Larmor's equation (Equation 2.143) gives the radiated power in the electron rest frame as

$$P' = \frac{2(e')^2 (a'_\perp)^2}{3c^3} = \frac{2e^2 (a'_\perp)^2}{3c^3} \tag{5.24}$$

because $e = e'$ (electric charge is a relativistic invariant) follows directly from Maxwell's relativistically correct equations.

The magnetic acceleration $a_\perp = (a_y^2 + a_z^2)^{1/2}$ of the electron in the frame of an observer at rest in the Galaxy can be derived by applying the chain rule for derivatives to the differential Lorentz coordinate transform (Equations 5.16 and 5.17).

$$a_y \equiv \frac{dv_y}{dt} = \frac{dv_y}{dt'}\frac{dt'}{dt} = \frac{1}{\gamma}\frac{dv'_y}{dt'}\frac{dt'}{dt} = \frac{a'_y}{\gamma^2}. \tag{5.25}$$

Similarly, $a_z = a'_z/\gamma^2$ so

$$a_\perp = \frac{a'_\perp}{\gamma^2}. \tag{5.26}$$

Thus

$$P' = \frac{2e^2 (a'_\perp)^2}{3c^3} = \frac{2e^2 a_\perp^2 \gamma^4}{3c^3}. \tag{5.27}$$

The next step is to transform from the radiated power $P' = dE'/dt'$ in the electron frame to $P = dE/dt$, the power measured by an observer at rest in the Galaxy, by applying the chain rule to the mass–energy Equation 5.18:

$$P \equiv \frac{dE}{dt} = \frac{dE}{dt'}\frac{dt'}{dt} = \frac{dE}{dE'}\frac{dE'}{dt'}\frac{dt'}{dt} = \gamma P' \gamma^{-1} = P'; \tag{5.28}$$

that is, the *power is a relativistic invariant*. Consequently,

$$P = P' = \frac{2e^2 a_\perp^2 \gamma^4}{3c^3} \qquad (a_\parallel = 0). \tag{5.29}$$

To calculate $a_\perp$, combine force balance in a circular orbit

$$a_\perp \equiv \frac{dv_\perp}{dt} = \omega_B v_\perp \tag{5.30}$$

with Equation 5.21 for ω_B to get

$$a_\perp = \frac{eBv_\perp}{\gamma m_e c} = \frac{eBv\sin\alpha}{\gamma m_e c}, \tag{5.31}$$

where the constant angle α between the electron velocity $\vec{v}$ and the magnetic field $\vec{B}$ is called the **pitch angle**.

Inserting $a_\perp$ from Equation 5.31 into Equation 5.29 gives the power radiated by a single electron moving with pitch angle α:

$$P = \frac{2e^2}{3c^3}\gamma^2\frac{e^2 B^2}{m_e^2 c^2}v^2\sin^2\alpha. \tag{5.32}$$

This power is usually written in terms of the **Thomson cross section** of an electron, σ_T. The Thomson cross section is the classical radiation-scattering cross section of a charged particle. If a plane wave of electromagnetic radiation passes a free charged particle initially at rest, the electric field of that radiation will accelerate the particle, which in turn will radiate power in all directions according to Larmor's equation. This process is called scattering rather than absorption because the total power in electromagnetic radiation is unchanged—all of the power extracted from the incident plane wave is reradiated at the same frequency but in other directions. It is a straightforward exercise to show that the geometric area that would intercept this amount of incident power from the plane wave is

$$\boxed{\sigma_T \equiv \frac{8\pi}{3}\left(\frac{e^2}{m_e c^2}\right)^2.} \tag{5.33}$$

Numerically,

$$\sigma_T = \frac{8\pi}{3}\left[\frac{(4.80\times 10^{-10}\text{ statcoul})^2}{9.11\times 10^{-28}\text{ g }(3.00\times 10^{10}\text{ cm s}^{-1})^2}\right] \approx 6.65\times 10^{-25}\text{ cm}^2. \tag{5.34}$$

The reason for using the Thomson cross section will become clear in the discussion of inverse-Compton scattering of radiation by the same cosmic rays that are producing synchrotron radiation (Section 5.5.1).

It is also conventional to eliminate the B^2 in Equation 5.32 in favor of the **magnetic energy density**

$$\boxed{U_B = \frac{B^2}{8\pi}.} \tag{5.35}$$

Then

$$P = \left[\frac{8\pi}{3}\left(\frac{e^2}{m_e c^2}\right)^2\right]2\left(\frac{B^2}{8\pi}\right)c\gamma^2\frac{v^2}{c^2}\sin^2\alpha \tag{5.36}$$

simplifies to

$$\boxed{P = 2\sigma_T\beta^2\gamma^2 c\, U_B\sin^2\alpha.} \tag{5.37}$$

The synchrotron power radiated by a single electron depends only on physical constants, the square of the electron kinetic energy (via γ^2), the magnetic energy density U_B, and the pitch angle α.

The relativistic electrons in radio sources can have **lifetimes** of thousands to millions of years before losing their ultrarelativistic energies via synchrotron radiation or other processes. During their lifetimes they are scattered repeatedly by magnetic-field fluctuations and charged particles in their environment, and the distribution of their pitch angles gradually becomes random and isotropic. The **average synchrotron power** $\langle P \rangle$ per electron in an ensemble of electrons having the same Lorentz factor γ and isotropically distributed pitch angles α is

$$\langle P \rangle = 2\sigma_{\rm T} \beta^2 \gamma^2 c \, U_B \langle \sin^2 \alpha \rangle, \tag{5.38}$$

where $\langle \sin^2 \alpha \rangle$ is the average over all pitch angles:

$$\langle \sin^2 \alpha \rangle \equiv \frac{\int \sin^2 \alpha \, d\Omega}{\int d\Omega} = \frac{1}{4\pi} \int \sin^2 \alpha \, d\Omega \tag{5.39}$$

$$= \frac{1}{4\pi} \int_{\phi=0}^{2\pi} \int_{\alpha=0}^{\pi} \sin^2 \alpha \sin \alpha \, d\alpha \, d\phi = \frac{1}{4\pi} 2\pi \frac{4}{3} \tag{5.40}$$

$$= \frac{2}{3}. \tag{5.41}$$

Thus the **average synchrotron power** per relativistic electron in a source with an isotropic pitch-angle distribution is

$$\boxed{\langle P \rangle = \frac{4}{3} \sigma_{\rm T} \beta^2 \gamma^2 c \, U_B .} \tag{5.42}$$

For all $\gamma \gg 1$, the factor $\beta^2 = 1 - \gamma^{-2} \approx 1$ can be ignored. Relativistic effects multiply the average radiated power by a factor γ^2 compared with the nonrelativistic ($\gamma = 1$) Larmor equation.

5.3 SYNCHROTRON SPECTRA

5.3.1 The Synchrotron Spectrum of a Single Electron

Why does synchrotron radiation appear at frequencies much higher than $\omega_B = \omega_G/\gamma$? First, **relativistic aberration** beams the dipole pattern of Larmor radiation in the electron frame sharply along the direction of motion in the observer's frame as v approaches c (Figure 5.3). Relativistic photon beaming follows directly from the relativistic **velocity addition equations** (Equations C.22 through C.26) that relate the photon velocity component v_x in the unprimed frame of the observer to the photon velocity component v'_x in the primed frame and the velocity $v = \beta c$ of the primed frame in which the electron is instantaneously at rest:

$$v_x \equiv \frac{dx}{dt} = \frac{dx}{dt'}\frac{dt'}{dt} = \gamma \left(\frac{dx'}{dt'} + v\frac{dt'}{dt'} \right)\left(\frac{dt}{dt'} \right)^{-1} \tag{5.43}$$

$$= \gamma(v'_x + v)\left[\gamma \left(1 + \frac{\beta}{c}\frac{dx'}{dt'} \right) \right]^{-1}, \tag{5.44}$$

$$\boxed{v_x = (v'_x + v)\left(1 + \frac{\beta v'_x}{c} \right)^{-1}.} \tag{5.45}$$

In the y-direction,

$$v_y \equiv \frac{dy}{dt} = \frac{dy}{dt'}\frac{dt'}{dt} = \frac{dy'}{dt'}\left(\frac{dt}{dt'} \right)^{-1}, \tag{5.46}$$

$$\boxed{v_y = \frac{v'_y}{\gamma}\left(1 + \frac{\beta v'_x}{c} \right)^{-1}.} \tag{5.47}$$

Consider the synchrotron photons emitted with speed c at an angle θ' from the x'-axis. Let v'_x and v'_y be the projections of the photon speed onto the x'- and y'-axes. Then

$$\cos\theta' = \frac{v'_x}{c}, \qquad \sin\theta' = \frac{v'_y}{c}. \tag{5.48}$$

In the observer's frame the same photons have

$$\cos\theta = \frac{v_x}{c}, \qquad \sin\theta = \frac{v_y}{c}. \tag{5.49}$$

Inserting the velocity Equations 5.45 and 5.47 into the angle Equations 5.48 and 5.49 yields relations connecting θ and θ':

$$\cos\theta = \left(\frac{v'_x + v}{1 + \beta v'_x/c} \right)\frac{1}{c} = \left(\frac{c\cos\theta' + v}{1 + \beta c\cos\theta'/c} \right)\frac{1}{c} = \frac{\cos\theta' + \beta}{1 + \beta\cos\theta'} \tag{5.50}$$

and

$$\sin\theta = \frac{v'_y}{c\gamma(1 + \beta v'_x/c)} = \frac{\sin\theta'}{\gamma(1 + \beta\cos\theta')}. \tag{5.51}$$

In the frame moving with the electron, the Larmor equation implies a power pattern proportional to $\cos^2\theta'$ with nulls at $\theta' = \pm\pi/2$. In the observer's frame, these nulls are offset by much smaller angles

$$\theta = \pm\arcsin(1/\gamma). \tag{5.52}$$

An observer at rest sees the radiation confined to a very narrow beam of width $2/\gamma$ between nulls, as shown in Figure 5.3. For example, a 10 GeV electron has $\gamma \approx 2 \times 10^4$ so $2/\gamma \approx 10^{-4}$ rad ≈ 20 arcsec! Although the electron is emitting

Figure 5.3. Relativistic aberration transforms the dipole power pattern of Larmor radiation in the electron rest frame (dotted curve) into a narrow searchlight beam in the observer's frame. The solid curve is the transformed pattern for $\gamma = 5$. The observed angle between the nulls of the forward beam falls to $\Delta\theta = 2\arcsin(1/\gamma)$, which approaches $\Delta\theta = 2/\gamma$ in the limit $\gamma \gg 1$.

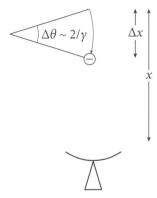

Figure 5.4. The beamed radiation from an ultrarelativistic electron is visible only while the electron's velocity points within $\pm 1/\gamma$ of the line of sight ($\Delta\theta \approx 2/\gamma$). During that time Δt the electron moves a distance $\Delta x = v\Delta t$ toward the observer, almost keeping up with the radiation that travels a distance $c\,\Delta t$. As a result, the observed pulse duration is shortened by a factor $(1 - v/c)$.

continuously, the observer sees a short pulse of radiation only from the tiny fraction

$$\frac{2}{2\pi\gamma} = \frac{1}{\pi\gamma} \tag{5.53}$$

of the electron orbit where the electron is moving almost directly toward the observer.

The duration $\Delta t_{\rm p}$ of the observed pulse is even shorter than the time Δt the electron needs to cover $1/(\pi\gamma)$ of its orbit because the electron is moving almost directly toward the observer with a speed approaching c (Figure 5.4) when it is observable. As it travels toward the observer, the electron nearly keeps up with the radiation that it emits:

$$\Delta t_{\rm p} = t(\text{end of observed pulse}) - t(\text{start of observed pulse}) \tag{5.54}$$

$$= \frac{\Delta x}{v} + \frac{(x - \Delta x)}{c} - \frac{x}{c}. \tag{5.55}$$

The first term in this equation represents the time taken by the electron to cover the distance Δx, the second is the light travel time from the electron position at the *end* of the pulse seen by the observer, and the third is the light travel time from the electron

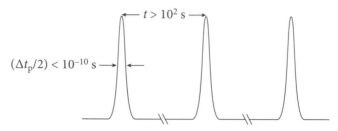

Figure 5.5. Synchrotron radiation is a very spiky series of widely spaced narrow pulses. The numerical values on this plot of power versus time correspond to an electron with $\gamma \sim 10^4$ in a magnetic field $B \sim 10\,\mu$G.

position at the *beginning* of the pulse seen by the observer. Note that the observed pulse duration

$$\Delta t_{\rm p} = \frac{\Delta x}{v} - \frac{\Delta x}{c} = \frac{\Delta x}{v}\left(1 - \frac{v}{c}\right) \ll \frac{\Delta x}{v} = \Delta t \qquad (5.56)$$

is much less than the time Δt the electron needs to move a distance Δx because, in the observer's frame, the electron nearly keeps up with its own radiation. In the limit $v \to c$,

$$\left(1 - \frac{v}{c}\right) = \left(1 - \frac{v}{c}\right)\frac{1 + v/c}{1 + v/c} = \frac{1 - v^2/c^2}{1 + v/c} \approx \frac{\gamma^{-2}}{2} = \frac{1}{2\gamma^2} \qquad (5.57)$$

so

$$\Delta t_{\rm p} = \frac{\Delta t}{2\gamma^2} = \frac{\Delta x}{v}\frac{1}{2\gamma^2} = \frac{\Delta\theta}{\omega_B}\frac{1}{2\gamma^2}. \qquad (5.58)$$

Recall that $\Delta\theta \approx 2/\gamma$ (Figure 5.4) so

$$\Delta t_{\rm p} = \frac{2}{\gamma\omega_B 2\gamma^2} = \frac{1}{\gamma^3\omega_B} = \frac{1}{\gamma^2\omega_G} \qquad (5.59)$$

is the *full* observed duration of the pulse. Allowing for the motion of the electron parallel to the magnetic field replaces the total magnetic field by its perpendicular component $B_\perp = B\sin\alpha$, yielding

$$\Delta t_{\rm p} = \frac{1}{\gamma^2\omega_G \sin\alpha}, \qquad (5.60)$$

where α is the pitch angle of the electron. Thus a plot of the power received as a function of time is very spiky. If $\gamma \approx 10^4$ and $B \sim 10\,\mu$G, the half width of each pulse is $\Delta t_{\rm p}/2 < 10^{-10}$ s and the spacing between pulses is $\gamma/\nu_G > 10^2$ s (Figure 5.5).

The observed synchrotron power spectrum is the Fourier transform of this time series of pulses. The pulse train is the convolution of the individual pulse profile with the **shah function** (see Figure A.1 or Bracewell [15], a valuable reference book):

$$\mathrm{III}(t/\Delta t) \equiv \sum_{n=-\infty}^{\infty} \delta[(t/\Delta t)] - n), \qquad (5.61)$$

where each **delta function** δ is an infinitesimally narrow spike at integer $t/\Delta t = n$ and whose integral is unity. The convolution theorem (Equation A.15) states that the Fourier transform of the pulse train is the product of the Fourier transform of one pulse and the Fourier transform of the shah function.

The Fourier transform of the shah function is also the shah function (Figure A.1), so the similarity theorem (Equation A.11) implies that the Fourier transform of

$$\mathrm{III}\left(\frac{t\nu_{\mathrm{G}}}{\gamma}\right) \tag{5.62}$$

in the time domain is proportional to

$$\mathrm{III}\left(\frac{\nu\gamma}{\nu_{\mathrm{G}}}\right), \tag{5.63}$$

which is a nearly continuous series of spikes in the frequency domain. Adjacent spikes are separated in frequency by only

$$\Delta\nu = \frac{\nu_{\mathrm{G}}}{\gamma} < 10^{-3}\ \mathrm{Hz}. \tag{5.64}$$

Although this is not formally a continuous spectrum, the frequency shifts caused by even tiny fluctuations in electron energy, magnetic field strength, or pitch angle cause frequency shifts much larger than $\Delta\nu$, so the spectrum of synchrotron radiation is effectively continuous.

Thus the synchrotron spectrum of a single electron is fairly flat at low frequencies and tapers off at frequencies above

$$\nu_{\max} \approx \frac{1}{2\Delta t_{\mathrm{p}}} \approx \pi\gamma^{2}\nu_{\mathrm{G}}\sin\alpha \propto \gamma^{2}B_{\perp}. \tag{5.65}$$

It isn't necessary to know the Fourier transform of the pulse shape precisely to calculate the synchrotron spectra of celestial sources because real sources don't contain electrons with just one energy and one pitch angle in a uniform magnetic field. The actual energy distribution of cosmic rays in a real radio source is a very broad power law, broad enough to smear out the details of the spectrum from each electron-energy range. Just for the record, the synchrotron power spectrum of a single electron is

$$\boxed{P(\nu) = \frac{\sqrt{3}e^{3}B\sin\alpha}{m_{e}c^{2}}\left(\frac{\nu}{\nu_{c}}\right)\int_{\nu/\nu_{c}}^{\infty}K_{5/3}(\eta)d\eta,} \tag{5.66}$$

where $K_{5/3}$ is a modified Bessel function and ν_{c} is the **critical frequency** whose value is

$$\boxed{\nu_{c} = \frac{3}{2}\gamma^{2}\nu_{\mathrm{G}}\sin\alpha.} \tag{5.67}$$

For the full mathematical derivation of Equations 5.66 and 5.67, see Pacholczyk [77], a valuable reference for details of radiation processes.

The synchrotron power spectrum of a single electron is plotted in Figure 5.6. It has a logarithmic slope

$$\frac{d \log P(\nu)}{d \log \nu} \approx \frac{1}{3} \tag{5.68}$$

at low frequencies, a broad peak near the critical frequency ν_c, and falls off sharply at higher frequencies. One way to look at ν_c is

$$\nu_c = \left(\frac{3}{2} \sin \alpha\right) \left(\frac{E}{mc^2}\right)^2 \frac{eB}{2\pi m_e c} \propto E^2 B_\perp. \tag{5.69}$$

That is, the frequency at which each electron emits most strongly is proportional to the square of its energy multiplied by the strength of the perpendicular component of the magnetic field.

5.3.2 Synchrotron Spectra of Optically Thin Radio Sources

If a synchrotron source containing any arbitrary distribution of electron energies is optically thin ($\tau \ll 1$), then its spectrum is the superposition of the spectra from individual electrons and its flux density cannot rise faster than $\nu^{1/3}$ at any frequency ν. In other words, the (negative) spectral index $\alpha \equiv -d \log P_\nu / d \log \nu$ (be careful not to confuse this spectral index α with the electron pitch angle α) must always be greater than $-1/3$. Most astrophysical sources of synchrotron radiation have spectral indices near $\alpha \approx 0.75$ at frequencies where they are optically thin, and their high-frequency spectral indices reflect their electron energy distributions, not the spectra of individual electrons.

The energy distribution of cosmic-ray electrons in most synchrotron sources is roughly a power law:

$$n(E)dE \propto E^{-\delta}dE, \tag{5.70}$$

where $n(E)dE$ is the number of electrons per unit volume with energies E to $E + dE$. The energy range around $\gamma \sim 10^4$ is relevant to the production of radio radiation. Because $n(E)$ is nearly a power law over more than a decade of energy and the critical frequency ν_c is proportional to E^2, the synchrotron spectrum will reflect this power law over a frequency range of at least $10^2 = 100$. Consequently the detailed spectra of individual electrons can be ignored because they are smeared out in the source spectrum by the broad power-law energy distribution. The source spectrum can be calculated with good accuracy from the approximation that each electron radiates all of its average power (Equation 5.42)

$$P = -\frac{dE}{dt} = \frac{4}{3}\sigma_T \beta^2 \gamma^2 c U_B \tag{5.71}$$

at the single frequency

$$\nu \approx \gamma^2 \nu_G, \tag{5.72}$$

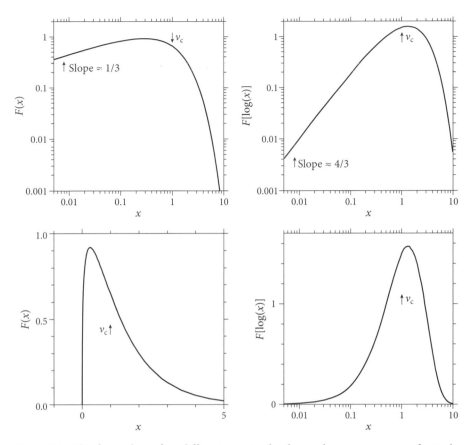

Figure 5.6. This figure shows four different ways to plot the synchrotron spectrum of a single electron in terms of $F(x) \equiv x \int_x^\infty K_{5/3}(\eta)d\eta$, where $x \equiv \nu/\nu_c$ is the frequency in units of the critical frequency ν_c. Although they all plot the same spectrum, they look quite different and emphasize or suppress information in different ways. (1) Simply plotting $F(x)$ versus x on linear axes (lower left panel) completely obscures the spectrum below the peak of $F(x)$ at $x \approx 0.29$. (2) Replotting on logarithmic axes (upper left panel) shows that the low-frequency spectrum has a logarithmic slope of $1/3$, but it obscures the fact that most of the power is emitted at frequencies near $x \sim 1$ because $F(x)$ is the spectral power per unit frequency, not per unit log(frequency). (3) The power per unit $\log x$ is $F(\log x) = \ln(10)x F(x)$, which is plotted on logarithmic axes in the upper right panel. It has a slope of $4/3$ at low frequencies, making it clearer that most of the power is emitted near $x \sim 1$, as required to justify the approximation (used in the next section) that all of the power is emitted at $x = 1$. (4) The lower right panel plots $F(\log x)$ with a linear ordinate but a logarithmic abscissa to expand the low-frequency spectrum lost in the lower left panel. It is clearly consistent with the approximation that all emission is near $x = 1$ but doesn't clearly show that the low-frequency spectrum is a power law. Note also that the peak of $F(\log x)$ is at $x \approx 1.3$, not $x \approx 0.29$. Areas under the curves in the two lower panels are proportional to the power radiated in given frequency ranges. Both lower panels show that about half of the power is emitted at frequencies below the critical frequency and half at higher frequencies.

which is very close to the critical frequency (Equation 5.67). Then the emission coefficient (Equation 2.26) of synchrotron radiation by an ensemble of electrons is

$$j_\nu d\nu = -\frac{dE}{dt} n(E) dE, \tag{5.73}$$

where

$$E = \gamma m_e c^2 \approx \left(\frac{\nu}{\nu_G}\right)^{1/2} m_e c^2. \tag{5.74}$$

Differentiating $E(\nu)$ in Equation 5.74 gives

$$dE \approx \frac{m_e c^2 \nu^{-1/2}}{2 \nu_G^{1/2}} d\nu \tag{5.75}$$

so

$$j_\nu \propto \left(\frac{4}{3} \sigma_T \beta^2 \gamma^2 c U_B\right)(E^{-\delta})\left(\frac{m_e c^2 \nu^{-1/2}}{2 \nu_G^{1/2}}\right). \tag{5.76}$$

Eliminating E in favor of ν/ν_G and then using $\nu_G \propto B$ yields an expression for j_ν in terms of ν and B only:

$$j_\nu \propto \left(\frac{\nu}{\nu_G}\right) B^2 \left(\frac{\nu}{\nu_G}\right)^{-\delta/2} (\nu \nu_G)^{-1/2} \propto \left(\frac{\nu}{B}\right) B^2 \left(\frac{\nu}{B}\right)^{-\delta/2} (\nu B)^{-1/2}. \tag{5.77}$$

This simplifies to

$$\boxed{j_\nu \propto B^{(\delta+1)/2} \nu^{(1-\delta)/2}.} \tag{5.78}$$

Thus the spectrum of optically thin synchrotron radiation from a power-law distribution $n(E) \propto E^{-\delta}$ of electrons is also a power law, and the spectral index $\alpha = -d\ln S/d\ln \nu$ depends only on δ:

$$\boxed{\alpha = \frac{\delta - 1}{2}.} \tag{5.79}$$

In our Galaxy and in many other synchrotron sources, $\alpha \approx 0.75$ near $\nu \approx 1$ GHz, so the radio spectra imply $\delta \approx 2.5$. This value of δ reflects the initial power-law energy slope δ_0 of cosmic rays accelerated in shocks, such as the shocks produced by supernova remnants expanding into the ambient interstellar medium, modified by loss processes that deplete the population of relativistic electrons. For example, the rate at which an electron loses energy to synchrotron radiation is proportional to E^2, so higher-energy electrons are depleted more rapidly. The critical frequency is also proportional to E^2, so synchrotron losses eventually steepen source spectra at higher frequencies. If relativistic electrons with initial power-law slope δ_0 are continuously injected into a synchrotron source, synchrotron losses will eventually steepen that slope to $\delta = \delta_0 + 1$ at high energies and the high-frequency spectral index will steepen by $\Delta\alpha = 1/2$. See Pacholczyk [77] for a more detailed discussion of energy losses and their spectral consequences.

5.3.3 Synchrotron Self-Absorption

The brightness temperatures of synchrotron sources cannot become arbitrarily large at low frequencies because for every emission process there is an associated absorption process. If the emitting particles are in local thermodynamic equilibrium (LTE), they have a Maxwellian energy distribution and the source is thermal. No thermal source can have a brightness temperature greater than the kinetic temperature of the emitting particles. If the energy distribution of relativistic electrons in a synchrotron source were a (relativistic) Maxwellian, the electrons would have a well-defined kinetic temperature, and **synchrotron self-absorption** would prevent the brightness temperature of the synchrotron radiation from exceeding the kinetic temperature of the emitting electrons. Most astrophysical synchrotron sources are nonthermal sources because the energy distribution of the relativistic electrons is a power law and there is no well-defined electron temperature. However, synchrotron self-absorption occurs for *any* electron energy distribution, and the low-frequency spectrum of an optically thick synchrotron source is a power law whose slope is $\alpha = -d \ln S / d \ln \nu = -5/2$. That result is derived below.

Electrons with energy $E = \gamma m_e c^2$ emit most of their synchrotron power near the critical frequency

$$\nu_c \sim \frac{\gamma^2 e B}{2\pi m_e c},\tag{5.80}$$

so the synchrotron emission at frequency ν comes primarily from electrons with Lorentz factors near

$$\gamma \approx \left(\frac{2\pi m_e c \nu}{e B} \right)^{1/2}.\tag{5.81}$$

In this approximation that only those electrons having one particular energy E contribute to the emission (and hence absorption) at each frequency ν, all other electrons *could* have a relativistic Maxwellian energy distribution to match without changing the resulting emission and absorption at that frequency. Consequently, a sufficiently bright synchrotron source *will* be optically thick, and its brightness temperature at any frequency cannot exceed the effective temperature of the electrons emitting at that frequency.

In an ultrarelativistic gas, the ratio of specific heats at constant pressure and at constant volume is $c_p/c_v = 4/3$, not the nonrelativistic 5/3, so the relation between electron energy E and temperature T_e is

$$E = 3kT_e, \quad \text{not} \quad \frac{3kT_e}{2}.\tag{5.82}$$

Thus the **effective temperature** of relativistic electrons with energy E can be defined as

$$T_e \equiv \frac{E}{3k} = \frac{\gamma m_e c^2}{3k},\tag{5.83}$$

even if the ensemble of electrons has a nonthermal energy distribution. Using Equation 5.81 to eliminate γ in favor of ν gives the effective temperature of those electrons producing most of the synchrotron radiation at frequency ν:

$$T_e \approx \left(\frac{2\pi m_e c \nu}{eB}\right)^{1/2} \frac{m_e c^2}{3k}. \tag{5.84}$$

Numerically,

$$\left(\frac{T_e}{K}\right) \approx 1.18 \times 10^6 \left(\frac{\nu}{Hz}\right)^{1/2} \left(\frac{B}{gauss}\right)^{-1/2}. \tag{5.85}$$

For example, the effective temperature of the relativistic electrons emitting synchrotron radiation at $\nu = 0.1\,\text{GHz} = 10^8\,\text{Hz}$ in a $B = 100\,\mu\text{gauss} = 10^{-4}$ gauss magnetic field is

$$\left(\frac{T_e}{K}\right) \approx 1.18 \times 10^6 \cdot (10^8)^{1/2}(10^{-4})^{-1/2} \approx 10^{12}. \tag{5.86}$$

At a sufficiently low frequency ν, the brightness temperature T_b of any synchrotron source will approach the effective electron temperature T_e of electrons emitting at that frequency and the source will become opaque. Equation 2.33 defines brightness temperature in the Rayleigh–Jeans limit:

$$T_b \equiv \frac{I_\nu c^2}{2k\nu^2}. \tag{5.87}$$

Setting $T_b \approx T_e$ and using Equation 5.85 to eliminate T_e in favor of ν and B gives

$$I_\nu \approx \frac{2kT_e\nu^2}{c^2} \propto \nu^{1/2}\nu^2 B^{-1/2} = \nu^{5/2}B^{-1/2}. \tag{5.88}$$

Thus at low frequencies the spectrum of a synchrotron self-absorbed and spatially homogeneous source is a power law of slope $5/2$:

$$S(\nu) \propto \nu^{5/2}, \tag{5.89}$$

independent of the slope δ of the electron-energy spectrum. The flux density of an opaque but truly thermal source (e.g., an H II region) is proportional to ν^2; the extra $\nu^{1/2}$ for synchrotron radiation comes from the fact that $T_e \propto \nu^{1/2}$ (Equation 5.85).

The full spectrum of a homogeneous cylindrical synchrotron source (Figure 5.7) is [77]

$$S \propto \left(\frac{\nu}{\nu_1}\right)^{5/2} \left\{1 - \exp\left[-\left(\frac{\nu}{\nu_1}\right)^{-(\delta+4)/2}\right]\right\}, \tag{5.90}$$

where ν_1 is the frequency at which $\tau = 1$. Real astrophysical sources are inhomogeneous, so synchrotron self-absorption always yields slopes much lower than $5/2$, as illustrated by Figure 5.8.

Substituting $T_b \approx T_e$ into Equation 5.85 yields an estimate of the **magnetic field strength** in a self-absorbed source whose brightness temperature has been measured

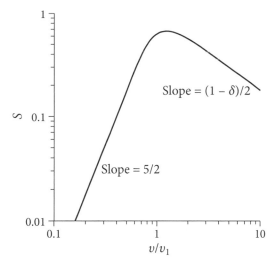

Figure 5.7. The spectrum of a homogeneous cylindrical synchrotron source in terms of the frequency ν_1 at which $\tau = 1$. Equation 5.90 shows that it approaches a power law with slope 5/2 at frequencies $\nu \ll \nu_1$ and slope $(1 - \delta)/2$ for $\nu \gg \nu_1$. Real astrophysical sources are inhomogeneous, so their low-frequency spectral slopes are smaller than 5/2 and their spectral peaks are not so sharp.

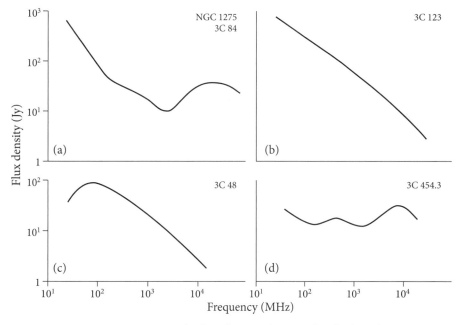

Figure 5.8. Representative spectra of radio galaxies and quasars [111]. The radio source 3C 84 in the nearby galaxy NGC 1275 contains a very compact nuclear component that is opaque below about 20 GHz. The radio galaxy 3C 123 is transparent at all plotted frequencies, and energy losses steepen its spectrum above a few GHz. The quasar 3C 48 is synchrotron self-absorbed only below 100 MHz, while the quasar 3C 454.3 contains structures of different sizes that become opaque at different frequencies.

at frequency ν:

$$\boxed{\left(\frac{B}{\text{gauss}}\right) \approx 1.4 \times 10^{12} \left(\frac{\nu}{\text{Hz}}\right)\left(\frac{T_b}{\text{K}}\right)^{-2}.}$$ (5.91)

For example, a self-absorbed radio source with observed brightness temperature $T_b \approx 10^{11}$ K at $\nu = 1$ GHz has a magnetic field strength

$$\left(\frac{B}{\text{gauss}}\right) \approx 1.4 \times 10^{12} \cdot 10^9 \cdot (10^{11})^{-2} \approx 0.1.$$

The spectra of celestial radio sources are more complex because real sources have nonuniform magnetic fields and electron energy distributions in geometrically complex structures. Representative spectra of powerful radio galaxies and quasars are illustrated in Figure 5.8.

Synchrotron radiation from cosmic-ray electrons accelerated by the supernova remnants of relatively massive ($M > 8M_\odot$) and short-lived ($T < 3 \times 10^7$ yr) stars dominates the radio continuum emission of the nearby starburst galaxy M82 (Color Plate 13) at frequencies $\nu < 30$ GHz (see the dot-dash line in Figure 2.24). Thermal emission (dashed line) from H II regions ionized primarily by even more massive ($M > 15M_\odot$) and shorter-lived stars is strongest between about 30 and 200 GHz. At frequencies well below 1 GHz, free–free absorption flattens the overall spectrum.

5.4 SYNCHROTRON SOURCES

5.4.1 Minimum Energy and Equipartition

The existence of a synchrotron source implies the presence of relativistic electrons with some energy density U_e and a magnetic field whose energy density is $U_B = B^2/(8\pi)$. What is the minimum total energy in relativistic particles and magnetic fields required to produce a synchrotron source of a given radio luminosity

$$L = \int_{\nu_{\min}}^{\nu_{\max}} L_\nu d\nu$$ (5.92)

over the frequency range conventionally bounded by $\nu_{\min} = 10^7$ Hz and $\nu_{\max} = 10^{11}$ Hz?

The energy density of relativistic electrons in the energy range $E_{\min}$ to $E_{\max}$ is

$$U_e = \int_{E_{\min}}^{E_{\max}} E n(E)\, dE,$$ (5.93)

where $n(E)dE$ is the number density of electrons in the energy range E to $E + dE$. Electrons with energy E emit most of the radiation seen at frequency $\nu \propto E^2 B$, so the electron energy corresponding to radiation at frequency ν satisfies

$$E \propto B^{-1/2}.$$ (5.94)

Thus the ratio of U_e to L can be written in terms of the energy limits:

$$\frac{U_e}{L} \propto \frac{\int_{E_{\min}}^{E_{\max}} E\, n(E)\, dE}{-\int_{E_{\min}}^{E_{\max}} (dE/dt) n(E)\, dE}, \tag{5.95}$$

where the synchrotron power emitted per electron is $(-dE/dt) \propto B^2 E^2$. For a power-law electron energy distribution $n(E) \propto E^{-\delta}$,

$$\frac{U_e}{L} \propto \frac{\int_{E_{\min}}^{E_{\max}} E^{1-\delta}\, dE}{B^2 \int_{E_{\min}}^{E_{\max}} E^{2-\delta}\, dE} \propto \frac{E^{2-\delta}\big|_{E_{\min}}^{E_{\max}}}{B^2 E^{3-\delta}\big|_{E_{\min}}^{E_{\max}}}. \tag{5.96}$$

The energy limits $E_{\min}$ and $E_{\max}$ are both proportional to $B^{-1/2}$ (Equation 5.94) so

$$\frac{U_e}{L} \propto \frac{(B^{-1/2})^{2-\delta}}{B^2(B^{-1/2})^{3-\delta}} = \frac{B^{-1+\delta/2}}{B^2 B^{-3/2+\delta/2}} = B^{-3/2} \tag{5.97}$$

Thus the electron energy density needed to produce a given synchrotron luminosity scales as

$$\boxed{U_e \propto B^{-3/2},} \tag{5.98}$$

while the magnetic energy density is

$$U_B \propto B^2. \tag{5.99}$$

The "invisible" cosmic-ray protons and heavier ions emit negligible synchrotron power but they still contribute to the total cosmic-ray particle energy. If the ion/electron energy ratio is η, then the total energy density in cosmic rays is $U_E = (1+\eta)U_e$ and the total energy density U of both cosmic rays and magnetic fields is

$$\boxed{U = (1+\eta)U_e + U_B.} \tag{5.100}$$

Cosmic rays collected near the Earth have $\eta \approx 40$, but the value of η in radio galaxies and quasars has not been measured.

The greatly differing dependences of U_e and U_B on B means that the total energy density U has a fairly sharp minimum near **equipartition**, the point at which $(1+\eta)U_e \approx U_B$ (Figure 5.9). The minimum of the total energy density U occurs at

$$\frac{dU}{dB} = \frac{d[(1+\eta)U_e + U_B]}{dB} = 0. \tag{5.101}$$

The logarithmic derivative of the electron energy density $U_e \propto B^{-3/2}$ is

$$\frac{dU_e}{dB} \cdot U_e^{-1} = -\left(\frac{3}{2}\right) B^{-5/2} B^{3/2} = -\frac{3}{2B}, \tag{5.102}$$

so

$$\frac{dU_e}{dB} = -\frac{3U_e}{2B}. \tag{5.103}$$

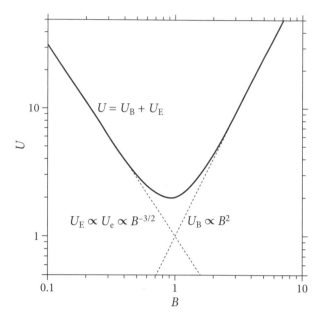

Figure 5.9. For a source of a given synchrotron luminosity, the particle energy density $U_E \equiv (1 + \eta)U_e$ is proportional to $B^{-3/2}$ and the magnetic energy density U_B is proportional to B^2. The total energy density $U = U_E + U_B$ has a fairly sharp minimum near equipartition of the particle and magnetic energy densities ($U_E \approx U_B$).

The logarithmic derivative of the magnetic-field energy density $U_B \propto B^2$ is

$$\frac{dU_B}{dB} \cdot U_B^{-1} = \frac{2B}{B^2} = \frac{2}{B}, \tag{5.104}$$

so

$$\frac{dU_B}{dB} = \frac{2U_B}{B}. \tag{5.105}$$

Inserting Equations 5.103 and 5.105 into the minimum-energy Equation 5.101 gives

$$-\frac{3(1 + \eta)U_e}{2B} + \frac{2U_B}{B} = 0. \tag{5.106}$$

The ratio of cosmic-ray particle energy density to magnetic field energy that minimizes the total energy is

$$\boxed{\frac{\text{particle energy density}}{\text{magnetic field energy density}} = \frac{(1 + \eta)U_e}{U_B} = \frac{4}{3}.} \tag{5.107}$$

This ratio is nearly unity, so **minimum energy** implies (near) **equipartition** of energy: the total cosmic-ray energy density (including the energy of the nonradiating ions) $(1 + \eta)U_e$ is nearly equal to the total magnetic energy density U_B.

Plate 1. The 100-m Green Bank Telescope (GBT). Image credit: NRAO/AUI/NSF.

Plate 2. The 305-m Arecibo Telescope. Image courtesy of the NAIC–Arecibo Observatory, a facility of the NSF.

Plate 3. The Westerbork Synthesis Radio Telescope (WSRT) is an east–west linear array of equatorially mounted dishes. Image courtesy of Adrian Renting.

Plate 4. The 1 km "D" configuration of the Jansky Very Large Array (VLA) of 27 25-m telescopes located on the plains of San Augustin in New Mexico at 2100 m elevation. Image credit: NRAO/AUI/NSF.

Plate 5. The Atacama Large Millimeter Array (ALMA) is on an extremely high (5000 m) and dry desert plain near Cerro Chajnator in Chile. Image credit: ALMA (ESO/NAOJ/NRAO), J. Guarda (ALMA).

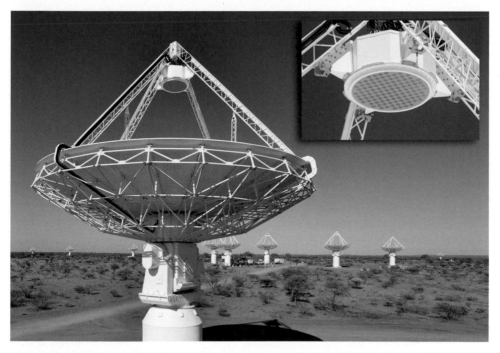

Plate 6. The CSIRO Australian Square Kilometre Array Pathfinder (ASKAP) with its multi-beam phased-array feed designed to survey the sky rapidly. Image credit: CSIRO and Natasha Hurley-Walker.

Plate 7. One tile of the MurchisonWidefield Array (MWA) [107]. Image credit: CSIRO.

Plate 8. The core of LOFAR (LOw Frequency Array for Radio astronomy). Image credit: ASTRON and Top-Foto, Assen.

Plate 9. Dust emission from the protoplanetary disk of HL Tau. Image credit: ALMA (NRAO/ESO/NAOJ); C. Brogan, B. Saxton (NRAO/AUI/NSF).

Plate 10. Composite image of the Crab Nebula. Blue indicates X-rays (from Chandra), green is optical (from the HST), and red is radio (from the VLA). Image credit: J. Hester (ASU), CXC, HST, NRAO, NSF, NASA.

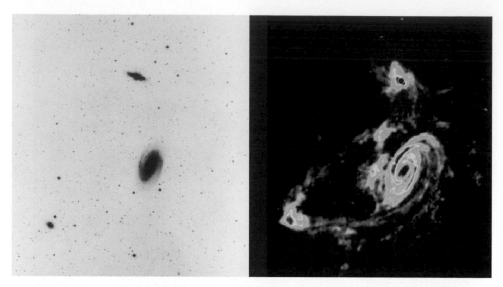

Plate 11. The 21-cm HI line highlights cold hydrogen tidally torn from the galaxies in the M81 group [118]. Image credit: NRAO/AUI/NSF Investigators: Min S. Yun, Paul T. P. Ho, and K. Y. Lo.

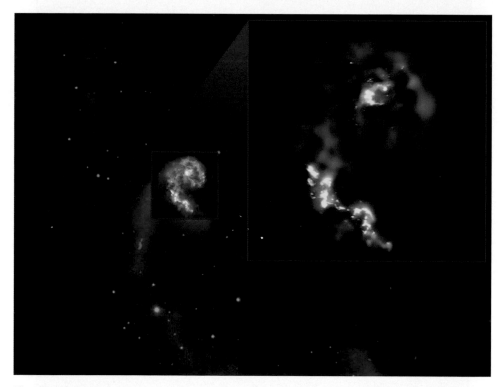

Plate 12. The interaction history of the Antennae Galaxies NGC 4038 and NGC 4039 is revealed by their long HI tidal tails (blue) and obscured star formation traced by CO emission (orange insert). Image credit: B. Saxton (NRAO/AUI/NSF) from data provided by ALMA (ESO/NAOJ/NRAO) and NASA/ESA.

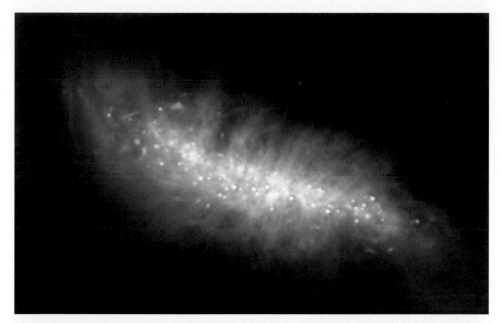

Plate 13. Radio continuum emission from M82. Image credit: Josh Marvil (NM Tech/NRAO), Bill Saxton (NRAO/AUI/NSF), Hubble (NASA/ESA/STScI).

Plate 14. The radio galaxy Hercules A (3C 348). Image credit: NASA, ESA, S. Baum and C. O'Dea (RIT), R. Perley and W. Cotton (NRAO/AUI/NSF), and the Hubble Heritage Team (STScI/AURA).

Plate 15. The radio source (red) in the galaxy cluster MS0735.6+7421 has displaced the X-ray emitting gas (blue) [71]. Image credit: NASA, ESA, CXC, STScI, B. McNamara, NRAO/AUI/NSF, and L. Birzan and team.

Plate 16. Cosmic microwave background fluctuations [105].

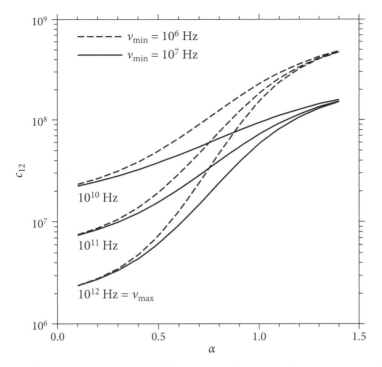

Figure 5.10. Plots of c_{12} in Gaussian CGS units as a function of (negative) spectral index $\alpha \equiv -d \log S/d \log \nu$ for $\nu_{\min} = 10^6$ Hz (dashed curves) and 10^7 Hz (solid curves) and $\nu_{\max} = 10^{10}$ Hz, 10^{11} Hz, and 10^{12} Hz.

It is not known whether most synchrotron sources are in equipartition, but radio astronomers often assume so because

1. it is physically plausible—systems with interacting components often tend toward equipartition;
2. extragalactic radio sources with high luminosities L and large volumes V such as Cyg A have enormous total energy $E = UV$ requirements even near equipartition; the "energy problem" is even worse otherwise;
3. it eliminates an unknown parameter and permits estimates of the relativistic particle energies and the magnetic field strengths of radio sources with measured luminosities and sizes.

Getting the actual numerical values of the particle and magnetic-field energy densities from the synchrotron emission coefficient is a straightforward but tedious algebraic chore (Wilson et al. [116, Section 10.10]). The results (from Pacholczyk [77, p. 171]) are summarized in Equations 5.109 and 5.110. The functions c_{12} and c_{13} in these equations absorb the integrations from frequency $\nu_{\min}$ to $\nu_{\max}$ and the physical constants in Gaussian CGS units. The values of c_{12} and c_{13} are plotted in Figures 5.10 and 5.11.

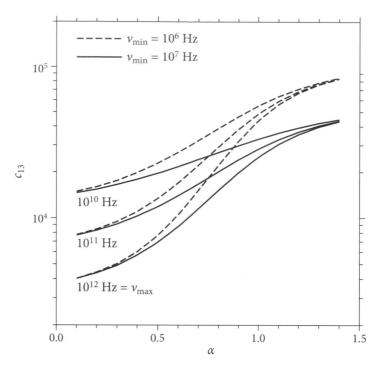

Figure 5.11. Plots of c_{13} in Gaussian CGS units as a function of (negative) spectral index $\alpha \equiv -d \log S/d \log \nu$ for $\nu_{\min} = 10^6$ Hz (dashed curves) and 10^7 Hz (solid curves) and $\nu_{\max} = 10^{10}$ Hz, 10^{11} Hz, and 10^{12} Hz.

For a spherical radio source with radius R and magnetic field strength B, the total magnetic energy is

$$E_B = U_B V = \frac{B^2}{8\pi} \frac{4\pi R^3}{3} = \frac{B^2 R^3}{6}. \tag{5.108}$$

The minimum-energy magnetic field strength for a source of radio luminosity L is

$$\boxed{B_{\min} = [4.5(1 + \eta)c_{12}L]^{2/7} R^{-6/7} \text{ gauss}} \tag{5.109}$$

and the corresponding total energy is

$$\boxed{E_{\min}(\text{total}) = c_{13}[(1 + \eta)L]^{4/7} R^{9/7} \text{ ergs.}} \tag{5.110}$$

The functions c_{12} and c_{13} in Gaussian CGS units are plotted in Figures 5.10 and 5.11.

The **synchrotron lifetime** of a source is defined as the ratio of total electron energy E_e to the energy loss rate L from synchrotron radiation:

$$\tau_s \equiv \frac{E_e}{L}. \tag{5.111}$$

It approximates the lifetime of a synchrotron source if the primary loss mechanism is synchrotron radiation; if other loss mechanisms (e.g., inverse-Compton scattering) are significant, the actual source lifetime will be shortened. The synchrotron lifetime can be written as

$$\boxed{\tau_s \approx c_{12} B_\perp^{-3/2}.}$$ (5.112)

5.4.2 The Eddington Luminosity Limit

The steady-state luminosity of an astronomical object of total mass M is limited by the requirement that the outward radiation pressure cannot exceed the pull of gravity. Otherwise, radiation pressure would expel the outer layers of a star or disrupt accretion onto a compact object such as a black hole or neutron star. If the atmosphere of the star or the infalling material is primarily ionized hydrogen, the free electrons will Thomson-scatter outflowing radiation. The Thomson-scattering cross section σ_T is given by Equation 5.33. Each electron being pushed away by radiation pressure will drag along one proton ($m_p \gg m_e$) with it to maintain charge neutrality. Balancing the forces from radiation and gravity on each electron/proton pair at distance r from the accreting object defines the **Eddington luminosity**:

$$\frac{L_E}{4\pi r^2} \frac{\sigma_T}{c} = \frac{GM(m_p + m_e)}{r^2} \approx \frac{GMm_p}{r^2}.$$ (5.113)

Both forces are proportional to r^{-2} so

$$L_E \approx \frac{4\pi GMm_p c}{\sigma_T}$$ (5.114)

is proportional to the mass M and independent of distance. In CGS units,

$$L_E = \frac{4\pi \cdot 6.67 \times 10^{-8} \text{ dyne cm}^2 \text{ g}^{-2} \cdot M \cdot 1.66 \times 10^{-24} \text{ g} \cdot 3 \times 10^{10} \text{ cm s}^{-1}}{6.65 \times 10^{-25} \text{ cm}^2},$$

$$L_E(\text{erg s}^{-1}) = 6.28 \times 10^4 \, M(\text{g}).$$ (5.115)

Normalized to "solar" units $L_\odot \approx 3.83 \times 10^{33}$ erg s^{-1} and $M_\odot \approx 1.99 \times 10^{33}$ g,

$$\left(\frac{L_E}{L_\odot}\right) \approx \frac{6.28 \times 10^4 \cdot 1.99 \times 10^{33} \text{ g}}{3.83 \times 10^{33} \text{ erg s}^{-1}} \left(\frac{M}{M_\odot}\right),$$ (5.116)

$$\boxed{\left(\frac{L_E}{L_\odot}\right) \approx 3.3 \times 10^4 \left(\frac{M}{M_\odot}\right).}$$ (5.117)

For example, as the mass of a main-sequence star approaches $M \approx 100 M_\odot$, its luminosity approaches its Eddington luminosity. Very massive stars often have radiation-driven winds, and stars more massive than $100 M_\odot$ may not be stable. The Eddington limit doesn't apply only to objects surrounded by ionized hydrogen, and analogs to the classical Eddington limit can be derived for other absorbers, such as **interstellar dust**. The ratio of the absorption cross section to mass for a typical dust grain is ~ 500 times larger than the ratio for ionized hydrogen, so the maximum luminosity-to-mass ratio of a dusty galaxy is ~ 500 times lower than given by Equation 5.117. Thus a supermassive black hole emitting at its Eddington limit for ionized hydrogen in the nucleus of a dusty galaxy may remove the dusty ISM if the

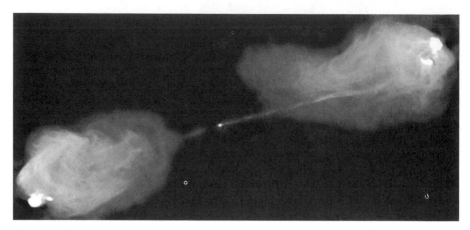

Figure 5.12. A high-resolution VLA image of the radio source Cyg A. The bright central component is thought to coincide with a supermassive black hole that accelerates the relativistic electrons along two jets terminating in lobes well outside the host galaxy. Image credit: NRAO/AUI/NSF Investigators: R. Perley, C. Carilli, & J. Dreher.

galaxy mass is less than $\sim$ 500 times the black-hole mass. Such radiative "feedback" processes may account for the observed mass ratio $\sim$ 500 of galaxy bulges and their central black holes [37].

5.4.3 Application to the Luminous Radio Galaxy Cyg A

Cyg A is a luminous double radio source (Figure 5.12) in a peculiar galaxy at a distance $d \approx 230$ Mpc. Its radio lobes have radii $R \approx 30$ kpc, and the total flux density of Cyg A is

$$S_\nu \approx 2000 \text{ Jy} \left(\frac{\nu}{\text{GHz}} \right)^{-0.8}.$$

To estimate the total radio luminosity of Cyg A, first convert the data from "astronomical" units to Gaussian CGS units:

$$R = 30 \text{ kpc} \left(\frac{10^3 \text{ pc}}{\text{kpc}} \right) \left(\frac{3.09 \times 10^{18} \text{ cm}}{\text{pc}} \right) \approx 9.0 \times 10^{22} \text{ cm},$$

$$S_\nu = 2000 \text{ Jy} \left(\frac{10^{-23} \text{ erg s}^{-1} \text{ Hz}^{-1} \text{ cm}^{-2}}{\text{Jy}} \right) \left(\frac{\nu}{10^9 \text{ Hz}} \right)^{-0.8}$$

$$= \frac{3.17 \times 10^{-13} \text{ erg s}^{-1} \text{ Hz}^{-1}}{\text{cm}^2} \left(\frac{\nu}{\text{Hz}} \right)^{-0.8},$$

$$d = 230 \text{ Mpc} \left(\frac{10^6 \text{ pc}}{\text{Mpc}} \right) \left(\frac{3.09 \times 10^{18} \text{ cm}}{\text{pc}} \right) \approx 7.1 \times 10^{26} \text{ cm}.$$

The spectral luminosity of Cyg A is

$$L_\nu \approx 4\pi d^2 S_\nu$$

$$\approx 4\pi \, (7.1 \times 10^{26} \text{ cm})^2 \left(\frac{3.17 \times 10^{-13} \text{ erg s}^{-1} \text{ Hz}^{-1}}{\text{cm}^2} \right) \left(\frac{\nu}{\text{Hz}} \right)^{-0.8}$$

$$\approx 2.0 \times 10^{42} \text{ erg s}^{-1} \text{ Hz}^{-1} \left(\frac{\nu}{\text{Hz}} \right)^{-0.8}.$$

The total radio luminosity of Cyg A in the frequency range 10^7 Hz to 10^{11} Hz is

$$L = \int_{10^7 \text{ Hz}}^{10^{11} \text{ Hz}} L_\nu \, d\nu \approx 2.0 \times 10^{42} \text{ erg s}^{-1} \text{ Hz}^{-1} \left(\frac{\nu^{0.2}}{0.2} \right) \Bigg|_{\nu=10^7 \text{ Hz}}^{\nu=10^{11} \text{ Hz}}$$

$$\approx 1.33 \times 10^{45} \text{ erg s}^{-1}.$$

In units of the bolometric solar luminosity $L_\odot \approx 3.83 \times 10^{33}$ erg s^{-1}, the radio luminosity of Cyg A is

$$\frac{L}{L_\odot} \approx \frac{1.33 \times 10^{45} \text{ erg s}^{-1}}{3.83 \times 10^{33} \text{ erg s}^{-1}} \approx 3.5 \times 10^{11}.$$

The *radio* power from Cyg A exceeds the *total* power produced by all of the stars in our Galaxy.

The energy source for this radio emission is a compact object at the center of the host galaxy. The Eddington limit (Equation 5.117) yields a lower limit to its mass M:

$$\left(\frac{M}{M_\odot} \right) \geq \frac{3.5 \times 10^{11}}{3.3 \times 10^4} \approx 10^7.$$

Note that the Eddington mass limit depends only on the instantaneous power emitted by the source, not on the total energy of the source, the source age, or any other indicator of its history.

The magnetic field strength $B_{\min}$ that minimizes the total energy in the relativistic particles and magnetic fields implied by the luminous synchrotron source can be estimated with Equation 5.109. Approximate Cyg A (Figure 5.12) by two spherical lobes of radius $R \approx 30$ kpc and luminosity $L/2$ each, where L is the total luminosity of Cyg A:

$$B_{\min} \approx [4.5(1 + \eta)c_{12}(L/2)]^{2/7} R^{-6/7}$$

$$\approx [(4.5 \cdot 3.9 \times 10^7 \cdot 1.33 \times 10^{45} \text{ erg s}^{-1}/2)^{2/7} (9 \times 10^{22} \text{ cm})^{-6/7}] (1 + \eta)^{2/7}.$$

The ion/electron energy ratio η has not been measured in extragalactic radio sources such as Cyg A. The cosmic rays accelerated by a supermassive black hole might be primarily electrons and positrons. Electrons and positrons have equally large charge/mass ratios, so an electron–positron plasma would have $\eta \approx 1$. If electrons and protons are accelerated to the same velocity (same γ), then the protons carry $m_p/m_e \sim 2 \times 10^3$ as much energy but emit almost nothing and $\eta \sim 2 \times 10^3$. Fortunately, $B_{\min} \propto (1 + \eta)^{2/7}$ is only weakly dependent on η—varying η from 1 to

2×10^3 changes $(1 + \eta)^{2/7}$ from about 1 to 9:

$$B_{\min} \approx 1.45 \times 10^{15} \cdot 2.1 \times 10^{-20} \cdot (1 \text{ to } 9) \text{ gauss} \tag{5.118}$$

$$\approx (30 \text{ to } 300) \times 10^{-6} \text{ gauss} \sim 10^{-4} \text{ gauss.} \tag{5.119}$$

The minimum total energy (Equation 5.110) of Cyg A is twice the energy of each lobe:

$$E_{\min} \approx 2(\text{lobes}) \cdot c_{13}[(1 + \eta)L]^{4/7} R^{9/7}$$

$$\approx 2 \cdot 2.0 \times 10^4 \left(\frac{1.33 \times 10^{45} \text{ erg s}^{-1}}{2} \right)^{4/7} (9 \times 10^{22} \text{ cm})^{9/7} (1 + \eta)^{4/7},$$

where $(1 + \eta)^{4/7}$ is in the range of about 1 to 80;

$$E_{\min} \approx 4 \times 10^4 \cdot 4.1 \times 10^{25} \cdot 3.26 \times 10^{29} \cdot (1 \text{ to } 80) \text{ ergs} \tag{5.120}$$

$$\approx 5.4 \times 10^{59} \cdot (1 \text{ to } 80) \text{ ergs} \sim 5 \times 10^{60} \text{ ergs.} \tag{5.121}$$

Such large calculated energies can be confirmed observationally for sources in clusters of galaxies. Color Plate 15 shows that the radio source (red) in the galaxy cluster MS07356+7421 has displaced the X-ray emitting gas (blue) in a large volume. The gas pressure can be derived from the intensity of its the X-ray emission, and the total energy required to displace the gas is the product of the volume and the pressure [71].

This enormous energy implies an independent lower limit to the mass of the central object powering the radio source. Mass cannot be converted to energy with more than 100% efficiency, so the minimum mass needed to produce $E_{\min}$ is

$$M \geq \frac{E_{\min}}{c^2} \approx \frac{5 \times 10^{60} \text{ ergs}}{(3 \times 10^{10} \text{ cm s}^{-1})^2} \approx 6 \times 10^{39} \text{ g,} \tag{5.122}$$

$$M \geq 6 \times 10^{39} \text{ g} \left(\frac{M_\odot}{1.99 \times 10^{33} \text{ g}} \right) \approx 3 \times 10^6 M_\odot. \tag{5.123}$$

This is a very conservative lower limit. Nuclear fusion can convert mass to energy with only $\sim 1\%$ efficiency, so $M > 3 \times 10^8 M_\odot$ would be required for nuclear fusion in stars. Accretion onto a spinning black hole can yield efficiencies up to $(1 - 3^{-1/2}) \approx 0.4$ in theory, implying $M > 10^7 M_\odot$. Many authors assume that mass accreted by astronomical black holes is converted to energy with about 10% efficiency; this yields $M > 3 \times 10^7 M_\odot$. The small size of the radio core measured by Very Long Baseline Interferometry (VLBI) and its observed flux variability on timescales of months to years combined with the large minimum masses estimated from the Eddington limit and the total energy of the radio lobes together make it difficult to avoid the conclusion that the compact, massive object powering the radio source is a **supermassive black hole** (SMBH). The adjective *supermassive* is used for black holes with $M > 10^6 M_\odot$, which is far more massive than the most massive stars, $M \sim 100 M_\odot$.

A lower limit to the age τ of the radio source Cyg A is the average synchrotron lifetime (Equation 5.111) of the relativistic electrons estimated by taking the ratio of

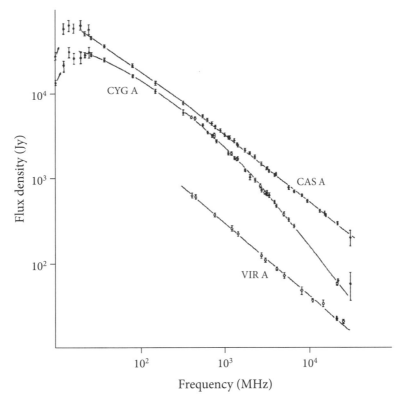

Figure 5.13. The radio spectrum of Cyg A (and Cas A, Vir A) from Baars et al. [6]. Note the spectral steepening above $\nu \sim 10^3$ MHz.

the electron energy to the observed synchrotron luminosity:

$$\tau \geq \tau_s \equiv \frac{E_e}{L} \geq \frac{E_{\min}/(1+\eta)}{L}, \tag{5.124}$$

$$\tau \geq \frac{5.4 \times 10^{59} \text{ erg} (1+\eta)^{4/7}}{1.33 \times 10^{45} \text{ erg s}^{-1} (1+\eta)} \approx 4 \times 10^{14} \text{ s} \cdot \eta^{-3/7} \sim 10^{14} \text{ s} \sim 3 \times 10^6 \text{ yr}. \tag{5.125}$$

Because each electron radiates energy at a rate proportional to E^2 and the critical frequency is proportional to E^2, the most energetic electrons emitting at the highest frequencies have the shortest lifetimes. The rapid depletion of high-energy electrons steepens the radio spectrum (Figure 5.13) of Cyg A frequencies higher than $\nu \sim 1$ GHz. Suppose that new relativistic electrons are continuously injected with a power-law energy distribution

$$N(E) \propto E^{-\delta_0} \tag{5.126}$$

into a radio source. After a long time, electrons emitting at frequencies higher than ν will be depleted by radiative losses $\propto E^2$ and these high-energy electrons will eventually reach an energy distribution $N(E) \propto E^{-(\delta_0+1)}$. Consequently, the

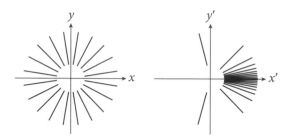

Figure 5.14. For a relativistic electron at rest in the "primed" frame moving with velocity v along the x-axis, the angle of incidence θ' of incoming photons will be much less than the corresponding angle θ in the rest frame of the observer (Equations 5.50 and 5.51). This figure shows the aberration of an isotropic radiation field (left) seen in a moving frame with $\gamma = (1 - v^2/c^2)^{-1/2} = 5$ (right).

(negative) spectral index will be

$$\alpha_0 = \frac{\delta_0 - 1}{2} \tag{5.127}$$

at low frequencies and approach

$$\alpha = (\delta_0 + 1 - 1)/2 = (\alpha_0 + 1/2) \tag{5.128}$$

at higher frequencies; that is, the high-frequency spectrum steepens by $\Delta\alpha = 1/2$.

If the observed frequency v of the spectral bend is high enough, the implied synchrotron lifetime of electrons with $v_c \sim v$ may be less than the time needed for new relativistic electrons to travel from the radio core to the emitting feature in a jet or lobe. This implies **in situ acceleration**—something outside the radio core (e.g., shocks in the jet) must replenish the supply of relativistic electrons. The radio spectrum of Vir A, the source in the galaxy M87, is straight to at least $v \sim 30$ GHz (Figure 5.13), and this synchrotron emission extends to optical frequencies, so many of the cosmic rays must be accelerated in the bright shocked regions, not just near the black hole.

5.5 INVERSE-COMPTON SCATTERING

The ambient radiation field is normally fairly isotropic in the rest frame of a synchrotron source. However, such a radiation field looks extremely anisotropic to each ultrarelativistic ($\gamma \gg 1$) electron producing the synchrotron radiation. **Relativistic aberration** (Section 5.3.1) causes nearly all ambient photons to approach within an angle $\sim \gamma^{-1}$ rad of head-on (Figure 5.14). Thomson scattering of this highly anisotropic radiation systematically reduces the electron kinetic energy and converts it into **inverse-Compton** (IC) radiation by upscattering radio photons to become optical or X-ray photons. Inverse-Compton "cooling" of the relativistic electrons also limits the maximum rest-frame brightness temperature of an incoherent synchrotron source to $T_b \approx 10^{12}$ K.

5.5.1 IC Power from a Single Electron

To derive the equations describing inverse-Compton scattering, first consider non-relativistic Thomson scattering in the rest frame of an electron. If the Poynting flux (power per unit area) of a plane wave incident on the electron is

$$\vec{S} = \frac{c}{4\pi} \vec{E} \times \vec{B} = \frac{c}{4\pi} |\vec{E}|^2, \tag{5.129}$$

the electric field of the incident radiation will accelerate the electron, and the accelerated electron will in turn emit radiation according to Larmor's equation. The net result is simply to scatter a portion of the incoming radiation, with no net transfer of energy between the radiation and the electron. The scattered radiation has power

$$P = |\vec{S}| \sigma_{\mathrm{T}}, \tag{5.130}$$

where

$$\sigma_{\mathrm{T}} \equiv \frac{8\pi}{3} \left(\frac{e^2}{m_e c^2} \right)^2 \approx 6.65 \times 10^{-25} \ \mathrm{cm}^2 \tag{5.131}$$

is called the **Thomson cross section** (Equation 5.33) of an electron. In other words, the electron will extract from the incident radiation the amount of power flowing through the area σ_{T} and reradiate that power over the doughnut-shaped pattern given by Larmor's equation. The scattered power can be rewritten as

$$\boxed{P = \sigma_{\mathrm{T}} c U_{\mathrm{rad}},} \tag{5.132}$$

where $U_{\mathrm{rad}} = |\vec{S}|/c$ is the energy density of the incident radiation.

Next consider radiation scattering by an ultrarelativistic electron. Equation 5.132 is valid only in the primed frame instantaneously moving with the electron:

$$P' = \sigma_{\mathrm{T}} c U'_{\mathrm{rad}}. \tag{5.133}$$

This nonrelativistic result needs to be transformed to the unprimed rest frame of an observer. Using the result $P = P'$ (Equation 5.28) gives

$$P = \sigma_{\mathrm{T}} c U'_{\mathrm{rad}}. \tag{5.134}$$

To transform U'_{rad} into U_{rad}, suppose that an electron moving with speed $v = v_x$ in the rest frame of the observer is hit successively by two low-energy photons approaching from an angle θ in the observer's frame (θ' in the electron frame) from the x-axis as shown in Figure 5.15. If the coordinates corresponding to the first and second photons hitting the electron (which is always located at $x' = y' = z' = 0$) are

$$(x_1, 0, 0, t_1) \text{ and } (x_2, 0, 0, t_2) \tag{5.135}$$

in the observer's frame, then the Lorentz transform Equation 5.12 gives the coordinates of these two events as

$$(\gamma v t'_1, 0, 0, \gamma t'_1) \text{ and } (\gamma v t'_2, 0, 0, \gamma t'_2), \tag{5.136}$$

as shown in Figure 5.15. In the observer's frame, the time Δt elapsed between the arrival of these two photons *at the plane (dashed line in Figure 5.15) normal to the*

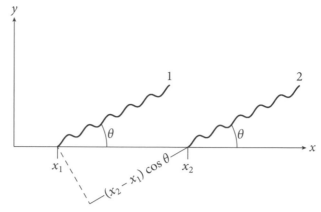

Figure 5.15. Two successive photons striking an electron moving to the right. The photons approach at angle θ from the x-axis, as seen in the unprimed observer's frame.

direction of propagation is

$$\Delta t = t_2 + \frac{(x_2 - x_1)}{c} \cos\theta - t_1 \tag{5.137}$$

$$= \gamma t_2' + \frac{(\gamma v t_2' - \gamma v t_1')}{c} \cos\theta - \gamma t_1' \tag{5.138}$$

$$= (t_2' - t_1')[\gamma(1 + \beta \cos\theta)], \tag{5.139}$$

where $\beta \equiv v/c$. The time between being hit by the two photons in the electron's frame is $\Delta t' = t_2' - t_1'$ so

$$\Delta t = \Delta t'[\gamma(1 + \beta \cos\theta)]. \tag{5.140}$$

The relativistic **Doppler equation** follows immediately from Equation 5.140. Let Δt be the time between the arrivals of two successive cycles of a wave whose frequency is $\nu = (\Delta t)^{-1}$ in the observer's frame and $\nu' = (\Delta t')^{-1}$ in the moving frame. Then

$$\nu^{-1} = (\nu')^{-1}[\gamma(1 + \beta \cos\theta)] \tag{5.141}$$

or

$$\boxed{\nu' = \nu[\gamma(1 + \beta \cos\theta)].} \tag{5.142}$$

In the electron's frame, the frequency ν' and energy $E' = h\nu'$ of each photon are multiplied by $[\gamma(1 + \beta \cos\theta)]$. Moreover, the *rate* at which successive photons arrive is multiplied by the same factor. If n_γ is the photon number density in the observer's frame, then $n_\gamma' = n_\gamma[\gamma(1 + \beta \cos\theta)]$. In the observer's frame, the radiation energy density is

$$U_{\rm rad} = n_\gamma h\nu. \tag{5.143}$$

In the electron's frame

$$U'_{\rm rad} = n'_\gamma h\nu' = n_\gamma [\gamma(1 + \beta \cos\theta)] h\nu [\gamma(1 + \beta\cos\theta)] = U_{\rm rad}[\gamma(1 + \beta\cos\theta)]^2. \tag{5.144}$$

Thus the transformation between $U_{\rm rad}$ and $U'_{\rm rad}$ depends on the angle θ between the direction of the photons and the direction of the electron motion.

The energy density in the electron's frame of a radiation field that is isotropic in the observer's frame is obtained by integrating Equation 5.144 over all directions:

$$U'_{\rm rad} = U_{\rm rad}\left(\frac{1}{4\pi}\int_{\phi=0}^{2\pi}\int_{\theta=0}^{\pi}[\gamma(1 + \beta\cos\theta)]^2 \sin\theta\, d\theta\, d\phi\right), \tag{5.145}$$

where ϕ is the azimuthal angle around the x-axis. Thus

$$U'_{\rm rad} = \frac{U_{\rm rad}\gamma^2}{2}\int_{\theta=0}^{\pi}(1 + \beta\cos\theta)^2 \sin\theta\, d\theta. \tag{5.146}$$

To evaluate this integral, substitute $z \equiv \cos\theta$ so $dz = -\sin\theta\, d\theta$ and eliminate β in favor of γ:

$$U'_{\rm rad} = \frac{U_{\rm rad}\gamma^2}{2}\int_{1}^{-1}(1 + \beta z)^2(-1)dz = U_{\rm rad}\gamma^2(1 + \beta^2/3) \tag{5.147}$$

$$= U_{\rm rad}\left[\gamma^2 + \frac{\gamma^2}{3} - \left(\frac{\gamma^2}{3} - \frac{\gamma^2\beta^2}{3}\right)\right] = U_{\rm rad}\left[\frac{4\gamma^2}{3} - \frac{1}{3}\gamma^2(1 - \beta^2)\right]. \tag{5.148}$$

Recall that $\gamma^2(1 - \beta^2) = 1$ so

$$U'_{\rm rad} = U_{\rm rad}\frac{4(\gamma^2 - 1/4)}{3}. \tag{5.149}$$

Substituting this result for $U'_{\rm rad}$ into Equation 5.134 yields

$$P = \frac{4}{3}\sigma_T c U_{\rm rad}(\gamma^2 - 1/4) \tag{5.150}$$

for the *total* power radiated after inverse-Compton upscattering of low-energy photons. The *initial* power of these photons was $\sigma_T c U_{\rm rad}$, so the *net* power added to the radiation field by inverse-Compton scattering is

$$P_{\rm IC} = \frac{4}{3}\sigma_T c U_{\rm rad}\left(\gamma^2 - \frac{1}{4}\right) - \sigma_T c U_{\rm rad} = \frac{4}{3}\sigma_T c U_{\rm rad}(\gamma^2 - 1). \tag{5.151}$$

Replacing $(\gamma^2 - 1)$ by $\beta^2\gamma^2$ gives the final result

$$\boxed{P_{\rm IC} = \frac{4}{3}\sigma_T c\beta^2\gamma^2 U_{\rm rad}} \tag{5.152}$$

for the **net inverse-Compton power** gained by the radiation field and lost by the electron. Dividing by the corresponding synchrotron power (Equation 5.42)

$$P_{\rm syn} = \frac{4}{3}\sigma_T c\beta^2\gamma^2 U_B \tag{5.153}$$

reveals the remarkably simple **ratio of IC to synchrotron radiation losses**:

$$\boxed{\frac{P_{IC}}{P_{syn}} = \frac{U_{rad}}{U_B}.}$$

(5.154)

The IC loss is proportional to the radiation energy density and the synchrotron loss is proportional to the magnetic energy density. Note that synchrotron and inverse-Compton losses have the same electron-energy dependence ($dE/dt \propto \gamma^2$), so their effects on radio spectra (Equation 5.128) are indistinguishable.

5.5.2 The IC Spectrum of a Single Electron

What is the spectrum of the inverse-Compton radiation? Suppose the ambient radiation field in the observer's frame contains only photons of frequency ν_0, and consider scattering by a single electron moving with ultrarelativistic velocity $+v$ along the x-axis. In the inertial frame moving with the electron, relativistic aberration causes most of the photons to approach nearly head-on. The relativistic Doppler Equation 5.142 gives the frequency ν_0' in the electron frame of a photon approaching near the x-axis ($\theta \ll 1$); it is

$$\nu_0' = \nu_0[\gamma(1 + \beta\cos\theta)] \approx \nu_0[\gamma(1 + \beta)].$$

(5.155)

In the electron frame, Thomson scattering produces radiation with the same frequency as the incident radiation: the scattered photons have $\nu' = \nu_0'$. In the observer's frame, relativistic aberration beams the scattered photons in the direction of the electron's motion, and the frequency ν of radiation scattered nearly along the $+x$-direction ($\theta \approx 0$) is given by the relativistic Doppler formula:

$$\nu = \nu'[\gamma(1 + \beta\cos\theta)] \approx \nu'[\gamma(1 + \beta)] \approx \nu_0[\gamma(1 + \beta)]^2.$$

(5.156)

In the ultrarelativistic limit $\beta \to 1$,

$$\boxed{\frac{\nu}{\nu_0} \approx 4\gamma^2.}$$

(5.157)

This is the **maximum frequency** of the upscattered radiation in the observer's frame.

Oblique collisions ($\theta > 0$) result in lower frequencies ν. For an isotropic radiation field in the observer's frame, the average energy $\langle E \rangle$ of scattered photons equals the average scattered power P_{IC} per electron divided by $\dot{N}_{IC}$, the number of photons scattered per second per electron. This rate is the scattered power divided by the photon energy in the observer's frame, or

$$\dot{N}_{IC} = \frac{\sigma_T c U_{rad}}{h\nu_0}.$$

(5.158)

Thus

$$\langle E \rangle = h\langle \nu \rangle = \frac{P_{IC}}{\dot{N}_{IC}} = \frac{4}{3}\sigma_T c \beta^2 \gamma^2 U_{rad} \left(\frac{h\nu_0}{\sigma_T c U_{rad}}\right)$$

(5.159)

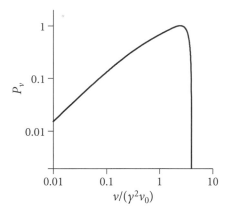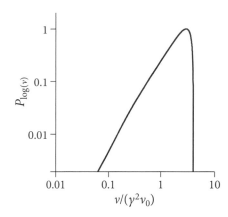

Figure 5.16. The inverse-Compton spectrum of electrons with energy γ irradiated by photons of frequency ν_0. The log-log plot of power per logarithmic frequency range (right) indicates more clearly just how peaked the spectrum is.

and the **average frequency** $\langle \nu \rangle$ of upscattered photons is

$$\frac{\langle \nu \rangle}{\nu_0} = \frac{4}{3}\gamma^2. \tag{5.160}$$

For example, isotropic radio photons at $\nu_0 = 1$ GHz, IC scattered by electrons having $\gamma = 10^4$, will be upscattered to an average frequency

$$\langle \nu \rangle = 10^9 \text{ Hz } \frac{4}{3}(10^4)^2 \approx 1.3 \times 10^{17} \text{ Hz}$$

corresponding to X-ray radiation. The principal astronomical effect of inverse-Compton scattering is to drain energy from cosmic-ray electrons that produce radio radiation and use it to produce X-ray radiation instead.

Because the maximum frequency (Equation 5.157) is only three times the average frequency (Equation 5.160), the IC spectrum must be sharply peaked near the average frequency. The detailed Compton-scattering spectrum resulting from an isotropic single-frequency radiation field has been calculated (Blumenthal and Gould [13]; see also Pacholczyk [77]). It is indeed sharply peaked just below the maximum $\nu/\nu_0 = 4\gamma^2$, as shown in Figure 5.16. This spectrum is even more peaked than the synchrotron spectrum of monoenergetic electrons. Therefore it is not necessary to use the detailed Compton-scattering spectrum of monoenergetic electrons to calculate the inverse-Compton spectrum of an astrophysical source containing a power-law distribution of relativistic electrons. If the electron-energy distribution is $n(E) \propto E^{-\delta}$, the inverse-Compton spectrum will also be a power law with spectral index

$$\alpha = \frac{\delta - 1}{2}, \tag{5.161}$$

which is the same spectral index that Equation 5.79 gives for synchrotron radiation emitted by the same power-law distribution of electron energies.

5.5.3 Synchrotron Self-Compton Radiation

Synchrotron self-Compton radiation results from inverse-Compton scattering of synchrotron radiation by the same relativistic electrons that produced the synchrotron radiation. Equation 5.154,

$$\frac{P_{\text{IC}}}{P_{\text{syn}}} = \frac{U_{\text{rad}}}{U_B},\tag{5.162}$$

implies that multiplying the density of relativistic electrons by some factor f multiplies both the synchrotron power and its contribution to U_{rad} by f, so the synchrotron self-Compton power is proportional to f^2.

The self-Compton radiation also contributes to U_{rad} and leads to significant second-order scattering as the synchrotron self-Compton contribution to U_{rad} approaches the synchrotron contribution in compact sources. This runaway positive feedback is a very sensitive function of the source brightness temperature, so inverse-Compton losses very strongly cool the relativistic electrons if the source brightness temperature exceeds $T_{\text{b}} \sim 10^{12}$ K in the rest frame of the source. Radio sources with brightness temperatures significantly higher than

$$\boxed{T_{\text{max}} \sim 10^{12} \text{ K}}\tag{5.163}$$

in the observer's frame are either Doppler boosted or not incoherent synchrotron sources (e.g., pulsars are coherent radio sources). The active galaxy Markarian 501 emits strong synchrotron self-Compton radiation and the radio emission approaches this rest-frame **brightness limit** for incoherent synchrotron radiation. The synchrotron and synchrotron self-Compton spectra of Mrk 501 are shown in Figure 5.17.

5.6 EXTRAGALACTIC RADIO SOURCES

5.6.1 Relativistic Bulk Motion

The results above apply only to radio-emitting plasmas that are not moving relativistically with respect to the observer. Bright radio-source **components** (discrete regions of enhanced brightness) are often seen to move with apparent transverse velocities exceeding the speed of light. This illusion of **superluminal velocities** can occur if the components are moving obliquely toward the observer with relativistic speeds, as shown in Figure 5.18.

Suppose the radio-emitting component is moving toward the observer with constant speed $v = \beta c$ at an angle θ from the line of sight. Consider two "events" in the moving component, the first occurring a distance r from the observer at time $t = 0$, and the second at time t. Radiation from the first and second events will be received at times

$$t_1 = r/c \quad \text{and} \quad t_2 = \frac{r - vt\cos\theta}{c} + t,\tag{5.164}$$

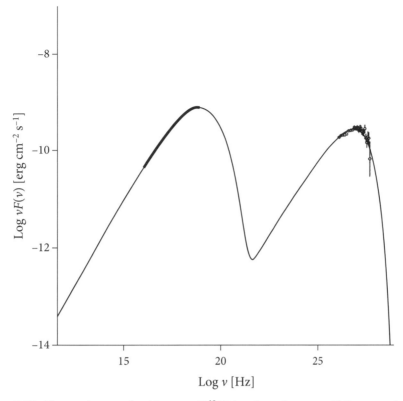

Figure 5.17. The synchrotron (peaking near 10^{19} Hz) and synchrotron self-Compton (peaking near 10^{27} Hz) spectra of Mrk 501 [60]. The thin curve shows the best-fit synchrotron self-Compton model, the thick points represent the X-ray data, and the γ-ray data are plotted as points with error bars. The ordinate νF_ν on this plot is proportional to flux density per logarithmic frequency range, so the relative heights of the two peaks indicate their relative contributions to $U_{\rm rad}$.

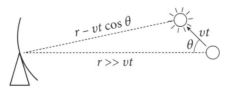

Figure 5.18. A source moving with speed $v < c$ at an angle $\theta < \pi/2$ from the line of sight may appear to be moving faster than c in projection onto the sky because the light travel time is reduced by $vt \cos\theta/c$ during time t.

respectively. The difference between these times is

$$t_2 - t_1 = t[1 - (v\cos\theta)/c]. \tag{5.165}$$

The *apparent* transverse velocity of the moving component is the actual transverse distance covered in time t divided by the *apparent* time interval $(t_2 - t_1)$:

$$v_\perp(\text{apparent}) = \frac{vt \sin \theta}{t_2 - t_1} = \frac{vt \sin \theta}{t[1 - (v \cos \theta)/c]}, \tag{5.166}$$

$$\boxed{\beta_\perp(\text{apparent}) = \frac{\beta \sin \theta}{1 - \beta \cos \theta}.} \tag{5.167}$$

For every speed β there is an angle θ_m that maximizes $\beta_\perp(\text{apparent})$. That angle satisfies

$$\frac{\partial \beta_\perp(\text{apparent})}{\partial \theta} = 0 = \frac{(1 - \beta \cos \theta_m)\beta \cos \theta_m - (\beta \sin \theta_m)^2}{(1 - \beta \cos \theta_m)^2}, \tag{5.168}$$

$$\beta \cos \theta_m - \beta^2 \cos^2 \theta_m - \beta^2 \sin^2 \beta = \beta \cos \theta_m - \beta^2 = 0. \tag{5.169}$$

Thus

$$\boxed{\cos \theta_m = \beta} \tag{5.170}$$

and

$$\boxed{\sin \theta_m = (1 - \cos^2 \theta_m)^{1/2} = (1 - \beta^2)^{1/2} = \gamma^{-1}.} \tag{5.171}$$

Inserting $\cos \theta = \beta$ and $\sin \theta = \gamma^{-1}$ into Equation 5.167 for $\beta_\perp(\text{apparent})$ yields the highest apparent transverse speed of a source whose actual speed is β:

$$\boxed{\max[\beta_\perp(\text{apparent})] = \frac{\beta(1 - \beta^2)^{1/2}}{1 - \beta^2} = \beta\gamma.} \tag{5.172}$$

Figure 5.19 shows five successive high-resolution radio images of the quasar 3C 279. The bright component at the left is taken to be the fixed radio core, and the bright spot at the right appears to have moved 25 light years across the plane of the sky between 1991 and 1998, for an apparently superluminal motion of 25 light years in 7 years: $\beta_\perp(\text{apparent}) \approx 3.6$. What is the minimum component speed β consistent with these images? What is the corresponding angle θ_m between that motion and the line of sight? From the results above,

$$\beta\gamma = \frac{\beta}{(1 - \beta^2)^{1/2}} \geq \beta_\perp(\text{apparent}), \tag{5.173}$$

so

$$\beta \geq \left[\frac{\beta_\perp^2(\text{apparent})}{1 + \beta_\perp^2(\text{apparent})} \right]^{1/2}, \tag{5.174}$$

$$\beta \geq \left[\frac{(25/7)^2}{1 + (25/7)^2} \right]^{1/2} \approx 0.96. \tag{5.175}$$

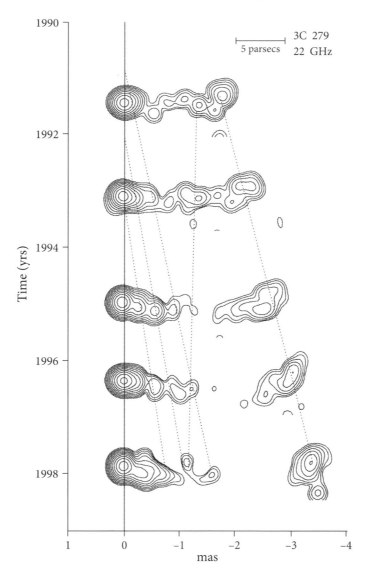

Figure 5.19. Apparently superluminal motion of the radio components in the quasar 3C 279 [81].

The corresponding θ_m is given by

$$\cos\theta_m = \beta \approx 0.96, \qquad (5.176)$$

$$\theta_m \approx 0.28 \text{ rad} \approx 16°. \qquad (5.177)$$

The relativistic Doppler formula (5.142) relates the frequency ν' emitted in the component frame to the observed frequency ν. Note that we have replaced θ by $(\pi - \theta)$ radians in the current analysis by calling it the angle between the line of sight and the velocity of an *approaching* component, so

$$\nu = \frac{\nu'}{\gamma(1 - \beta\cos\theta)}, \tag{5.178}$$

where $\theta = 0$ now corresponds to a radio component moving directly toward the observer. The quantity

$$\delta \equiv [\gamma(1 - \beta\cos\theta)]^{-1} = \frac{\nu}{\nu'} \tag{5.179}$$

is called the **Doppler factor**. If $\theta = \pi/2$, there is a **transverse Doppler shift**

$$\boxed{\delta = \frac{\nu}{\nu'} = \gamma^{-1}.} \tag{5.180}$$

The transverse Doppler shift has no nonrelativistic counterpart because the source has no component of velocity parallel to the line of sight; it exists only because moving clocks run slower by a factor γ. The Doppler factors associated with a given source speed β range from

$$\delta \geq (2\gamma)^{-1} \tag{5.181}$$

for directly receding ($\theta = \pi$ rad) sources to

$$\delta \leq 2\gamma \tag{5.182}$$

for directly approaching ($\theta = 0$) sources. For example, the ratio ν/ν' of the observed frequency to the emitted frequency in 3C 279 ($\beta = 0.96$, $\cos\theta = \beta$) is

$$\frac{\nu}{\nu'} = [\gamma(1 - \beta\cos\theta)]^{-1} = [\gamma(1 - \beta^2)]^{-1} = \gamma$$

$$= (1 - \beta^2)^{-1/2} \approx 3.6.$$

The observed flux density S of a relativistically moving component emitting isotropically in its rest frame depends critically on its Doppler factor δ. The exact amount of **Doppler boosting** caused by relativistic beaming is somewhat model dependent [100] but probably lies in the range

$$\boxed{\delta^{2+\alpha} < \frac{S}{S_0} < \delta^{3+\alpha},} \tag{5.183}$$

where S_0 would be the observed flux density if the source were stationary and $\alpha = -d\log S/d\log\nu$ is the (negative) spectral index. If $\gamma \sim 5$, then $0.1 < \delta < 10$ depending on the angle θ. Relativistic components approaching at angles $\theta < \gamma^{-1}$ can easily be boosted by factors $> 10^3$ compared with components moving in the sky plane or away from us. For example, the approaching jet of 3C 279 ($\delta = \nu/\nu' \approx 3.6$ and $\alpha \approx 0.7$) is Doppler boosted by

$$\delta^{2+\alpha} < \frac{S}{S_0} < \delta^{3+\alpha},$$

$$3.6^{2.7} < \frac{S}{S_0} < 3.6^{3.7},$$

$$32 < \frac{S}{S_0} < 114.$$

The receding counterjet is dimmed by a comparable factor, so the jet/counterjet observed flux-density ratio of 3C 279 is probably $> 10^3$.

Doppler boosting strongly favors approaching relativistic jets and components and discriminates against those with $\theta > \gamma^{-1}$ in flux-limited samples of compact radio sources. Radio quasars aren't isotropic candles spread throughout the universe; they are beamed flashlights. The brightest aren't always the most luminous; they are just pointing in our direction. For every flashlight we see, there are many others in the same volume of space that we don't see simply because they are not pointing at us.

The fact that the two lobes of very extended radio sources such as Cyg A (Figure 5.12) and 3C 348 (Color Plate 14) typically have flux ratios < 2 indicates that the lobes are moving outward with speeds $v \ll c$. The radio jets feeding nearly equal lobes often appear quite unequal, with one jet being very strong and the other undetectable. The jets of very luminous sources often terminate in bright **hotspots** in the lobes.

Because the jets feed the lobes, the lobe symmetry suggests that the jets are intrinsically similar, but the approaching jet is boosted while the receding **counterjet** is dimmed. Another feature of many radio jets is gaps near the core. If jets start out relativistic at the core and are inclined by more than $\theta \sim \gamma^{-1}$ from the line of sight, *both* will be Doppler dimmed. If they proceed with constant θ but gradually decelerate as they move away from the core, one or both may become visible beyond the point where $\gamma \sim \theta^{-1}$.

Extragalactic radio sources with jets and lobes can be divided into two morphological classes: (1) those, like 3C 31 (Figure 5.20), that appear to fade away at large distances from the center and (2) sources with edge-brightened lobes, like 3C 175 (Figure 5.21). Such sources are called **FR I** and **FR II** sources, respectively, after Fanaroff and Riley [38], who first made such classifications and noted that FR I sources are usually less luminous than FR II sources, with the dividing line being $L_\nu \sim 10^{24}$ W Hz^{-1} at 1.4 GHz. FR I sources generally have lower equipartition energy densities and hence lower equipartition pressures. The jets of FR I sources are fairly symmetric at distances greater than several kpc from the cores, suggesting that the low-luminosity jets are quickly decelerated to nonrelativistic speeds. The low-energy FR I jets are easily influenced by ambient matter. Low-luminosity galaxies moving through the intracluster medium of a cluster of galaxies frequently have bent head-tail radio morphologies similar to the wake of a moving boat.

5.6.2 Unified Models

The combination of orientation-dependent beaming and obscuration by dust has led to various **unified models** (Figure 5.22) of active galactic nuclei (AGN).

These models attribute some or all of the differences between observationally different objects to the inclinations of their jets relative to the line of sight. If the inclination is small, the base of the approaching jet will be strongly Doppler boosted, and the compact optical **broad-line region** and inner **accretion disk** will not be obscured by the larger dusty accretion torus lying in a plane normal to the jet. The observed radio emission will be dominated by a one-sided jet that may be variable in intensity and apparently superluminal. Thermal emission from the inner parts of the accretion disk may be visible as a **big blue bump** in the optical/UV spectrum, and

Figure 5.20. Inner jets of the radio galaxy 3C 31 [64]. If the jets are close to the plane of the sky and decelerate from relativistic speeds as they recede from the core, only the inner portions of the jets are Doppler dimmed. Image credit: NRAO/AUI/NSF. The simulation at http://www.cv.nrao.edu/~abridle/3c31free/3c31anim_const_sen_flame.htm shows how 3C 31 would appear at different angles θ between the jet and the line of sight. The best fit to the data occurs at $\theta = 52°$. Image credit: NRAO/AUI/NSF. Investigators: Alan Bridle & Robert Laing.

Doppler-broadened emission lines from the small (< 1 pc) broad-line region will not be obscured. If the optical AGN emission is much brighter than the starlight of the host galaxy, the object will be called a **quasi-stellar object** (QSO), but otherwise a **Seyfert I galaxy**. In extreme cases, optical synchrotron emission may dominate the big blue bump and emission lines. Objects with lineless power-law optical spectra are often called **BL Lac objects** after their prototype BL Lacertae, which was originally thought to be a Galactic star (hence the constellation name). If the inclination angle is larger than about 45 degrees, the optical core may be obscured by the dusty torus and highly relativistic radio jets may be Doppler dimmed, and we will see either a double-lobed radio galaxy or a **Seyfert II galaxy** (a Seyfert galaxy with only the narrow emission lines directly visible). The ongoing debate over unified models is not about *whether* relativistic beaming and dust obscuration affect the appearance of AGNs, but *how much*.

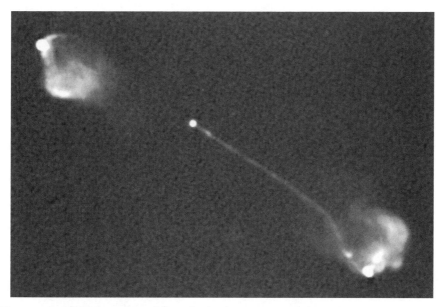

Figure 5.21. This VLA image of the radio-loud quasar 3C 175 shows the core, an *apparently* one-sided jet, and two radio lobes with hot spots of comparable flux densities. The jet is intrinsically two sided but relativistic, so Doppler boosting brightens the approaching jet and dims the receding jet. Both lobes and their hot spots are comparably bright and thus are not moving relativistically. Image credit: NRAO/AUI/NSF Investigators: Alan Bridle, David Hough, Colin Lonsdale, Jack Burns, & Robert Laing.

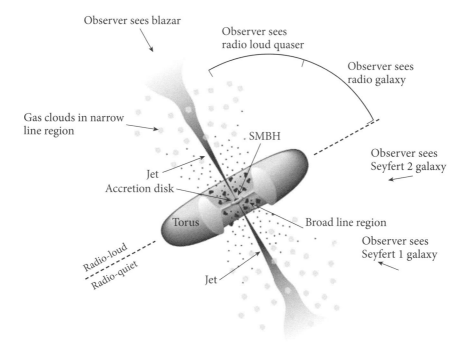

Figure 5.22. This cartoon shows the main features of a "unified model" for active galactic nuclei [109]. Image Credit: Robert Findlay.

5.6.3 Radio Emission from Normal Galaxies

The radio emission from a **normal galaxy** is not powered by an AGN. The continuum radio emission from normal galaxies is dominated by a combination of

1. free–free emission from HII regions ionized by massive ($M > 15M_\odot$) main-sequence stars, and
2. synchrotron radiation from cosmic-ray electrons, most of which were accelerated in the supernova remnants (SNRs) of massive ($M > 8M_\odot$) stars.

Stars more massive than $\sim 8M_\odot$ have main-sequence lifetimes $\tau < 3 \times 10^7$ yr, much less than the $> 10^{10}$ yr age of our Galaxy. Also, the synchrotron lifetimes of cosmic-ray electrons in the typical interstellar magnetic field are $\tau < 10^8$ yr. Thus the current radio continuum emission from normal galaxies is an extinction-free tracer of recent star formation, unconfused by emission from older stars. The radio emission is roughly coextensive with the locations of star formation, spanning the stellar disks of most spiral galaxies.

Massive stars form by gravitational collapse in dusty molecular clouds. The dust absorbs most of their visible and ultraviolet radiation, is heated to temperatures of several tens of K, and reemits the input energy at far-infrared (FIR) wavelengths $\lambda \sim 100\,\mu$m. The molecular clouds are not opaque at FIR wavelengths, so FIR luminosity is a good quantitative measure of the current star-formation rate. Remarkably, the radio luminosities of normal galaxies are very tightly correlated with their FIR luminosities.

The physical origin of this famous **FIR/radio correlation** (Figure 5.23) is poorly understood, particularly at low frequencies where most of the radio emission is synchrotron radiation. It is not surprising that the FIR and free–free radio fluxes would be correlated because both are roughly proportional to the ionizing luminosities of massive young stars. However, free–free emission accounts for only a small fraction of the total radio luminosity at low frequencies $\nu \ll 30$ GHz. The FIR/radio spectrum of the nearby starburst galaxy M82 (Figure 2.24) is typical.

At $\nu \approx 1$ GHz, about 90% of the radio flux is produced by synchrotron radiation, yet the FIR/radio luminosity ratio is confined to a very narrow range. If the star-formation rate (SFR) in a galaxy is fairly constant on timescales longer than 3×10^7 yr, then the number of young SNRs would be proportional to the present number of massive stars, so it is plausible that the current production rate of cosmic-ray electrons is proportional to the current star-formation rate. However, most of the synchrotron radiation from normal galaxies does not originate in the SNRs themselves, but rather comes from cosmic-ray electrons that have diffused into the interstellar medium (ISM). The power radiated by each electron is proportional to the magnetic energy density $U_B = B^2/(8\pi)$ in the ISM. The equipartition fields in normal galaxies range from very low values to $B \sim 5\,\mu$G in a typical spiral galaxy like ours, to $B \sim 100\,\mu$G in M82 and up to $B \sim 1000\,\mu$G in a particularly compact and luminous starburst galaxy such as Arp 220. Thus the power radiated by each cosmic-ray electron must vary by up to several orders of magnitude from one galaxy to another, yet all of these galaxies obey the same FIR/radio correlation.

The **calorimeter model** [112] was devised to explain how the FIR/radio ratio could be independent of U_B. The total radio *energy* radiated by each electron might

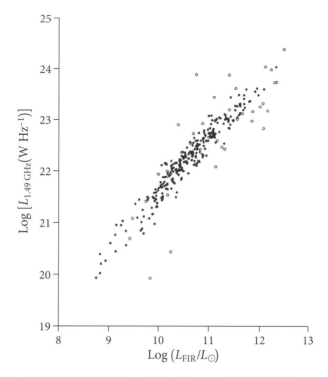

Figure 5.23. The FIR/radio (1.4 GHz) correlation for normal galaxies [26].

be independent of U_B if the lifetime of the electron is proportional to U_B^{-1}. Thus, a cosmic-ray electron in a strong magnetic field radiates a high power for a short time, while one in a weak magnetic field radiates a lower power for a proportionately longer time. For a given production rate of cosmic-ray electrons, the average synchrotron power will then be independent of U_B. The calorimeter model works well to explain the FIR/radio correlation so long as the *fraction* of the electron energy going into synchrotron radiation is about the same in all normal galaxies. However, there are many other energy-loss channels. One is inverse-Compton scattering off the cosmic microwave background, starlight, FIR radiation, etc. Another is diffusion out of the magnetic field of a galaxy—some electrons escape silently into intergalactic space. Electrons may also lose energy by colliding with atoms in the ISM.

Galaxy–galaxy collisions can trigger intense **starbursts** (star-formation episodes so intense that they will deplete the available ISM on timescales much shorter than 10^{10} yr) within several hundred parsecs of the centers of galaxies and produce compact central sources. The radiation energy density in a compact starburst galaxy such as Arp 220 can reach $U_{rad} \sim 3 \times 10^8$ erg cm^{-3}, yet Arp 220 still obeys the FIR/radio correlation. The fact that inverse-Compton cooling doesn't seriously deplete the population of synchrotron-emitting electrons sets a lower limit to the interstellar magnetic energy density $U_B = B^2/(8\pi) \sim U_{rad}$ via Equation 5.154. This limit is $B \sim 1000\ \mu$G in Arp 220 [31].

Despite its uncertain theoretical basis, the FIR/radio correlation makes radio continuum emission from normal galaxies a very useful, quantitative, and extinction-free indicator of the rate at which massive stars are being formed. The rate (measured in units of solar masses per year) at which stars with masses $M > 5M_\odot$ are formed in a galaxy can be estimated from the thermal (free–free) and nonthermal (synchrotron) spectral luminosities by the following equations [24]:

$$\left(\frac{L_T}{\text{W Hz}^{-1}}\right) \approx 5.5 \times 10^{20} \left(\frac{\nu}{\text{GHz}}\right)^{-0.1} \left[\frac{\text{SFR}(M > 5M_\odot)}{M_\odot \text{ yr}^{-1}}\right], \tag{5.184}$$

$$\left(\frac{L_{NT}}{\text{W Hz}^{-1}}\right) \approx 5.3 \times 10^{21} \left(\frac{\nu}{\text{GHz}}\right)^{-0.8} \left[\frac{\text{SFR}(M > 5M_\odot)}{M_\odot \text{ yr}^{-1}}\right]. \tag{5.185}$$

5.6.4 Extragalactic Radio-Source Populations and Cosmological Evolution

Surveys of discrete radio sources have been made over large areas of the sky and at many frequencies ranging from 38 MHz to 857 GHz. The most extensive **sky survey** is the NRAO VLA Sky Survey (NVSS) [30], which covers the whole sky north of **declination** (latitude on the celestial sphere) $\delta = -40°$ and detected nearly 2×10^6 sources stronger than $S \approx 2.3$ mJy at 1.4 GHz. Extremely sensitive sky surveys covering much smaller areas have reached flux densities $S \sim 5\,\mu\text{Jy}$. Sources detected by **blind surveys** covering representative areas of sky give us an unbiased statistical sample of the radio-source population.

The distribution of discrete sources on the sky is extremely isotropic, as shown in Figure 5.24. This isotropy indicates that nearly all radio sources in a flux-limited sample are extragalactic—the center of our Galaxy is barely visible as the curved band at the left of the upper panel. A similar plot of the brightest galaxies selected at optical or near-infrared wavelengths is much clumpier than the radio plots because galaxies cluster on scales ~ 10 Mpc in size. The reason for this difference is that the strongest extragalactic radio sources are much farther away than the optically brightest galaxies. Radio galaxies are at least as clustered as optical galaxies, but the average distance between radio galaxies is so much greater than 10 Mpc that their clustering can be detected only by sensitive statistical tests. Indeed, the distribution of radio sources on the sky is so uniform that the small ($< 1\%$) dipole anisotropy in source density caused by Doppler boosting from the Earth's motion relative to the frame defined by distant galaxies has been detected. The velocity of the Earth deduced from this anisotropy is consistent with the motion deduced from the corresponding anisotropy in the cosmic microwave background radiation produced by the hot big bang [11].

Only a small fraction ($\sim 1\%$) of radio sources in a flux-limited sample are closer than about 100 Mpc. By identifying those sources with bright galaxies and determining their Hubble distances, we can determine the space density of nearby radio sources as a function of radio spectral luminosity; this is called the **local luminosity function**. The local luminosity function can be further refined by specifying separately the space densities of radio sources powered by AGN and those powered by star-forming galaxies containing H II regions, SNRs, etc. (Figure 5.25).

In a given volume of space, radio sources in star-forming galaxies outnumber radio galaxies containing AGN by an order of magnitude. However, the rarer AGN

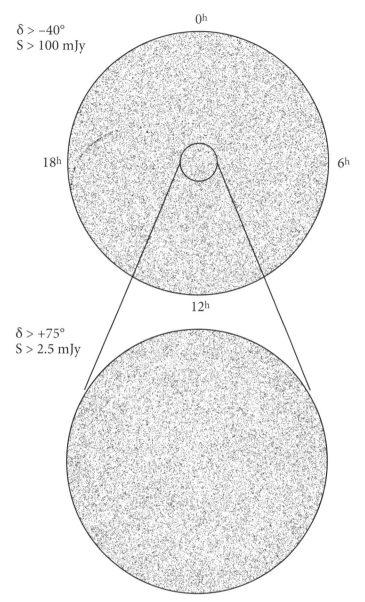

Figure 5.24. These two equal-area plots show the sky distribution of discrete sources stronger than $S = 100$ mJy at 1.4 GHz in the 82% of the sky north of $\delta = -40°$ (upper panel) and the sky distribution of sources stronger than $S = 2.5$ mJy at 1.4 GHz in the 1.70% of the sky north of $\delta = +75°$ (lower panel).

produce all of the most luminous sources, so they account for slightly over half of all radio emission produced by discrete sources.

If we assume that the comoving space density of radio sources in the expanding universe is independent of time, we can use the local luminosity function to calculate the total number of radio sources per steradian of sky as a function of flux density.

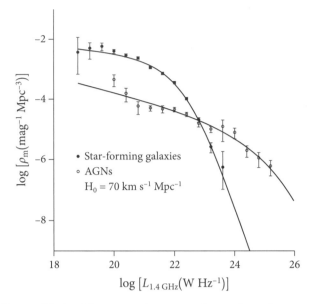

Figure 5.25. The 1.4 GHz local luminosity functions of normal star-forming galaxies (filled symbols) and of AGN (open symbols) [28].

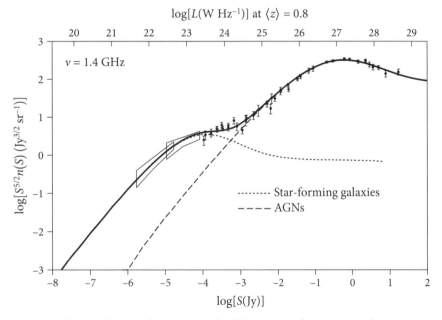

Figure 5.26. The 1.4 GHz Euclidean-normalized luminosity function ϕ and source count $S^{5/2}n(S)$ [29] are consistent with strong ($\sim 10\times$) luminosity evolution of all extragalactic radio sources. The dashed and dotted curves indicate the contributions of AGNs and star-forming galaxies, respectively, to the total source count. Consequently, radio-loud AGNs in radio galaxies and quasars (dashed curve in Figure 5.26) account for most sources stronger than $S \sim 0.1$ mJy at 1.4 GHz, and the numerous but less luminous star-forming galaxies (dotted curve in Figure 5.26) dominate the microJy radio-source population.

The resulting **source counts** are usually tabulated in differential form: $n(S)dS$ is the number of sources per sr with flux densities between S and $S + dS$. In a static Euclidean universe, the flux density of any source at distance r is proportional to r^{-2}, and the volume enclosed by a sphere is proportional to r^3, so the integral number $N(> S)$ of sources stronger than any given flux density S should be proportional to $S^{-3/2}$ and the differential number $n(S)$ per unit flux density should be $n(S) \propto S^{-5/2}$. Plotting the normalized differential source count $n(S) \times S^{5/2}$ as a function of S should yield a horizontal line in a static Euclidean universe.

The actual plot for sources selected at 1.4 GHz is shown in Figure 5.26. The source counts are not consistent with a static Euclidean universe, so most radio sources cannot be "local" extragalactic sources. In an expanding universe with a constant comoving source density, distant sources will be Doppler dimmed and the normalized source counts should decline monotonically at low flux densities. This is not the case either; the normalized counts have a clear maximum near $S \sim 500$ mJy. This peak indicates that radio sources must be evolving on cosmological timescales; that is, their comoving space density varies with time and was higher at some time in the past. The discovery of radio-source evolution was used as evidence against the steady-state model of the universe [99], years before the discovery of the cosmic microwave background radiation decisively confirmed the hot big-bang model.

Detailed models consistent with the local luminosity function, radio source counts, and redshift distributions of radio sources identified with galaxies and quasars have been constructed to measure the amount of evolution. The results are actually quite simple: **cosmological evolution** is so strong that most radio sources in flux-limited samples have redshifts near the median $\langle z \rangle \sim 0.8$. To radio astronomers, the universe looks like a nearly hollow spherical shell centered on the Earth. This observation does not conflict with the Copernican principle, which states that the Earth is not in a special position at the center of the universe; it requires only that the universe evolve with time. Most radio sources seen today have distances of 5 to 10×10^9 light years, and their dominance reflects the higher AGN and star-formation activity of 5 to 10 Gyr ago. For sources in a thin shell, there is little correlation between flux density and average distance; rather, flux density is more closely correlated with absolute luminosity as shown by labels on the lower and upper abscissae in Figure 5.26.

6 Pulsars

6.1 PULSAR PROPERTIES

Pulsars are magnetized **neutron stars** that appear to emit periodic short pulses of radio radiation with periods between 1.4 ms and 8.5 s. The radical proposal that neutron stars even exist was made with trepidation by Baade & Zwicky in 1934 [5]: "With all reserve we advance the view that a supernova represents the transition of an ordinary star into a new form of star, the *neutron star*, which would be the endpoint of stellar evolution. Such a star may possess a very small radius and an extremely high density. As neutrons can be packed much more closely than ordinary nuclei and electrons, the 'gravitational packing' energy in a cold neutron star may become very large, and, under certain circumstances, may far exceed the ordinary nuclear packing fractions."

The name **pulsar** blends "pulse" and "star," but pulsars are not pulsating stars. Like lighthouses, they continuously emit rotating beams of radiation and appear to flash each time the beam sweeps across the observer's line of sight. The pulse periods are quite stable because they equal the rotation periods of massive neutron stars. Even though their radio emission mechanism is not well understood, pulsars have become uniquely valuable astrophysical tools:

1. Neutron stars are physics laboratories sampling extreme conditions—deep gravitational potentials $GM/(rc^2) \sim 1$, densities $\rho \sim 10^{14}$ g cm^{-3} exceeding those in atomic nuclei, and magnetic field strengths as high as $B \sim 10^{14}$ or even 10^{15} gauss—not reproducible on Earth.
2. Pulse periods can be timed with fractional errors as small as 10^{-16}. Accurate **pulsar timing** permits exquisitely sensitive measurements of quantities such as the power of gravitational radiation emitted by a pulsar in a binary system, neutron-star masses, general relativistic effects in strong gravitational fields, orbital perturbations from binary companions as light as planets, accurate pulsar positions and proper motions, and potentially the distortions of interstellar space produced by long-wavelength gravitational radiation from the mergers of supermassive black holes throughout the universe.

Lorimer and Kramer [69] and Lyne and Graham-Smith [70] have written excellent reference books about pulsars and their astrophysical applications.

6.1.1 Discovery

Pulsars were discovered [50] serendipitously in 1967 on chart recordings (Figure 6.1) obtained during a low-frequency ($\nu = 81$ MHz) search for compact extragalactic radio sources that **scintillate** in the interplanetary plasma, just as stars twinkle in the Earth's atmosphere.

This history of this important discovery is a warning against overprocessing data before looking at them, ignoring unexpected signals, and failing to explore the observational "parameter space" covered by the data—here the relevant parameter being time. The strong pulsar PSR B0329+54 was clearly visible in 1954 survey data from the Jodrell Bank 250-foot radio telescope [70], but nobody paid any attention to it at the time. The X-ray pulsar in the Crab Nebula (Color Plate 10) was present in data taken several months before the discovery of radio pulsars, but only after the radio pulsar in the Crab Nebula was announced were the X-ray pulses extracted [39]. As radio instrumentation and data-processing software become more sophisticated, more data are "cleaned up" automatically before they reach the astronomer. Matched filtering that brings out the expected signal usually suppresses the unexpected. Thus, clipping circuits or software remove the strong impulses that are usually caused by terrestrial interference, and integrators smooth out fluctuations shorter than the integration time. Most pulses seen by radio astronomers are just artificial interference from radar, electric cattle fences, etc., and short pulses from sources at interstellar distances imply unexpectedly high brightness temperatures $T_b > 10^{25}$ K, which is much higher than the $\sim 10^{12}$ K upper limit for incoherent electron-synchrotron radiation imposed by synchrotron self-Compton cooling (Section 5.5.3).

Cambridge University graduate student Jocelyn Bell recognized that pulsars are astronomical sources where others had failed because she noticed that some "scruffy" pulses in her chart-recorder data (Figure 6.1) didn't look like other forms of interference and they reappeared exactly once per sidereal day, indicating an origin outside the Solar System. Fortunately, she and her supervisor, Antony Hewish, "decided initially not to computerize the output because until we were familiar with the behavior of our telescope and receivers we thought it better to inspect the data visually, and because a human can recognize signals of different character whereas it is difficult to program a computer to do so." Read http://www.bigear.org/vol1no1/burnell.htm for the full discovery story in her own words.

6.1.2 Neutron Star Masses and Densities

The sources of the pulses were originally unknown, and even intelligent transmissions by "LGM" (Little Green Men) were seriously suggested as explanations for pulsars. Their short periods imply very compact sources such as white dwarf stars, black holes, and neutron stars; their stable periods rule out black holes. Astronomers were familiar with slowly varying or pulsating emission from stars, but the natural period of a radially pulsating star depends on its mean density ρ and is typically days, not seconds. There is a comparable lower limit to the rotation period P of a gravitationally bound star, set by the requirement that the centrifugal acceleration at its equator not exceed the gravitational acceleration. If a nearly spherical star of

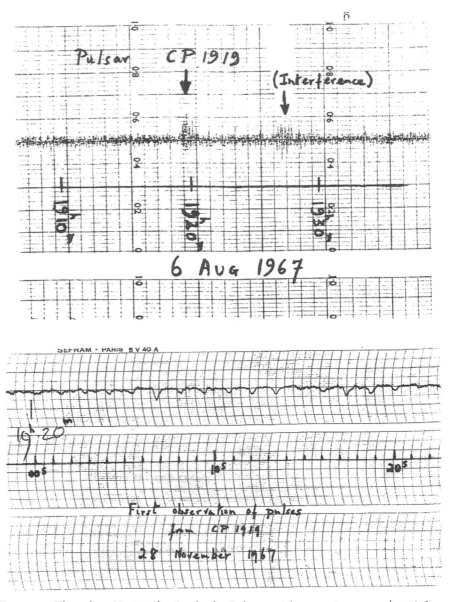

Figure 6.1. The pulsar CP1919 (for Cambridge Pulsar at **right ascension** $\alpha = 19^{\mathrm{h}}\,19^{\mathrm{m}}$) first appeared as "scruff" barely distinguishable from interference on a scintillation survey chart record (*top*). On a higher-speed chart recording of CP1919 (*bottom*), dips in the upper trace spaced by $P \approx 1.3$ s showed that the "scruff" was actually a series of periodic pulses. Image credit: Jocelyn Bell Burnell and Antony Hewish.

mass M and radius R rotates with angular velocity $\Omega \equiv 2\pi/P$,

$$\Omega^2 R < \frac{GM}{R^2} \tag{6.1}$$

implies

$$P^2 > \left(\frac{4\pi R^3}{3}\right)\frac{3\pi}{GM}.$$ (6.2)

In terms of the mean density

$$\rho \equiv M\left(\frac{4\pi R^3}{3}\right)^{-1},$$ (6.3)

$$P > \left(\frac{3\pi}{G\rho}\right)^{1/2}$$ (6.4)

and

$$\rho > \frac{3\pi}{GP^2}.$$ (6.5)

Equation 6.5 gives a conservative lower limit to the mean density because a rapidly spinning star is oblate, which increases the centrifugal acceleration and decreases the gravitational acceleration at its equator.

The first pulsar discovered has a period $P = 1.3$ s, so its mean density is at least

$$\rho > \frac{3\pi}{GP^2} = \frac{3\pi}{6.67 \times 10^{-8}\ \text{dyne cm}^2\ \text{gm}^{-2}(1.3\ \text{s})^2} \approx 10^8\ \text{g cm}^{-3}.$$

This limit is just consistent with the known densities of **white dwarf** stars. But soon the faster ($P = 0.033$ s) pulsar in the Crab Nebula was discovered, and its period implies a density much higher than any stable white-dwarf star supported by electron-degeneracy pressure. Also, the **Crab Nebula** (Color Plate 10) is the remnant of a supernova recorded by Chinese astronomers as a "guest star" in 1054 AD, so the discovery of this pulsar confirmed the suggestion by Baade and Zwicky [5] that neutron stars are the compact remnants of supernovae. The fastest known pulsar has $P = 1.4 \times 10^{-3}$ s implying $\rho > 10^{14}$ g cm^{-3}, the density of atomic nuclei.

A star whose mass is greater than the **Chandrasekhar mass**

$$M_{\text{Ch}} \sim \left(\frac{\hbar c}{G}\right)^{3/2}\frac{1}{m_{\text{p}}^2} \approx 1.4 M_\odot$$ (6.6)

cannot be supported by electron degeneracy pressure and will collapse to become a neutron star. Equation 6.2 implies the maximum radius

$$R < \left(\frac{GMP^2}{4\pi^2}\right)^{1/3}$$ (6.7)

of a $P = 1.4 \times 10^{-3}$ s pulsar with mass $M \approx 1.4 M_\odot$ is

$$R < \left[\frac{6.67 \times 10^{-8}\ \text{dyne cm}^2\ \text{g}^{-2} \cdot 1.4 \cdot 2.0 \times 10^{33}\ \text{g} \cdot (1.4 \times 10^{-3}\ \text{s})^2}{4\pi^2}\right]^{1/3}$$

$$\approx 2 \times 10^6\ \text{cm} = 20\ \text{km}.$$ (6.8)

These numbers motivate the *definition* of the **canonical neutron star** as being a uniform density sphere with mass $M \approx 1.4 M_\odot$, radius $R \approx 10$ km, and moment of inertia $I = 2MR^2/5 \approx 10^{45}$ g cm^2. "Canonical" is perhaps too strong a term for what is really no more than a convention for the typical properties of neutron stars; individual neutron stars can have different masses and radii. The extreme density and pressure turn most of the star into a neutron superfluid that is a superconductor at temperatures up to $T \sim 10^9$ K. The masses of individual pulsars have been measured with varying degrees of accuracy, and many are close to the canonical $1.4 M_\odot$. The highest accurately measured pulsar masses are $M \approx 2.0 M_\odot$, and they rule out all currently proposed hyperon or boson condensate equations of state and "free" quarks [34]. Any neutron star of significantly higher mass ($M \sim 3 M_\odot$ in standard models) must collapse and become a black hole.

6.1.3 Magnetic Fields

The Sun and many other stars are known to possess approximately dipolar magnetic fields. Stellar interiors are fully ionized and hence good electrical conductors. Charged particles are constrained to move along magnetic field lines, and magnetic field lines are tied to the charged particles. When a star collapses from a radius $\sim 10^6$ km to ~ 10 km, its cross-sectional area a is divided by $\sim 10^{10}$, its magnetic flux $\Phi \equiv \int \vec{B} \cdot \hat{n}\, da$ (where $\hat{n}$ is the unit vector normal to each infinitesimal surface area da) is conserved, and the magnetic field strength is multiplied by $\sim 10^{10}$. An initial magnetic field strength $B \sim 100$ G becomes $B \sim 10^{12}$ G after collapse, so young neutron stars should have very strong dipolar fields. The best models of the core-collapse process show that a dynamo effect can generate even stronger magnetic fields. Such dynamos may be able to produce the 10^{14}–10^{15} G fields observed in **magnetars**, which are neutron stars having such strong magnetic fields that their radiation is powered by magnetic field decay. Conservation of angular momentum during collapse increases the rotation rate by about the same factor, 10^{10}, yielding initial rotation periods P_0 in the millisecond range. Thus young neutron stars should contain rapidly rotating magnetic dipoles (Figure 6.2).

6.1.4 Magnetic Dipole Radiation

If a rotating **magnetic dipole** is inclined by some **inclination angle** $\alpha > 0$ from the rotation axis, it emits electromagnetic radiation at the rotation frequency. Rewriting the Larmor formula (Equation 2.143) in terms of power radiated by a rotating electric dipole gives

$$P_{\rm rad} = \frac{2q^2 \dot{v}^2}{3c^3} = \frac{2}{3}\frac{(q\ddot{r}\sin\alpha)^2}{c^3} = \frac{2}{3}\frac{(\ddot{p}_\perp)^2}{c^3}, \tag{6.9}$$

where $p = qr$ is the electric dipole moment and $p_\perp = p\sin\alpha$ is its component perpendicular to the rotation axis. J. J. Thomson's derivation of the Larmor formula in terms of electric field lines (Section 2.7) is also valid for magnetic field lines. The Gaussian CGS units for magnetic and electric field are the same, so the power of

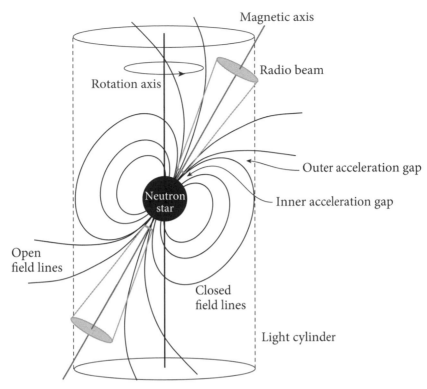

Figure 6.2. The traditional magnetic dipole model of a pulsar. Electrons and positrons from particle cascades are accelerated in one or more of the "gap" regions of the magnetosphere. They stream along the open magnetic field lines and emit coherent radio emission, and in the highest-energy pulsars potentially X-rays and γ-rays as well. The details of these dynamic, relativistic plasma processes are not well understood. Figure credit: Lorimer and Kramer [69].

magnetic dipole radiation is

$$P_{\text{rad}} = \frac{2}{3} \frac{(\ddot{m}_\perp)^2}{c^3}, \tag{6.10}$$

where $m_\perp = m \sin \alpha$ is the perpendicular component of the magnetic dipole moment. For a uniformly magnetized sphere with radius R and surface magnetic field strength B, the magnitude of the magnetic dipole moment is [55]

$$m = B R^3. \tag{6.11}$$

If the inclined magnetic dipole rotates with angular velocity $\Omega = 2\pi/P$, then

$$P_{\text{rad}} = \frac{2}{3} \frac{m_\perp^2 \Omega^4}{c^3} = \frac{2}{3c^3} (B R^3 \sin \alpha)^2 \left(\frac{2\pi}{P}\right)^4, \tag{6.12}$$

where P is the pulsar period. This electromagnetic radiation will appear at the *very* low radio frequency $\nu = P^{-1} < 1$ kHz, so low that it cannot propagate through the surrounding ionized nebula or ISM. Magnetic dipole radiation extracts rotational kinetic energy from the neutron star and causes the pulsar period to increase with time. The absorbed radiation deposits energy in the surrounding nebula, the Crab Nebula (Color Plate 10) being a prime example.

6.1.5 Spin-Down Luminosity

The rotational kinetic energy E of a spinning object is related to its moment of inertia I by

$$E = \frac{I\Omega^2}{2} = \frac{2\pi^2 I}{P^2}. \tag{6.13}$$

The **moment of inertia** of a small mass element about any rotation axis is its mass multiplied by the square of its radial distance r from the rotation axis. The moment of inertia of a sphere with radius R, mass M, and uniform density $\rho = 3M/(4\pi R^3)$ spinning about its z-axis is obtained by summing over its mass elements:

$$I = \int_{-R}^{R} \left(\int_0^{(R^2-z^2)^{1/2}} \rho r^2 \, 2\pi r \, dr \right) dz, \tag{6.14}$$

where r is the distance from the rotation axis and z is the height in cylindrical coordinates centered on the sphere. Then

$$I = \pi\rho \int_0^R (R^2 - z^2)^2 dz = \frac{8\pi\rho R^5}{15} = \frac{2MR^2}{5}. \tag{6.15}$$

The moment of inertia of a canonical neutron star is

$$I = \frac{2MR^2}{5} \approx \frac{2 \cdot 1.4 \cdot 2.0 \times 10^{33} \text{ g} \cdot (10^6 \text{ cm})^2}{5} \approx 10^{45} \text{ gm cm}^2. \tag{6.16}$$

The rotational kinetic energy of a canonical neutron star with the rotation period $P = 0.033$ s of the Crab pulsar is

$$E = \frac{2\pi^2 I}{P^2} \approx \frac{2\pi^2 \cdot 10^{45} \text{ g cm}^2}{(0.033 \text{ s})^2} \approx 1.8 \times 10^{49} \text{ ergs.}$$

As magnetic dipole radiation extracts rotational energy, it slowly increases the period of a pulsar:

$$\dot{P} \equiv \frac{dP}{dt} > 0. \tag{6.17}$$

Note that the **period derivative** $\dot{P}$ is a dimensionless (seconds per second) pure number. Combining the observed period P and period derivative $\dot{P}$ yields an estimate of the rate $\dot{E}$ at which the rotational energy is changing. The quantity

$$-\dot{E} \equiv -\frac{dE_{\rm rot}}{dt} = -\frac{d}{dt}\left(\frac{1}{2}I\Omega^2\right) = -I\Omega\dot{\Omega} \tag{6.18}$$

is called the **spin-down luminosity**. It is *not* a measured luminosity; it is the measured loss rate of rotational energy, which is *presumed* to equal the luminosity of magnetic dipole radiation. The spin-down luminosity is usually expressed in terms of the pulse period P:

$$\Omega = \frac{2\pi}{P} \qquad \text{so} \qquad \dot{\Omega} = 2\pi(-P^{-2}\dot{P}), \tag{6.19}$$

and Equation 6.18 becomes

$$\boxed{-\dot{E} = \frac{4\pi^2 I \dot{P}}{P^3}.} \tag{6.20}$$

The Crab pulsar has $P = 0.033$ s and $\dot{P} = 10^{-12.4}$. If $I = 10^{45}$ g cm^2 (Equation 6.16), its spin-down luminosity is

$$-\dot{E} = \frac{4\pi^2 I \dot{P}}{P^3} = \frac{4\pi^2 \cdot 10^{45} \text{ g cm}^2 \cdot 10^{-12.4} \text{ s s}^{-1}}{(0.033 \text{ s})^3} \approx 4 \times 10^{38} \text{ erg s}^{-1} \approx 10^5 L_\odot.$$

If $P_{\text{rad}} \approx -\dot{E} \approx 10^5 L_\odot$, the luminosity of the low-frequency ($\nu = P^{-1} \approx 30$ Hz) magnetic dipole radiation from the Crab pulsar is comparable with the *entire* radio output of our Galaxy! It even exceeds the Eddington luminosity limit (Section 5.4.2) of the neutron star, which is possible because the energy *source* is not accretion. It greatly exceeds the average radio pulse luminosity of the Crab pulsar, $\sim 10^{30}$ erg s^{-1}. The long wavelength ($\lambda = c/\nu \sim 10^7$ m) magnetic-dipole radiation is absorbed by and heats up the surrounding Crab Nebula (Color Plate 10), making it the "megawave oven" counterpart of a kitchen microwave oven. The observed bolometric luminosity of the Crab Nebula is comparable with the spin-down luminosity, supporting the model that rotational kinetic energy is converted to magnetic dipole radiation that is absorbed by the surrounding ionized nebula and later reradiated at radio through X-ray wavelengths.

6.1.6 Minimum Magnetic Field Strength

If $-\dot{E} \approx P_{\text{rad}}$, Equations 6.12 and 6.20 can be combined to yield a lower limit to the magnetic field strength $B > B \sin \alpha$ at the neutron star surface, where α is the unknown **inclination angle** between the rotation and magnetic axes:

$$P_{\text{rad}} = -\dot{E}, \tag{6.21}$$

$$\frac{2}{3c^3}(B R^3 \sin \alpha)^2 \left(\frac{4\pi^2}{P^2}\right)^2 = \frac{4\pi^2 I \dot{P}}{P^3}, \tag{6.22}$$

$$B^2 = \frac{3c^3 I P \dot{P}}{2 \cdot 4\pi^2 R^6 \sin^2 \alpha}, \tag{6.23}$$

$$B > \left(\frac{3c^3 I}{8\pi^2 R^6}\right)^{1/2} (P\dot{P})^{1/2}. \tag{6.24}$$

Inserting the constants for the canonical pulsar in CGS units yields for the first factor

$$\left[\frac{3 \cdot (3 \times 10^{10} \text{ cm s}^{-1})^3 \cdot 10^{45} \text{ g cm}^2}{8\pi^2 (10^6 \text{ cm})^6} \right]^{1/2} \approx 3.2 \times 10^{19}, \qquad (6.25)$$

so the **minimum magnetic field** strength at the surface of a canonical pulsar is

$$\boxed{\left(\frac{B}{\text{gauss}} \right) > 3.2 \times 10^{19} \left(\frac{P \dot{P}}{\text{s}} \right)^{1/2}.} \qquad (6.26)$$

It is sometimes called the **characteristic magnetic field** of a pulsar.

The minimum magnetic field strength of the Crab pulsar ($P = 0.033$ s, $\dot{P} = 10^{-12.4}$) is

$$\left(\frac{B}{\text{gauss}} \right) > 3.2 \times 10^{19} \left(\frac{0.033 \text{ s} \cdot 10^{-12.4}}{\text{s}} \right)^{1/2} = 4 \times 10^{12}.$$

This is an amazingly strong magnetic field. Its energy density is

$$U_B = \frac{B^2}{8\pi} > 6 \times 10^{23} \text{ erg cm}^{-3}. \qquad (6.27)$$

Just 10^3 cm^3 (about the volume of a grapefruit) of this magnetic field contains over 6×10^{19} J $= 6 \times 10^{16}$ kW s $\approx 2 \times 10^9$ kW yr of energy, the annual output of a large nuclear power station.

6.1.7 Characteristic Age

If the spin-down luminosity equals the magnetic dipole radiation luminosity and ($B \sin \alpha$) doesn't change significantly with time, a pulsar's age τ can be estimated from $P \dot{P}$ on the further assumption that the pulsar's initial period P_0 was much shorter than its current period. Solving Equation 6.23 for $P \dot{P}$ shows that

$$P \dot{P} = \frac{8\pi^2 R^6 (B \sin \alpha)^2}{3c^3 I} \qquad (6.28)$$

also doesn't change with time. Rewriting the identity $P \dot{P} = P \dot{P}$ as $P \, dP = P \dot{P} \, dt$ and integrating over the pulsar's age τ gives

$$\int_{P_0}^{P} P \, dP = \int_0^{\tau} (P \dot{P}) \, dt = P \dot{P} \int_0^{\tau} dt \qquad (6.29)$$

because $P \dot{P}$ is constant. Integrating Equation 6.29 gives

$$\frac{P^2 - P_0^2}{2} = P \dot{P} \tau. \qquad (6.30)$$

In the limit $P_0^2 \ll P^2$, the **characteristic age** of a pulsar *defined* by

$$\tau \equiv \frac{P}{2\dot{P}}$$

(6.31)

should be close to the actual age of the pulsar. The characteristic age depends only on the observables P and $\dot{P}$; it does not depend on the unknown radius R, moment of inertia I, or perpendicular magnetic field $B\sin\alpha$. However, a newly formed neutron star with small P_0 may be quite oblate and will initially be slowed down by emitting quadrupole gravitational radiation. This can cause the characteristic age to be somewhat larger than the true age for young pulsars.

For example, the characteristic age of the Crab pulsar ($P = 0.033$ s, $\dot{P} = 10^{-12.4}$) is

$$\tau = \frac{P}{2\dot{P}} = \frac{0.033 \text{ s}}{2 \cdot 10^{-12.4}} \approx 4.1 \times 10^{10} \text{ s} \approx \frac{4.1 \times 10^{10} \text{ s}}{10^{7.5} \text{ s yr}^{-1}} \approx 1300 \text{ yr.}$$

It is slightly larger than the actual age of the Crab pulsar, which is known to be just under 1000 yr because the Crab supernova was observed in 1054 AD.

6.1.8 Braking Index

If magnetic dipole radiation is solely responsible for pulsar spin down, then $P_{\text{rad}} = -\dot{E}$ and Equations 6.12 and 6.18 together imply

$$\dot{\Omega} \propto \Omega^3.$$

(6.32)

The pulsar **braking index** n is defined by

$$\dot{\Omega} \propto \Omega^n = C\Omega^n,$$

(6.33)

where C is the constant of proportionality. The braking index is completely determined by the observables P, $\dot{P}$, and $\ddot{P}$, so comparing its observed value with the expected $n = 3$ provides a useful check on pulsar spin-down models. The time derivative of Equation 6.33 is

$$\ddot{\Omega} = Cn\Omega^{n-1}\dot{\Omega} = n\left(\frac{C\Omega^n}{\Omega}\right)\dot{\Omega} = n\left(\frac{\dot{\Omega}}{\Omega}\right)\dot{\Omega} = \frac{n\dot{\Omega}^2}{\Omega},$$

(6.34)

so

$$n = \Omega\ddot{\Omega}/\dot{\Omega}^2.$$

(6.35)

Converting from angular velocities to periods with

$$\Omega = 2\pi P^{-1}, \quad \dot{\Omega} = -2\pi P^{-2}\dot{P}, \quad \text{and}$$

$$\ddot{\Omega} = -2\pi P^{-2}\ddot{P} + \dot{P}4\pi P^{-3}\dot{P} = 2\pi\left(\frac{\ddot{P}}{P^2} + \frac{2\dot{P}^2}{P^3}\right)$$

yields

$$n = \left(\frac{2\pi}{P}\right) 2\pi \left(\frac{\ddot{P}}{P^2} + \frac{2\dot{P}^2}{P^3}\right)(-2\pi P^{-2}\dot{P})^{-2}, \qquad (6.36)$$

$$\boxed{n = 2 - \frac{P\ddot{P}}{\dot{P}^2},} \qquad (6.37)$$

the braking index in terms of the observables P, $\dot{P}$, and $\ddot{P}$.

Braking indices in the range $1.4 \leq n < 3$ have been observed and used to investigate alternative spin-down mechanisms and to make more sophisticated estimates of pulsar ages and initial spin periods [59].

6.1.9 The Lives of Pulsars

Pulsars are born in supernovae and appear in the upper left corner of the pulsar **P$\dot{P}$ diagram** (Figure 6.3). If B is conserved and they age as described above, they gradually move to the right and down, along lines of constant B and crossing lines of constant characteristic age. Pulsars with characteristic ages $< 10^5$ yr are often found in or near recognizable supernova remnants (SNRs). Older pulsars are not, either because their SNRs have faded to invisibility or because the supernova explosions expelled the pulsars with enough speed that they have since escaped from their parent SNRs. The bulk of the pulsar population is older than 10^5 yr but much younger than the Galaxy ($\sim 10^{10}$ yr). The observed distribution of pulsars in the P$\dot{P}$ diagram indicates that something changes as pulsars age. One controversial possibility is that the magnetic fields of old pulsars must decay on timescales $\sim 10^7$ yr, causing old pulsars to move almost straight down in the P$\dot{P}$ diagram until either their magnetic field is too weak or their spin rate is too slow to produce radio emission via the normal, and as yet still highly uncertain, emission mechanism. **Rotating Radio Transients (RRATs)** are pulsars that emit so sporadically that they are more easily detected in searches for single pulses rather than for periodic pulse trains. RRATs with measurable periods usually have $P > 1$ s (Figure 6.3). Some RRATs may be old pulsars on the verge of becoming radio quiet.

Almost all short period ($P < 0.1$ s) pulsars having fairly weak magnetic field strengths ($B < 10^{11}$ G) are in binary systems, as evidenced by periodic orbital variations in their observed pulse periods (Figure 6.4). These **recycled pulsars** have been spun up by accreting mass and angular momentum from their stellar companions, to the point that they emit radio pulses despite their relatively low magnetic field strengths $B \sim 10^8$ G. (Apparently, currents in the accreting plasma "bury" the magnetic field of the neutron star itself.) The magnetic fields of neutron stars funnel the ionized accreting material onto the magnetic polar caps, which become so hot that they, as well as the hot accretion disk, emit X-rays. We observe these systems as the **Low Mass X-ray Binaries (LMXBs)**. As the neutron stars rotate, their inclined magnetic polar caps can appear and disappear from view, causing periodic fluctuations in X-ray flux, and making some of these systems detectable as X-ray pulsars.

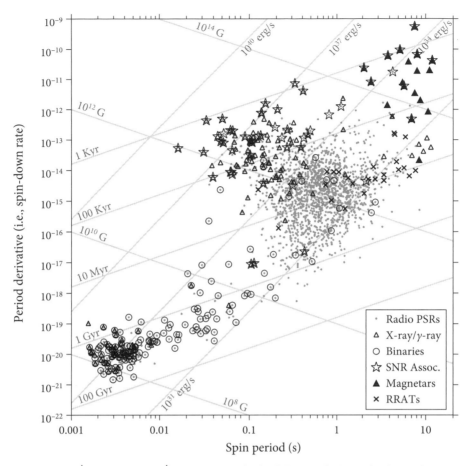

Figure 6.3. PṖ Diagram. The PṖ diagram is useful for following the lives of pulsars, playing a role similar to the Hertzsprung–Russell diagram for ordinary stars. It encodes a tremendous amount of information about the pulsar population and its properties in terms of the observables P (the spin period) and $\dot{P}$ (the time derivative of the spin period). Each pulsar's characteristic age τ (Equation 6.31), minimum magnetic field strength B (Equation 6.26), and spin-down luminosity $-\dot{E}$ (Equation 6.20) is determined by its location on the PṖ diagram, as indicated by the contour lines for τ, B, and $-\dot{E}$. Young pulsars in the upper middle of the diagram are often associated with supernova remnants (SNRs) and emit high-energy radiation. They slow down and evolve over millions of years to become one of the large population of $P \sim 1$ s "slow" pulsars, until their rate of rotation is too slow to power the radio emission mechanism, and they disappear from view. Rotating Radio Transients (RRATs) are pulsars that sporadically emit single pulses instead of continuous pulse trains. The millisecond pulsars in the lower left are mostly in binary systems and were "recycled" via accretion from their companion stars.

Millisecond pulsars (MSPs) with low mass ($M \sim 0.1$–$1M_\odot$) white-dwarf companions typically have orbits with small eccentricities owing to strong tidal dissipation during the accretion phase which spins up the pulsars. The **eccentricity** e of an elliptical orbit is defined as the ratio of the separation of the foci to the length of the major axis. It ranges between $e = 0$ for a circular orbit and $e = 1$ for a parabolic

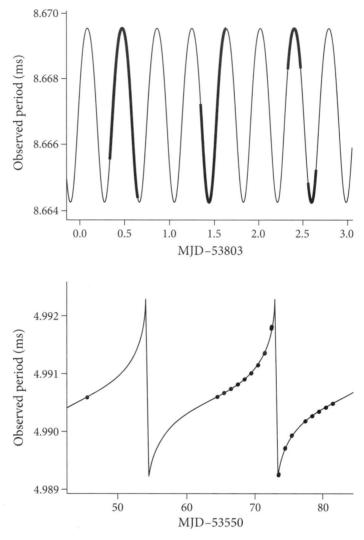

Figure 6.4. Examples of Doppler variations observed in binary systems containing pulsars. Top: The Doppler variations of the globular cluster MSP J1748−2446N in Terzan 5. This pulsar is in a nearly circular orbit (eccentricity $e = 4.6 \times 10^{-5}$) with a companion of minimum mass 0.47 $M_\odot$. The difference between the semimajor and semiminor axes for this orbit is only 51±4 cm! The thick lines show the periods as measured during GBT observations. Bottom: Similar Doppler variations from the highly eccentric binary MSP J0514−4002A in the globular cluster NGC 1851. This pulsar has one of the most eccentric orbits known ($e = 0.888$) and a massive white-dwarf or neutron-star companion.

orbit. Pulsars with extremely eccentric orbits usually have neutron-star companions, indicating that these companions exploded as asymmetric supernovae and nearly disrupted the binary system. Stellar interactions in globular clusters cause a much higher fraction of recycled pulsars per unit mass than in the Galactic disk. These interactions can result in very strange systems such as pulsar–main-sequence-star binaries and MSPs in highly eccentric orbits. In both cases, the original low-mass companion star that recycled the pulsar was ejected in an interaction and replaced by another star.

About 15% of millisecond pulsars are isolated. They were probably recycled via the standard scenario in binary systems, but the energetic MSPs eventually ablated their companions away. Recently, many new "**black widow**" and "**redback**" systems have been detected which seem to be doing exactly this. They exhibit eclipses of their radio MSP emission, likely owing to free–free absorption by the ionized gas in the systems being blown off the companion stars.

6.1.10 Emission Mechanisms

The radio pulses originate in the pulsar magnetosphere. Because the neutron star is a spinning magnetic dipole, it acts as a **unipolar generator**. The total **Lorentz force** acting on a charged particle is

$$\vec{F} = q\left(\vec{E} + \frac{\vec{v} \times \vec{B}}{c}\right). \tag{6.38}$$

Charges in the magnetic equatorial region redistribute themselves by moving along closed magnetic field lines until they build up an electrostatic field large enough to cancel the magnetic force and give $|\vec{F}| = 0$. The induced voltage is about 10^{16} V in MKS units. However, the corotating field lines emerging from the polar caps cross the **light cylinder** (Figure 6.2), an imaginary cylinder centered on the pulsar and aligned with the rotation axis at whose radius the corotating speed equals the speed of light, so these field lines cannot close. Electrons in the polar cap are magnetically accelerated to very high energies along the open but curved field lines, where the acceleration resulting from the curvature causes them to emit **curvature radiation** that is strongly polarized in the plane of curvature. As the radio beam sweeps across the line of sight, the plane of polarization is observed to rotate by up to 180 degrees, a purely geometrical effect. Normal "slow" pulsars usually have quite large amounts of typically linear polarization. There are several relativistic effects in the rapidly rotating magnetospheres of millisecond pulsars that complicate this emission-geometry-based polarization, yet millisecond pulsars are typically highly polarized as well. High-energy photons produced by curvature radiation interact with the magnetic field and lower-energy photons to produce electron–positron pairs that radiate additional high-energy photons. The final results of this cascade process are bunches of charged particles that emit at radio wavelengths. Pulsars "die" in the lower right corner of the $P\dot{P}$ diagram (Figure 6.3) when they have sufficiently low B and high P that the curvature radiation near the polar surface is no longer capable of generating particle cascades.

The extremely high brightness temperatures of pulsars are explained by **coherent radiation**. The electrons do not radiate as independent charges e; instead bunches of N electrons in volumes whose dimensions are less than a wavelength emit in phase as charges Ne. Larmor's formula indicates that the power radiated by a charge q is proportional to q^2, so the radiation intensity can be N times brighter than incoherent radiation from the same total number N of electrons. Because the coherent volume is smaller at shorter wavelengths, most pulsars have extremely steep radio spectra. Typical pulsar spectral indices (defined using the positive convention, see Equation 4.55) are $\alpha \sim -1.7$ (i.e., $S \propto \nu^{-1.7}$), although some can be much steeper ($\alpha < -3$), and a handful are almost flat ($\alpha > -0.5$).

6.2 PULSARS AND THE INTERSTELLAR MEDIUM

With their sharp and short-duration pulse profiles allowing continuous "on–off" differential measurements, small sizes, and high brightness temperatures, pulsars are unique probes of the interstellar medium (ISM). This section closely follows the discussion in Lorimer and Kramer [69], and Equations 6.39 through 6.42 are derived in Appendix D.

The electrons in the ISM make up a cold plasma whose **refractive index** is

$$\mu = \left[1 - \left(\frac{\nu_p}{\nu}\right)^2\right]^{1/2},$$

(6.39)

where ν is the frequency of the radio waves, ν_p is the **plasma frequency**

$$\nu_p = \left(\frac{e^2 n_e}{\pi m_e}\right)^{1/2} \approx 8.97 \text{ kHz} \left(\frac{n_e}{\text{cm}^{-3}}\right)^{1/2},$$

(6.40)

and n_e is the electron number density. For a typical ISM value $n_e \sim 0.03 \text{ cm}^{-3}$, $\nu_p \sim 1.5$ kHz. If $\nu < \nu_p$ then μ is imaginary and radio waves cannot propagate through the plasma.

For propagating radio waves, $\mu < 1$ and the **group velocity**

$$v_g = \mu c$$

(6.41)

of pulses is less than the vacuum speed of light. For most radio observations $\nu_p \ll \nu$, so

$$v_g \approx c\left(1 - \frac{\nu_p^2}{2\nu^2}\right).$$

(6.42)

A broadband pulse moves through a plasma more slowly at lower frequencies than at higher frequencies. If the distance to the source is d, the **dispersion delay** t at frequency ν is

$$t = \int_0^d \frac{dl}{v_g} - \frac{d}{c}$$

(6.43)

$$= \frac{1}{c} \int_0^d \left(1 + \frac{\nu_p^2}{2\nu^2}\right) dl - \frac{d}{c} = \left(\frac{e^2}{2\pi m_e c}\right) \nu^{-2} \int_0^d n_e \, dl.$$

(6.44)

In astronomically convenient units the dispersion delay is

$$\left(\frac{t}{\text{sec}}\right) \approx 4.149 \times 10^3 \left(\frac{\text{DM}}{\text{pc cm}^{-3}}\right) \left(\frac{\nu}{\text{MHz}}\right)^{-2}, \tag{6.45}$$

where

$$\text{DM} \equiv \int_0^d n_e \, dl \tag{6.46}$$

in units of pc cm^{-3} is called the **dispersion measure** and represents the integrated column density of electrons between the observer and the pulsar.

Because pulsar observations almost always cover a wide bandwidth, uncorrected differential delays across the band cause **dispersive smearing** of the pulsed signal when integrated across the band. If uncorrected, the delays would smear out the integrated pulse in time and make most pulsars undetectable (see Figure 6.5). For pulsar searches, the DM is unknown *a priori* and is a search parameter much like the pulsar spin frequency. This extra search dimension is one of the primary reasons that pulsar searches are computationally intensive. For high-precision timing observations of pulsars with known DMs, if Nyquist-sampled voltage data (often called "baseband" data) are available, the dispersion may be completely removed from the data by a technique known as **coherent dedispersion**. The dispersed data are de-convolved using the complex conjugate of the complex transfer function of the ISM that caused the dispersion in the first place.

Dispersion measures can be used to provide distance estimates to pulsars. Crude distances can be calculated for pulsars near the Galactic plane by assuming that $n_e \sim 0.03$ cm^{-3}. However, several sophisticated models of the Galactic electron-density distribution now exist (e.g., NE2001; Cordes and Lazio [33]) that provide much better ($\Delta d/d \sim 30\%$ or less) distance estimates.

The ionized ISM affects pulsar signals in several other ways besides dispersion. Inhomogeneities in the turbulent ISM result in both diffractive and refractive scintillation, or time- and frequency-dependent flux-density variations of the pulsar signal much like the "twinkling" of starlight by Earth's turbulent atmosphere. **Diffractive scintillations** occur over typical timescales of minutes to hours and radio bandwidths of kHz to hundreds of MHz, and they can cause more than order-of-magnitude flux-density fluctuations. **Refractive scintillations** tend to be less than a factor of ~ 2 in amplitude and occur on timescales of weeks.

A phenomenon related to scintillation is pulse broadening caused by **scattering** of radio waves. ISM inhomogeneities cause multipath propagation of a pulsar signal, whereby some rays travel longer physical distances because they do not follow straight lines to the observer. Such rays are delayed in time relative to those traveling more direct paths, and so cause a strongly frequency-dependent (typically $\propto \nu^{-4}$) exponential-like scattering tail of the pulse (see Figure 6.6). This scatter broadening

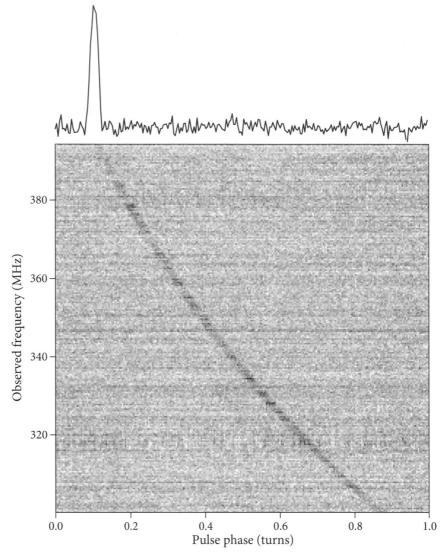

Figure 6.5. Pulsar dispersion. The gray scale shows the uncorrected dispersive delay ($\propto \nu^{-2}$) as a function of frequency from an observation of the recently discovered pulsar J1400+50. If the dispersion measure (DM) were higher or the pulse period shorter, then the pulses would *wrap* in pulse phase (potentially many times) because the data are folded (i.e., averaged) modulo the pulse period. The "de-dispersed" band-integrated pulse profile is shown at the top.

can greatly decrease both the observed pulsed flux density from a pulsar and its timing precision.

Finally, pulsars have broadband continuum spectra, so if there are gas clouds along our line of sight, pulsars can be used to probe the ISM via absorption by spectral lines of HI or molecules. Such absorption spectra can be used to estimate pulsar distances.

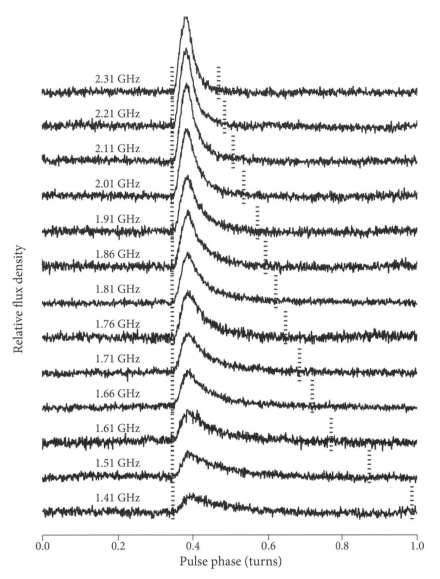

Figure 6.6. Pulse broadening caused by scattering. These data show the strong frequency dependence (typically v^{-3} to $v^{-4.4}$ depending on the nature of the scattering media) of pulse broadening due to multipath propagation as exhibited by the Galactic Center magnetar J1745−2900 [79]. At each frequency the vertical dotted lines span the pulse phase range containing most of the pulse flux.

6.3 PULSAR TIMING

Pulsars are intrinsically interesting and exotic objects, but much of the best science based on pulsar observations has come from their use as *tools* via pulsar timing. **Pulsar timing** is the regular monitoring of the rotation of the neutron star by tracking (nearly exactly) the arrival times of the radio pulses. The key point to remember is that *pulsar timing unambiguously accounts for every single rotation of the neutron star over long periods (years to decades) of time.* This unambiguous and very precise tracking of rotation phase allows pulsar astronomers to probe the interior physics of neutron stars, make extremely accurate astrometric measurements, uniquely test gravitational theories in the strong-field regime, and possibly within the next few years, directly detect **gravitational waves (GWs)** (GWs; propagating distortions of space–time) from supermassive black hole binaries.

For pulsar timing, astronomers "fold" (average) the data from many pulses modulo the instantaneous pulse period P. The instantaneous pulse frequency is $f = 1/P$, and the instantaneous **pulse phase** ϕ is defined by $d\phi/dt = f$. Pulse phase is usually measured in *turns* of 2π radians, so $0 < \phi < 1$.

Averaging over many pulses yields an **average pulse profile**. Although the shapes of individual pulses vary considerably because pulsar emission is intrinsically a noise process, the shape of the average profile is quite stable.

For timing, the average pulse profile is correlated with a template or model profile so that its phase offset can be determined. When multiplied by the instantaneous pulse period, that phase yields a time offset that can be added to a high-precision reference point on the profile (for example, the left edge of the profile based on the recorded time of the first sample of the observation) to create the **time of arrival (TOA)**.

The precision with which a TOA can be determined is approximately equal to the duration of a sharp pulse feature (e.g., the leading edge) divided by the signal-to-noise ratio of the average profile. It is usually expressed in terms of the width W_f of the pulse features in units of the pulse period P, and the signal-to-noise ratio (S/N) such that $\sigma_{\text{TOA}} \propto W_f P/(S/N)$. Strong, fast pulsars with narrow pulse profiles provide the most accurate arrival times.

In the nearly inertial frame of the Solar System **barycenter** (center of mass), the rotation period of a pulsar is nearly constant, so the time-dependent phase $\phi(t)$ of a pulsar can be approximated by the Taylor expansion

$$\phi(t) = \phi_0 + f(t - t_0) + \frac{1}{2}\dot{f}(t - t_0)^2 + \cdots, \qquad (6.47)$$

where ϕ_0 and t_0 are arbitrary reference phases and times for each pulsar. The critical constraint for pulsar timing is that *the observed rotational phase difference between each of the TOAs must contain an integer number of rotations.* Each TOA corresponds to a different time t, so the correct fitting parameters (e.g., f and $\dot{f}$) must result in a phase change between each pair of TOAs i and j that is an integer number of turns, or $\Delta\phi_{ij} = n$ turns. Because all measurements are made with regard to the integrated pulse *phase* rather than the instantaneous pulse period, the precision with which astronomers can make long-term timing measurements can be quite extraordinary.

With what precision can timing determine the spin frequency f of a pulsar? Because $f = d\phi/dt$ when ϕ is measured in turns, the precision is based on how precisely a change in phase $\Delta\phi$ can be measured over some time interval ΔT. Typically, ΔT is the length of time (up to several tens of years for many pulsars now) over which a pulsar's phase has been tracked through regular monitoring. $\Delta\phi$ is determined principally by the individual TOA precisions, although for some types of measurements a statistical component is important as well because precision improves as the number of measurements $N^{-1/2}$ if the random errors are larger than the systematic errors. For example, the original millisecond pulsar B1937+21 has pulse period $P \approx 0.00156$ s and the TOA precision is $\sigma_{\text{TOA}} \sim 1\mu$s, which corresponds to a phase error of $\Delta\phi \sim 6 \times 10^{-4}$ turns. This pulsar has been timed for $\Delta T > 25$ years, so

$$\Delta f \sim \frac{\Delta\phi}{\Delta T} = \frac{6 \times 10^{-4}}{25 \,\text{yrs} \cdot 10^{7.5} \,\text{s yr}^{-1}} = 8 \times 10^{-13} \,\text{Hz}. \tag{6.48}$$

That pulsar's spin frequency is 642 Hz, so the absolute frequency error Δf implies that the spin rate is known to 15 significant figures!

Many corrections have to be applied to the observed TOAs before $\phi(t)$ can be expressed as a Taylor series (Equation 6.47). The arrival of a pulse at an observatory on Earth at **topocentric** (topocentric means measured from a fixed point on the Earth's surface) time t_t can be corrected to the time t in the nearly inertial Solar System **barycentric frame**, which we assume to be nearly the same as the time in the frame comoving with the pulsar. Note that the measured pulse rates will differ from the actual pulse rates in the pulsar frame by the Doppler factor resulting from the unknown line-of-sight pulsar velocity as well as unknown relativistic corrections from within the pulsar system itself.

The principal terms in the timing equation are

$$t = t_t - t_0 + \Delta_{\text{clock}} - \Delta_{\text{DM}} + \Delta_{\text{R}\odot} + \Delta_{\text{E}\odot} + \Delta_{\text{S}\odot} + \Delta_{\text{R}} + \Delta_{\text{E}} + \Delta_{\text{S}}. \tag{6.49}$$

As before, t_0 is a reference epoch, Δ_{clock} represents a clock correction that accounts for differences between the observatory clocks and terrestrial time standards, and Δ_{DM} is the frequency-dependent dispersion delay caused by the ISM. The other Δ terms are delays from within the Solar System and, if the pulsar is in a binary, from within its orbit. The **Roemer delay** $\Delta_{\text{R}\odot}$ is the classical light travel time across the Earth's orbit. Its magnitude is $\sim 500 \cos\beta$ s, where β is the **ecliptic latitude** of the pulsar (the angle between the pulsar and the **ecliptic plane** containing the Earth's orbit around the Sun), and Δ_{R} is the corresponding delay across the orbit of a pulsar in a binary or multiple system. The **Einstein delay** Δ_{E} accounts for the time dilation from the moving pulsar (and observatory) and the gravitational redshift caused by the Sun and planets or the pulsar and any companion stars. The **Shapiro delay** Δ_{S} is the extra time required by the pulses to travel through the curved space–time containing the Sun, planets, and pulsar companions. Errors in any of these parameters, as well as other parameters such as f, $\dot{f}$, and proper motion, give very specific systematic signatures in plots of **timing residuals** (see Figure 6.7), which are simply the differences between the observed TOAs and the predicted TOAs based on the current timing model parameters.

A good example showing how pulsar timing can be extremely useful is timing-based pulsar astrometry. Pulsar positions on the sky are determined by timing a pulsar over the course of a year as the Earth orbits the Sun and tracking the changing Roemer delay.

The Roemer delay Δ_{R} of a pulsar at ecliptic coordinates λ (ecliptic longitude) and β (ecliptic latitude) is

$$\Delta_{\text{R}} \simeq 500 \,\text{s} \, \cos(\beta) \cos(\theta(t) + \lambda), \tag{6.50}$$

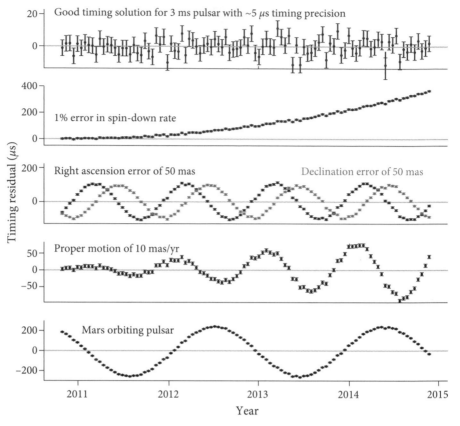

Figure 6.7. Pulsar timing residuals. The top panel shows a "good" timing solution for a fairly average millisecond pulsar with an rms timing precision of about 5 μs over 4 years. The four remaining panels show how the timing residuals are affected by various timing parameter errors. From top to bottom, an error of 1% in the spin-down rate of the pulsar (causing a quadratic drift in pulsar phase), a position error in either right ascension or declination of only 50 mas (an annual sinusoid reflecting the Earth's motion), a pulsar proper motion of 10 mas/yr (an annual sinusoid growing linearly with time), or the presence of a planet with the mass and orbital period of Mars around the pulsar.

where $\theta(t)$ is the orbital phase of the Earth with respect to the vernal equinox (the intersection of the ecliptic plane and the celestial equator). Equation 6.50 is only approximate because the Earth's orbit is not quite circular.

If there is an error in our position estimate, the individual position error components $\Delta\lambda$ and $\Delta\beta$ cause a differential time delay $\Delta\tau$ to be present in the timing residuals with respect to the correct Roemer delay:

$$\Delta\tau \simeq 500\,\text{s}\,[\cos(\beta + \Delta\beta)\cos(\theta(t) + \lambda + \Delta\lambda) - \cos\beta\cos(\theta(t) + \lambda)]. \quad (6.51)$$

If the position errors are small enough that $\sin x \sim x$, $\cos x \sim 1$, and $\Delta\beta\,\Delta\lambda \sim 0$, we can use trigonometric angle-sum identities and then simplify to get

$$\Delta\tau \simeq -500\,\text{s}\,[\Delta\lambda\cos\beta\sin(\theta(t) + \lambda) + \Delta\beta\sin\beta\cos(\theta(t) + \lambda)]. \quad (6.52)$$

Applying the trig identity $\sin(\theta(t) + \phi) = \cos\phi \sin\theta(t) + \sin\phi \cos\theta(t)$ to the equation for $\Delta\tau$, we see that

$$A\cos\phi = -500\,\mathrm{s}\,\Delta\lambda\cos\beta, \tag{6.53}$$

$$A\sin\phi = -500\,\mathrm{s}\,\Delta\beta\sin\beta, \tag{6.54}$$

and therefore

$$\Delta\lambda = -\frac{A\cos\phi}{500\,\mathrm{s}\,\cos\beta}, \tag{6.55}$$

$$\Delta\beta = -\frac{A\sin\phi}{500\,\mathrm{s}\,\sin\beta}, \tag{6.56}$$

where A and ϕ are the amplitude (in time units because it corresponds to light-travel time delay) and phase of an error sinusoid that we would observe in the timing residuals from the pulsar (see the middle panel in Figure 6.7).

If the pulsar is located near the ecliptic plane ($\beta \sim 0$), then $\cos\beta \sim 1$ and there is maximum timing leverage, and therefore minimum error, in the determination of λ. However, $\sin\beta \sim 0$ and so the errors on β are huge. A similar problem occurs near the ecliptic pole for λ. VLBI positions give better astrometric accuracy for pulsars near the ecliptic plane or ecliptic pole.

In a timing fit for position, the amplitude A of the error sinusoid in the timing residuals as described above will be determined to an absolute precision ΔA approximately equal to the TOA uncertainty divided by the square-root of the number of observations made of the pulsar (so long as there is at least one year of timing data).

For **binary pulsars**, the time delays across the neutron-star orbit allow for similar high-precision measurements of the pulsar orbital parameters. The binary pulsar Roemer delays comprise up to five **Keplerian parameters** describing elliptical orbits: the projected semimajor axis $x \equiv (a_1 \sin i)/c$, the longitude of periastron ω, the time of periastron passage T_0, the orbital period P_b, and the orbital eccentricity e.

Example. How precisely can we measure the position of a millisecond pulsar at an ecliptic latitude of $\beta \sim 45°$, with typical TOA errors of $2\,\mu$s, and where there have been 16 observations over a year?

With $2\,\mu$s TOAs and 16 independent observations, we should be able to constrain the amplitude of the sinusoidal position error amplitude A to be within about $\Delta A \sim 2\,\mu\mathrm{s}/\sqrt{16} = 5 \times 10^{-7}$ s.

Because both $\cos\beta = \sin\beta \sim 0.71$, our errors on both λ and β should be similar:

$$\Delta\lambda \sim \Delta\beta \sim \frac{\Delta A}{500\,\mathrm{s}\cos\beta} = \frac{5 \times 10^{-7}\,\mathrm{s}}{500\,\mathrm{s}\cdot 0.71} = 1.4 \times 10^{-9}\,\mathrm{radians}.$$

There are 206,265 arcseconds (as) per radian, so those errors correspond to errors in both directions of only $\sim$290 μas! Even normal pulsars with slow spin periods typically provide astrometric precisions of 0.1 as or better.

Table 1 | Physical parameters for PSR J1614-2230

Parameter	Value
Ecliptic longitude (λ)	245.78827556(5)°
Ecliptic latitude (β)	−1.256744(2)°
Proper motion in λ	9.79(7) mas yr^{-1}
Proper motion in β	−30(3) mas yr^{-1}
Parallax	0.5(6) mas
Pulsar spin period	3.1508076534271(6) ms
Period derivative	9.6216(9) × 10^{-21} s s^{-1}
Reference epoch (MJD)	53,600
Dispersion measure*	34.4865 pc cm^{-3}
Orbital period	8.6866194196(2) d
Projected semimajor axis	11.2911975(2) light s
First Laplace parameter (e sin ω)	1.1(3) × 10^{-7}
Second Laplace parameter (e cos ω)	−1.29(3) × 10^{-6}
Companion mass	0.500(6)$M_\odot$
Sine of inclination angle	0.999894(5)
Epoch of ascending node (MJD)	52,331.1701098(3)
Span of timing data (MJD)	52,469–55,330
Number of TOAs†	2,206 (454, 1,752)
Root mean squared TOA residual	1.1 μs
Right ascension (J2000)	16 h 14 min 36.5051(5) s
Declination (J2000)	−22° 30′ 31.081(7)′′
Orbital eccentricity (e)	1.30(4) × 10^{-6}
Inclination angle	89.17(2)°
Pulsar mass	1.97(4)$M_\odot$
Dispersion-derived distance‡	1.2 kpc
Parallax distance	>0.9 kpc
Surface magnetic field	1.8 × 10^8 G
Characteristic age	5.2 Gyr
Spin-down luminosity	1.2 × 10^{34} erg s^{-1}
Average flux density* at 1.4 GHz	1.2 mJy
Spectral index, 1.1–1.9 GHz	−1.9(1)
Rotation measure	−28.0(3) rad m^{-2}

Figure 6.8. An example millisecond pulsar timing ephemeris for the two-solar-mass neutron star system J1614−2230. Arrival times for this pulsar can be measured at the microsecond level, allowing extremely precise measurements of astrometric, spin, and orbital parameters, as well as a high-precision measurement of the relativistic Shapiro delay, which provided the masses of the neutron star and its white-dwarf companion. The numbers in parentheses for each parameter are the errors in the last digits of those parameters. From Demorest et al. [34].

Relativistic binaries, particularly those with compact and elliptical orbits, may allow the measurement of up to five **post-Keplerian (PK) parameters**: the rate of periastron advance $\dot{\omega}$ (elliptical orbits do not close in relativistic theories), the orbital period decay $\dot{P}_b$ caused by the emission of gravitational radiation, the relativistic γ term describing time dilation and gravitational redshift, and the Shapiro delay terms r (range) and s (shape). An example of a full "timing solution," listing high-precision spin, astrometric, binary, and two post-Keplerian parameters (r, s), is shown in Figure 6.8.

In any theory of gravity, the five PK parameters are functions only of the pulsar mass m_1, the companion mass m_2, and the standard five Keplerian orbital parameters. For general relativity, the formulas are

$$\dot{\omega} = 3 \left(\frac{P_b}{2\pi} \right)^{-5/3} (T_\odot M)^{2/3} (1 - e^2)^{-1}, \tag{6.57}$$

$$\gamma = e \left(\frac{P_b}{2\pi} \right)^{1/3} T_\odot^{2/3} M^{-4/3} m_2 (m_1 + 2m_2), \tag{6.58}$$

$$\dot{P}_b = -\frac{192\pi}{5} \left(\frac{P_b}{2\pi} \right)^{-5/3} \left(1 + \frac{73}{24} e^2 + \frac{37}{96} e^4 \right) (1 - e^2)^{-7/2} T_\odot^{5/3} m_1 m_2 M^{-1/3}, \tag{6.59}$$

$$r = T_\odot m_2, \tag{6.60}$$

$$s = x \left(\frac{P_b}{2\pi} \right)^{-2/3} T_\odot^{-1/3} M^{2/3} m_2^{-1}. \tag{6.61}$$

In these equations, $T_\odot \equiv G M_\odot / c^3 = 4.925490947 \, \mu s$ is the solar mass in time units (which is known much more precisely than either G or $M_\odot$ individually), m_1, m_2, and $M \equiv m_1 + m_2$ are in solar masses, and $s \equiv \sin i$ (where i is the orbital inclination). If any two of these PK parameters are measured, the masses of the pulsar and its companion can be determined. If more than two are measured, each additional PK parameter yields a different test of a gravitational theory.

For the famous case of the Hulse–Taylor binary pulsar B1913+16, high-precision measurements of $\dot{\omega}$ and γ were first made to determine the masses of the two neutron stars accurately. The Nobel-prize-winning measurement came with the eventual detection of $\dot{P}_b$, which implied that the orbit was decaying in accordance with general relativity's predictions for the the emission of gravitational radiation.

More recently, the double-pulsar system J0737−3039 was discovered, which contains two pulsar clocks in a more compact orbit (2.4 hrs compared to 7.7 hrs for PSR B1913+16), allowing the measurement of all five PK parameters as well as the pulsar mass ratio R and relativistic spin precession, giving a total of five tests of general relativity. Kramer et al. [61] showed that general relativity is correct at the 0.05% level and measured the masses of the two neutron stars to within 1 part in 10^4, and the measurements continue to improve (see Figure 6.9).

The double neutron-star systems have indirectly confirmed the existence of gravitational waves by matching the predicted orbital decays as orbital energy is carried away by gravitational radiation. But over the last decade, one of the driving efforts in pulsar astronomy has become the direct detection of GWs using **pulsar timing arrays (PTAs)**. A PTA is an array of MSPs spread over the sky rather than an array of telescopes, and the goal is to use that pulsar array to detect correlated signals in the timing residuals of dozens of MSPs caused by the distortions of interstellar space from nanohertz (i.e., periods of years) GWs passing through our Galaxy.

Because GW emission and propagation is a quadrupolar process in general relativity, in contrast to the dipolar emission of electromagnetic waves from an accelerating electron, GWs cause specific angular correlations in the timing residuals between pairs of pulsars on the sky. Pulsars close together on the sky will be similarly affected by a passing GW, whereas those much farther apart on the sky will be uncorrelated or even negatively correlated by the same GW. That angular pattern is known as the **Hellings and Downs curve** [49], and detecting it would be the key to confirming that correlated signals in timing residuals are caused by GWs and not by other effects such as clock or planetary ephemeris errors.

The most likely sources of detectable nanohertz GWs are supermassive black hole binaries (with total masses of 10^8–10^{10} $M_\odot$) and years-long orbital periods after

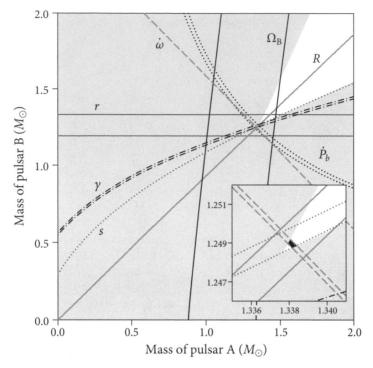

Figure 6.9. A recent PSR J0737−3039 mass versus mass diagram showing lines corresponding to the post-Keplerian parameters measured for the system. In this case seven(!) parameters have been measured, including the mass ratio R, since both neutron stars are (or were) pulsar clocks. The dark region in the inset plot is the only portion of parameter space consistent with general relativity, effectively testing it at the 0.05% level [19]. The parameter Ω_B is a measurement of the relativistic spin precession of the "B" pulsar in the system from detailed modeling of the eclipses of the "A" pulsar by the magnetosphere of "B" [17].

their parent galaxies have merged. A single massive nearby system, or an ensemble of more distant, less massive systems (making a "stochastic background" of GWs), will cause tens-of-nanosecond spatially and temporally correlated systematics in PTA timing residuals. This level of long-term timing precision for the best MSPs is now being achieved.

Three PTA experiments have been working on this endeavor: NANOGrav in North America and the Parkes and European PTAs in Australia and Europe, respectively. Together, they are collaborating in the International Pulsar Timing Array (IPTA), and huge progress toward a detection has been made recently. Current PTA limits, based on upper limits for appropriately correlated low-frequency timing residuals, are beginning to constrain models of galaxy mergers throughout the universe. And given continued improvements in pulsar timing capability and many new high-precision MSPs from recent surveys, a direct detection of GWs seems possible or even likely within the next five years.

7 Spectral Lines

7.1 INTRODUCTION

Spectral lines are narrow ($\Delta \nu \ll \nu$) emission or absorption features in the spectra of gaseous and ionized sources. Examples of radio spectral lines include recombination lines of ionized hydrogen and heavier atoms, rotational lines of polar molecules such as carbon monoxide (CO), and the $\lambda = 21$ cm hyperfine line of interstellar HI.

Spectral-line emission and absorption are intrinsically quantum phenomena. Classical *particles* and *waves* are idealized concepts like infinitesimal points or perfectly straight lines in geometry; they don't exist in the real world. Some things are nearly waves (e.g., radio waves) and others are nearly particles (e.g., electrons), but all share characteristics of both particles and waves. Unlike idealized waves, real radio waves do not have a continuum of possible energies. Instead, electromagnetic radiation is quantized into photons whose energy is proportional to frequency: $E = h\nu$. Unlike idealized particles, real particles of momentum p are associated with waves whose **De Broglie wavelength** is $\lambda = h/p$. An electron's stable orbit about the nucleus of an atom shares a property with standing waves: its circumference must equal an integer number of wavelengths. Planck's constant $h \approx 6.62607 \times 10^{-27}$ erg s in these two equations is a quantum of **action** whose dimensions are (mass $\times$ length$^2 \times$ time^{-1}), the same as (energy $\times$ time) or (angular momentum) or (length $\times$ momentum). Although h has dimensions of energy $\times$ time, physically acceptable solutions (the wave functions and their derivatives must be finite and continuous) to the time-independent Schrödinger equation exist only for discrete values of the total energy, so spectral lines have definite frequencies resulting from transitions between discrete energy states. A second quantum effect important to spectral lines, particularly at radio wavelengths where $h\nu \ll kT$, is *stimulated emission* (Section 7.3.1). Fortunately, the fundamental characteristics of radio spectral lines from interstellar atoms and molecules can be derived from fairly simple applications of quantum mechanics and thermodynamics.

Spectral lines are powerful diagnostics of physical and chemical conditions in astronomical objects. Their rest frequencies identify the specific atoms and molecules involved, and their Doppler shifts measure radial velocities. These velocities yield the redshifts and Hubble distances of extragalactic sources, plus rotation curves and radial mass distributions for resolved galaxies. Collapse speeds, turbulent velocities, and thermal motions contribute to line broadening in Galactic sources.

Temperatures, densities, and chemical compositions of HII regions, dust-obscured dense molecular clouds, and diffuse interstellar gas are also constrained by spectral-line data. Radio spectral lines have some unique characteristics:

1. Their "natural" line widths are much smaller than Doppler-broadened line widths, so gas temperatures and very small changes in radial velocity can be measured.
2. Stimulated emission is important because $h\nu \ll kT$. This causes line opacities to vary as T^{-1} and favors the formation of natural masers.
3. The ability of radio waves to penetrate dust in our Galaxy and in other galaxies allows the detection of line emission emerging from dusty molecular clouds, protostars, and molecular disks orbiting AGNs.
4. Frequency (inverse time) can be measured with much higher precision than wavelength (length), so very sensitive searches for small changes in the fundamental physical constants over cosmic timescales are possible.

Although the interstellar medium (ISM) of our Galaxy is dynamic, it tends toward a rough pressure equilibrium because mass motions with speeds up to the speed of sound try to reduce pressure gradients. Temperatures equilibrate more slowly, so there are wide ranges of temperature T and particle number density n consistent with a given pressure p and the **ideal gas law**

$$\boxed{p = nkT.} \tag{7.1}$$

Typical ISM pressures lie in the range $p/k = nT \sim 10^3$–10^4 cm^{-3} K [57]. Radiative cooling by spectral-line emission depends strongly on temperature, so most of the ISM exists in several distinct *phases* having comparable pressures but quite different temperatures:

1. cold (10s of K) dense molecular clouds
2. cool ($\sim 10^2$ K) neutral HI gas
3. warm ($\sim 5 \times 10^3$ K) neutral HI gas
4. warm ($\sim 10^4$ K) ionized HII gas
5. hot ($\sim 10^6$ K) low-density ionized gas (in *bubbles* formed by expanding supernova remnants, for example)

All but the hottest phase are sources of radio spectral lines.

7.2 RECOMBINATION LINES

7.2.1 Recombination Line Frequencies

The semiclassical **Bohr atom** (Figure 7.1) contains a nucleus of protons and neutrons around which one or more electrons move in circular orbits. The nuclear mass M is always much greater than the sum of the electron masses m_e, so the nucleus is nearly at rest in the center-of-mass frame. The wave functions of the electrons have

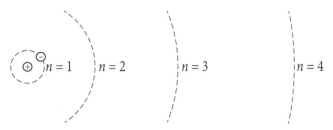

Figure 7.1. The radius of the nth Bohr orbit is proportional to n^2, so radio-emitting hydrogen atoms with $n \sim 100$ are $\sim 10^4$ times larger than ordinary hydrogen atoms in the $n = 1$ ground state.

De Broglie wavelengths

$$\lambda = \frac{h}{p} = \frac{h}{m_e v}, \tag{7.2}$$

where p is the electron's momentum and v is its speed. Only those orbits whose circumferences equal an integer number n of wavelengths correspond to standing waves and are permitted. Thus the **Bohr radius** a_n of the nth permitted electron orbit satisfies the quantization rule

$$2\pi a_n = n\lambda = \frac{nh}{m_e v}, \tag{7.3}$$

where the number n is called the **principal quantum number**. The requirement that

$$a_n = \frac{nh}{2\pi m_e v} = \frac{n\hbar}{m_e v} \tag{7.4}$$

implies that the orbital angular momentum $L = m_e v a_n = n\hbar$ is an integer multiple of the **reduced Planck's constant** $\hbar \equiv h/(2\pi)$. The relation between a_n and v is determined by the balance of Coulomb and centrifugal forces on electrons in circular orbits. For a hydrogen atom,

$$\frac{e^2}{a_n^2} = \frac{m_e v^2}{a_n}. \tag{7.5}$$

Equations 7.4 and 7.5 can be combined to eliminate v and solve for a_n in terms of n and physical constants:

$$\boxed{a_n = \frac{n^2 \hbar^2}{m_e e^2}.} \tag{7.6}$$

Numerically, the Bohr radius of a hydrogen atom whose electron is in the nth electronic energy level is

$$a_n = \frac{\hbar^2}{m_e e^2} n^2 = \frac{[6.63/(2\pi) \times 10^{-27} \text{ erg s}]^2}{9.11 \times 10^{-28} \text{ g} \cdot (4.8 \times 10^{-10} \text{ statcoul})^2} n^2$$

$$\approx 0.53 \times 10^{-8} \text{ cm} \cdot n^2.$$

The Bohr radius of a hydrogen atom in its **ground electronic state** ($n = 1$) is only $a_1 \approx 0.53 \times 10^{-8}$ cm, but the diameter of a highly excited ($n \approx 100$) radio-emitting hydrogen atom in the ISM can be remarkably large: $2a_{100} \approx 10^{-4}$ cm $= 1 \, \mu$m, which is bigger than most viruses!

The electron in a Bohr atom can fall from the level ($n + \Delta n$) to n, where Δn and n are any natural numbers (1, 2, 3, ...), by emitting a photon whose energy equals the energy difference ΔE between the initial and final levels. Such spectral lines are called **recombination lines** because formerly free electrons recombining with ions quickly cascade to the ground state by emitting such photons. Astronomers label each recombination line using the name of the element, the final level number n, and successive letters in the Greek alphabet to denote the level change Δn: α for $\Delta n = 1$, β for $\Delta n = 2$, γ for $\Delta n = 3$, etc. For example, the recombination line produced by the transition between the $n = 92$ and $n = 91$ levels of a hydrogen atom is called the H91α line.

The total electronic energy E_n is the sum of the kinetic (T) and potential (V) energies of the electron in the nth circular orbit:

$$E_n = T + V = -T = V/2 = -\frac{e^2}{2a_n} = -e^2 \left(\frac{m_e e^2}{2n^2 \hbar^2} \right) = -\left(\frac{m_e e^4}{2\hbar^2} \right) \frac{1}{n^2}. \quad (7.7)$$

The electronic energy change ΔE going from level ($n + \Delta n$) to level n is equal to the energy $h\nu$ of the emitted photon:

$$\Delta E = \frac{m_e e^4}{2\hbar^2} \left[\frac{1}{n^2} - \frac{1}{(n + \Delta n)^2} \right] = h\nu, \quad (7.8)$$

so the photon frequency is

$$\nu = \left(\frac{2\pi^2 m_e e^4}{h^3 c} \right) c \left[\frac{1}{n^2} - \frac{1}{(n + \Delta n)^2} \right]. \quad (7.9)$$

The factor in large parentheses is called the **Rydberg constant** R_∞, where the subscript refers to the limit of infinite nuclear mass M:

$$R_\infty \equiv \left(\frac{2\pi^2 m_e e^4}{h^3 c} \right) = 1.09737312 \ldots \times 10^5 \text{ cm}^{-1}. \quad (7.10)$$

The dimensions of R_∞ are length^{-1}, and the product $R_\infty c$ is the **Rydberg frequency**:

$$R_\infty c = 3.28984 \ldots \times 10^{15} \text{ Hz}. \quad (7.11)$$

Allowing for the relatively large but finite nuclear mass $M \gg m_e$ and repeating the analysis above in the atomic center-of-mass frame yields the same frequency formula with R_∞ replaced by R_M:

$$\nu = R_M c \left[\frac{1}{n^2} - \frac{1}{(n + \Delta n)^2} \right], \qquad \text{where} \qquad R_M \equiv R_\infty \left(1 + \frac{m_e}{M} \right)^{-1}, \quad (7.12)$$

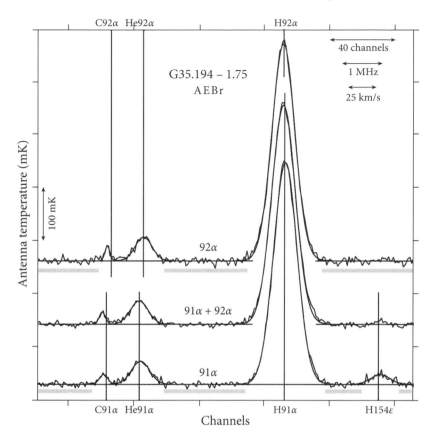

Figure 7.2. Observed recombination-line spectra from the 91α and 92α transitions of hydrogen, helium, and carbon observed in an HII region [84].

The hydrogen nucleus is a single proton of mass $m_p \approx 1836.1 m_e$ so $M(\text{H}) \approx 1836.1 m_e$ and $R_M c$ for a hydrogen atom is

$$R_M c = 3.28984 \times 10^{15}\,\text{Hz}\left(1 + \frac{1}{1836.1}\right)^{-1} = 3.28805 \times 10^{15}\,\text{Hz}.$$

Thus the frequency of the photon produced by the H109α transition ($n + \Delta n = 110$ to $n = 109$) is

$$\nu(\text{H}109\alpha) = 3.28805 \times 10^{15}\,\text{Hz}\left(\frac{1}{109^2} - \frac{1}{110^2}\right) \approx 5.0089 \times 10^9\,\text{Hz}.$$

The mass of a neutron is about equal to the mass of a proton, so the ^{4}He nucleus consisting of two protons and two neutrons has mass $M(^4\text{He}) \approx 4M(\text{H})$, the isotope of carbon with six protons and six neutrons has $M(^{12}\text{C}) \approx 12M(\text{H})$, and so on. Electrons recombining onto singly ionized atoms with any number N_p of protons and ($N_p - 1$) electrons orbit in the potential produced by a net charge of one proton, so the recombination lines of heavier atoms are very similar to those of hydrogen, but at the slightly higher frequencies (Figure 7.2) given by Equation 7.12. For example,

Figure 7.3. The $\Delta n = 1$ radio recombination lines of singly ionized atoms, shown here as vertical bars, are closely spaced in frequency.

the primordial abundance of the rare helium isotope ^{3}He is important because it depends on the photon/baryon ratio in the early universe. The abundance of ^{3}He in Galactic HII regions has been measured via radio recombination-line emission and indicates that baryons account for only a few percent of the critical density needed to close the universe.

The strongest radio recombination lines are produced by transitions with $\Delta n \ll n$, so the approximation

$$\left[\frac{1}{n^2} - \frac{1}{(n+\Delta n)^2} \right] \approx \frac{(n+\Delta n)^2 - n^2}{n^2(n+\Delta n)^2} = \frac{n^2 + 2n\Delta n + (\Delta n)^2 - n^2}{n^2[n^2 + 2n\Delta n + (\Delta n)^2]} \approx \frac{2n\Delta n}{n^4}$$

$$= \frac{2\Delta n}{n^3} \qquad (7.13)$$

yields simpler (but not extremely accurate) approximations for radio recombination line frequencies

$$\nu \approx \frac{2(R_M c)\Delta n}{n^3}, \qquad (7.14)$$

and for the frequency separation $\Delta \nu = \nu(n) - \nu(n+1)$ between adjacent lines

$$\boxed{\frac{\Delta \nu}{\nu} \approx \frac{3}{n}.} \qquad (7.15)$$

Adjacent high-n (low-ν) radio recombination lines have such small fractional frequency separations (Figure 7.3) that two or more transitions can often be observed simultaneously and averaged, to reduce the observing time needed to reach a given signal-to-noise ratio.

The H109α line was first detected by P. Mezger in 1965, despite (incorrect) theoretical predictions that pressure broadening would smear out the lines in frequency and make them undetectable. It is true that atomic collisions in the interstellar medium significantly disturb the energy levels of large atoms, but this disturbance is about the same for adjacent energy levels, so the *differential* disturbance that alters the line frequency is actually much smaller. His advice: "Don't abandon an observation just because you have been told that it will fail."

7.2.2 Recombination Line Strengths

The **spontaneous emission rate** is the average rate at which an isolated atom emits photons. Rigorous quantum-mechanical calculations of spontaneous emission rates are complicated, but a fairly good classical approximation can be derived by noting that radio photons are emitted by atoms with $n \gg 1$ and invoking the **correspondence principle**, Bohr's hypothesis that systems with large quantum numbers behave almost classically. The time-averaged radiated power $\langle P \rangle$ for classical transitions is given by Larmor's formula (Equation 2.143) for an electric dipole with dipole moment ea_n:

$$\langle P \rangle = \frac{2e^2}{3c^3}(\omega^2 a_n)^2 \langle \cos^2(\omega t)\rangle = \frac{2e^2}{3c^3}(2\pi\nu)^4 \frac{a_n^2}{2}. \tag{7.16}$$

The photon emission rate (s^{-1}) equals the average power emitted by one atom divided by the energy of each photon. The spontaneous emission rate for transitions from level n to level $(n-1)$ is denoted by $A_{n,n-1}$:

$$A_{n,n-1} = \frac{\langle P \rangle}{h\nu}, \tag{7.17}$$

where

$$\nu \approx \frac{2R_\infty c \Delta n}{n^3} \tag{7.18}$$

in the limit $\Delta n \ll n$ (Equation 7.14). In that limit, $A_{n+1,n} \approx A_{n,n-1}$ also.

The atomic radius (Equation 7.6) is

$$a_n \approx \frac{n^2 h^2}{4\pi^2 m_e e^2}, \tag{7.19}$$

so

$$A_{n+1,n} \approx \frac{\langle P \rangle}{h\nu} \approx \frac{2e^2}{3c^3}\left(\frac{16\pi^4 \nu^3}{h}\right)\frac{a_n^2}{2} \tag{7.20}$$

$$\approx \frac{16\pi^4}{3}\frac{e^2}{c^3 h}\left(\frac{2R_\infty c}{n^3}\right)^3\left(\frac{n^2 h^2}{4\pi^2 m_e e^2}\right)^2 \tag{7.21}$$

$$\approx \frac{16\pi^4}{3}\frac{e^2}{c^3 h}\left(\frac{4\pi^2 m_e e^4}{h^3}\right)^3\left(\frac{h^2}{4\pi^2 m_e e^2}\right)^2 \frac{1}{n^5}, \tag{7.22}$$

$$\boxed{A_{n+1,n} \approx \left(\frac{64\pi^6 m_e e^{10}}{3c^3 h^6}\right)\frac{1}{n^5}.} \tag{7.23}$$

Evaluating the constants yields the spontaneous emission rate of hydrogen atoms:

$$A_{n+1,n} \approx \left[\frac{64\pi^6 \cdot 9.11 \times 10^{-28}\,\text{g} \cdot (4.8 \times 10^{-10}\,\text{statcoul})^{10}}{3 \cdot (3 \times 10^{10}\,\text{cm s}^{-1})^3 \cdot (6.63 \times 10^{-27}\,\text{erg s})^6}\right]\frac{1}{n^5}, \tag{7.24}$$

$$A_{n+1,n} \approx 5.3 \times 10^9 \left(\frac{1}{n^5}\right) \text{s}^{-1}. \qquad (7.25)$$

For example, the 5.0089 GHz H109α transition rate is $A_{110,109} \approx 0.3\,\text{s}^{-1}$.

The associated **natural line width** or **intrinsic line width** follows from the uncertainty principle: $\Delta E \, \Delta t \approx \hbar$. Substituting $h\Delta\nu$ for ΔE and $A_{n+1,n}^{-1}$ for Δt for each energy level involved in the transition and summing these two uncertainties yields

$$\Delta\nu \sim A_{n+1,n}/\pi \sim 0.1\,\text{Hz}. \qquad (7.26)$$

Natural broadening is negligibly small at the large n that produce radio-frequency photons. Collisions of the emitting atoms cause **collisional broadening**, where the amount of collisional broadening is a small fraction of the collision rate when $\Delta n \ll n$. Except for very large n, collisional broadening is also small and the actual **line profile** (normalized intensity as a function of frequency) is primarily determined by Doppler shifts reflecting the **radial velocities** v_r of the emitting atoms. The sign convention is $v_r > 0$ for sources moving away from the observer. Radial velocities may be microscopic (from the thermal motions of individual atoms) or macroscopic (from large-scale turbulence, flows, or rotation). In the nonrelativistic limit $v_r \ll c$, the Doppler equation (Equation 5.142) relating the observed frequency ν to the line rest frequency ν_0 reduces to

$$\nu \approx \nu_0 \left(1 - \frac{v_r}{c}\right), \qquad (7.27)$$

so nonrelativistic radial velocities can be estimated from

$$v_r \approx \frac{c\,(\nu_0 - \nu)}{\nu_0}. \qquad (7.28)$$

The thermal component of the line profile from a recombination-line source in LTE is determined by the Maxwellian speed distribution (Equation B.49) of atoms with mass M and temperature T. The speed in any one coordinate of an isotropic distribution is $3^{-1/2}$ of the total speed in three dimensions, so the Gaussian

$$f(v_r) = \left(\frac{M}{2\pi kT}\right)^{1/2} \exp\left(-\frac{Mv_r^2}{2kT}\right) \qquad (7.29)$$

is the normalized ($\int f(v_r)\,dv_r = 1$) radial velocity distribution. The normalized line profile (Figure 7.4) $\phi(\nu)$ for thermal emission is

$$|\phi(\nu)d\nu| = f(v_r)dv_r, \qquad (7.30)$$

$$\phi(\nu) = \left(\frac{M}{2\pi kT}\right)^{1/2} \exp\left[-\frac{M}{2kT}\frac{c^2(\nu-\nu_0)^2}{\nu_0^2}\right]\left|\frac{dv_r}{d\nu}\right|, \qquad (7.31)$$

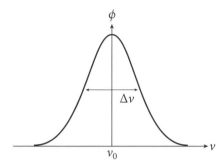

Figure 7.4. The parameters of the normalized ($\int \phi(v)\, dv = 1$) line profile $\phi(v)$ are the center frequency v_0, the FWHM line width Δv, and the profile height $\phi(v_0)$ at the center frequency.

$$\phi(v) = \frac{c}{v_0}\left(\frac{M}{2\pi kT}\right)^{1/2} \exp\left[-\frac{Mc^2}{2kT}\frac{(v - v_0)^2}{v_0^2}\right]. \tag{7.32}$$

This is a Gaussian line profile. Its full width between half-maximum points (FWHM) Δv is the solution of

$$\exp\left[-\frac{Mc^2}{2kT}\frac{(\Delta v/2)^2}{v_0^2}\right] = \frac{1}{2}, \tag{7.33}$$

$$\frac{Mc^2}{2kT}\frac{\Delta v^2}{4v_0^2} = \ln 2, \tag{7.34}$$

$$\Delta v = \left[\frac{8\ln(2)\,k}{c^2}\right]^{1/2}\left(\frac{T}{M}\right)^{1/2} v_0. \tag{7.35}$$

For example, the FWHM of the H109α line ($v_0 = 5.0089$ GHz) in a quiescent (no macroscopic motions) H II region with temperature $T \approx 10^4$ K is

$$\Delta v \approx \left[\frac{8\ln 2 \cdot 1.38 \times 10^{-16}\ \text{erg K}^{-1}}{(3 \times 10^{10}\ \text{cm s}^{-1})^2}\right]^{1/2}\left(\frac{10^4\ \text{K}}{1836 \cdot 9.11 \times 10^{-28}\ \text{g}}\right)^{1/2}$$
$$\times\, 5.0089 \times 10^9\ \text{Hz}$$
$$\approx 3.6 \times 10^5\ \text{Hz}.$$

Notice that the thermal line width Δv is much larger than the natural line width $A_{110,109} \approx 0.3$ Hz.

Normalization (requiring $\int \phi(v)dv = 1$) implies that the value of ϕ at the line center ($v = v_0$) is

$$\phi(v_0) = \frac{c}{v_0}\left(\frac{M}{2\pi kT}\right)^{1/2} = \frac{c}{\Delta v}\left(\frac{8\ln 2\, kT}{Mc^2}\frac{M}{2\pi kT}\right)^{1/2}, \tag{7.36}$$

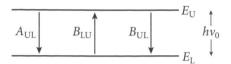

Figure 7.5. The three **Einstein coefficients** for a two-level system: A_{UL} for spontaneous emission, B_{LU} for absorption, and B_{UL} for stimulated emission.

$$\phi(\nu_0) = \left(\frac{\ln 2}{\pi}\right)^{1/2} \frac{2}{\Delta \nu}. \tag{7.37}$$

For a given integrated (over frequency) line strength, the line strength per unit frequency at any one frequency (e.g., at ν_0) is inversely proportional to the line width $\Delta \nu$. Integrated line strengths are frequently specified in the astronomically convenient units of Jy km s^{-1}, where 1 km s$^{-1} \approx \nu_0/3.00 \times 10^5$.

7.3 LINE RADIATIVE TRANSFER

7.3.1 Einstein Coefficients

The **spontaneous emission coefficient** A_{UL} is the average photon emission rate (s^{-1}) for an "undisturbed" atom or molecule transitioning from an upper (U) to a lower (L) energy state. The spectral-line radiative transfer problem also involves the **absorption coefficient** B_{LU} and the **stimulated emission coefficient** B_{UL} (Figure 7.5). Einstein showed that both the absorption and stimulated emission coefficients can be calculated from the spontaneous emission coefficient.

Consider any two energy levels E_U and E_L of a quantum system such as a single atom or molecule. The photon emitted or absorbed during a transition between the upper and lower states will have energy

$$E = E_U - E_L \tag{7.38}$$

and contribute to a spectral line with rest frequency $\nu_0 = E/h$. The energy levels actually have small but finite widths, so the spectral line has some narrow line profile $\phi(\nu)$ centered on $\nu = \nu_0$ and conventionally normalized such that $\int_0^\infty \phi(\nu)\, d\nu = 1$. A system in the lower energy state may absorb a photon of frequency $\nu \approx \nu_0$ and transition to the upper state. The rate (s^{-1}) for this process is proportional to the **profile-weighted mean radiation energy density**

$$\bar{u} \equiv \int_0^\infty u_\nu(\nu)\, \phi(\nu)\, d\nu \tag{7.39}$$

of the surrounding radiation field, so the Einstein absorption coefficient B_{LU} is defined to make the product

$$B_{LU}\bar{u} \tag{7.40}$$

equal the average rate (s^{-1}) at which photons are absorbed by a single atomic or molecular system in its lower energy state.

Einstein realized that there must be a third process in addition to spontaneous emission and absorption. It is **stimulated emission**, in which a photon of energy $E = h\nu_0$ stimulates the system in the upper energy state to emit a second photon with the same energy and direction. The rate for this process is also proportional to $\bar{u}$, so by analogy with Equation 7.40, the Einstein stimulated-emission coefficient B_{UL} is defined to make the product

$$\boxed{B_{UL}\bar{u}} \tag{7.41}$$

equal the average rate (s^{-1}) of stimulated photon emission by a single quantum system in its upper energy state. **Beware** that some authors use $\bar{I}$ instead of $\bar{u}$ in Equation 7.41 to define a stimulated emission coefficient that is $4\pi/c$ times the B_{UL} given by Equation 7.41 [97].

Stimulated emission is sometimes called **negative absorption**. Negative absorption is not familiar in everyday life because it is much weaker in room-temperature objects at visible wavelengths, where $h\nu/(kT) \gg 1$, but negative absorption competes effectively with ordinary absorption at radio wavelengths where $h\nu/(kT) \ll 1$.

Consider a macroscopic collection of many atoms or molecules in full thermo-dynamic equilibrium (TE) with the surrounding radiation field. TE is a **stationary state**. If there are (n_U, n_L) atoms or molecules per unit volume in the (upper, lower) energy states, then the average rate of photon creation by both spontaneous emission and stimulated emission must balance the average rate of photon destruction by absorption:

$$\boxed{n_U A_{UL} + n_U B_{UL}\bar{u} = n_L B_{LU}\bar{u}.} \tag{7.42}$$

In TE, the ratio of n_U to n_L is fixed by the **Boltzmann equation**

$$\frac{n_U}{n_L} = \frac{g_U}{g_L} \exp\left[-\frac{(E_U - E_L)}{kT}\right] = \frac{g_U}{g_L} \exp\left(-\frac{h\nu_0}{kT}\right), \tag{7.43}$$

where g_U and g_L are the numbers of distinct physical states having energies E_U and E_L, respectively. The quantities g_U and g_L are called the **statistical weights** of those energy states. Examples of statistical weights include the following:

1. Hydrogen atoms have $g_n = 2n^2$, where $n = 1,\ 2,\ 3,\ \dots$ is the principal quantum number. The number $2n^2$ is the product of the 2 electron spin states and n^2 orbital angular momentum states in the nth electronic energy level.
2. Rotating linear molecules (e.g., carbon monoxide, CO) have $g = 2J + 1$, where $J = 0,\ 1,\ 2,\ \dots$ is the angular-momentum quantum number. For each J, there are $2J + 1$ possible values of the z-component of the angular momentum: $J_z = -J, -(J-1), \dots, -1, 0, 1, \dots, (J-1), J$.
3. Hydrogen atoms have two hyperfine energy levels whose difference yields the $\lambda = 21$ cm ($\nu_0 = 1420.406\dots$ MHz) HI line: $g_U = 3$ and $g_L = 1$.

Solving Equation 7.42 for the profile-weighted mean energy density $\bar{u}$ of black-body radiation connects the properties of the quantum system (atom or molecule) to the radiation, just as Kirchhoff's law (Equation 2.30) did for continuum radiation:

$$\bar{u} = \frac{n_U A_{UL}}{n_L B_{LU} - n_U B_{UL}} = \frac{A_{UL}}{(n_L/n_U)B_{LU} - B_{UL}}. \tag{7.44}$$

For full TE at temperature T, Equations 7.43 and 7.44 imply

$$\bar{u} = A_{UL}\left[\frac{g_L}{g_U}\exp\left(\frac{h\nu_0}{kT}\right)B_{LU} - B_{UL}\right]^{-1} \tag{7.45}$$

for matter and Equation 2.92 implies

$$\bar{u} = \frac{4\pi}{c}\int_0^\infty B_\nu(T)\,\phi(\nu)d\nu \tag{7.46}$$

for radiation. Inserting the Planck radiation law (Equation 2.86) for $B_\nu(T)$ near $\nu = \nu_0$ gives

$$\bar{u} \approx \frac{4\pi}{c}\frac{2h\nu_0^3}{c^2}\left[\exp\left(\frac{h\nu_0}{kT}\right) - 1\right]^{-1}. \tag{7.47}$$

Equations 7.45 and 7.47 for $\bar{u}$ must agree:

$$A_{UL}\left[\frac{g_L}{g_U}\exp\left(\frac{h\nu_0}{kT}\right)B_{LU} - B_{UL}\right]^{-1} = \frac{4\pi}{c}\frac{2h\nu_0^3}{c^2}\left[\exp\left(\frac{h\nu_0}{kT}\right) - 1\right]^{-1} \tag{7.48}$$

for *all* temperatures T, so the equation

$$\frac{A_{UL}}{B_{UL}}\left[\frac{g_L}{g_U}\frac{B_{LU}}{B_{UL}}\exp\left(\frac{h\nu_0}{kT}\right) - 1\right]^{-1} = \frac{8\pi h\nu_0^3}{c^3}\left[\exp\left(\frac{h\nu_0}{kT}\right) - 1\right]^{-1} \tag{7.49}$$

implies *both*

$$\boxed{\frac{g_L}{g_U}\frac{B_{LU}}{B_{UL}} = 1} \tag{7.50}$$

and

$$\boxed{\frac{A_{UL}}{B_{UL}} = \frac{8\pi h\nu_0^3}{c^3}.} \tag{7.51}$$

Equations 7.50 and 7.51 are called the **equations of detailed balance**. The two equations relate the three quantities A_{UL}, B_{LU}, and B_{UL}, so all three can be computed if only one (e.g., the spontaneous emission coefficient A_{UL}) is known. Equations 7.50 and 7.51 also prove that B_{UL} cannot be zero; that is, stimulated emission *must* occur.

Note that these Equations 7.50 and 7.51 are valid for any *microscopic* physical system because they relate the coefficients A_{UL}, B_{UL}, and B_{LU}, characteristic of individual atoms or molecules for which the *macroscopic* statistical concepts of

TE or LTE are meaningless. Although TE was assumed for their derivation, the dependences on temperature T and frequency ν dropped out for a line at a single frequency ν_0. Thus Equations 7.50 and 7.51 also apply to *all* macroscopic systems, whether or not they are in TE or even LTE. (Recall the derivation of Kirchhoff's law $j_\nu(T)/\kappa(T) = B_\nu(T)$ (Equation 2.30), which also made use of full TE but also relates the emission and absorption coefficients of any matter in LTE at temperature T, independent of the actual ambient radiation field.)

7.3.2 Radiative Transfer and Detailed Balance

The two Equations 7.50 and 7.51 for detailed balance relate the three Einstein coefficients and allow the spectral-line radiative transfer problem to be solved in terms of the spontaneous emission coefficient A_{UL} alone. The radiative transfer equation (2.27) is

$$\frac{dI_\nu}{ds} = -\kappa I_\nu + j_\nu, \qquad (7.52)$$

where I_ν is the specific intensity, κ is the *net* fraction of photons absorbed (the difference between ordinary absorption and negative absorption) per unit length, and j_ν is the volume emission coefficient.

The ordinary opacity coefficient is the fraction of spectral brightness removed per unit length by absorption from the lower level to the upper level. At frequencies near the line center frequency ν_0, the photon energy is $h\nu_0$, the number of absorbers per unit volume is n_{L}, the number of absorptions per unit area per unit time is $n_{\mathrm{L}}B_{\mathrm{LU}}$, the fraction of photons absorbed per unit area per unit length is $n_{\mathrm{L}}B_{\mathrm{LU}}/c$, and the photon energy loss per unit length at frequency ν is

$$\frac{dI_\nu}{ds} = -\kappa I_\nu = -\left(\frac{h\nu_0}{c}\right)n_{\mathrm{L}}B_{\mathrm{LU}}\phi(\nu)I_\nu. \qquad (7.53)$$

Stimulated emission is best treated as negative absorption because, like ordinary absorption and unlike spontaneous emission, its strength is proportional to I_ν. The derivation of Equation 7.53 can be repeated to give

$$\frac{dI_\nu}{ds} = -\kappa I_\nu = \left(\frac{h\nu_0}{c}\right)n_{\mathrm{U}}B_{\mathrm{LU}}\phi(\nu)I_\nu. \qquad (7.54)$$

Adding Equations 7.53 and 7.54 yields the **net absorption coefficient**

$$\kappa = \left(\frac{h\nu_0}{c}\right)(n_{\mathrm{L}}B_{\mathrm{LU}} - n_{\mathrm{U}}B_{\mathrm{UL}})\phi(\nu). \qquad (7.55)$$

The spontaneous emission coefficient is the spectral brightness (power per unit frequency per steradian) added per unit volume by spontaneous transitions from the upper to lower energy levels. As above, the line photon energy is approximately $h\nu_0$, the number *density* in the upper energy level is n_{U}, and the photon emission rate per unit volume is $n_{\mathrm{U}}A_{\mathrm{UL}}$. These photons are emitted isotropically over 4π sr, so

$$\frac{dI_\nu}{ds} = j_\nu = \left(\frac{h\nu_0}{4\pi}\right)n_{\mathrm{U}}A_{\mathrm{UL}}\phi(\nu). \qquad (7.56)$$

Inserting the net absorption coefficient (Equation 7.55) and the spontaneous emission coefficient (Equation 7.56) into Equation 7.52 specifies the full spectral-line equation of radiative transfer:

$$\frac{dI_\nu}{ds} = -\left(\frac{h\nu_0}{c}\right)(n_L B_{LU} - n_U B_{UL})\phi(\nu)I_\nu + \left(\frac{h\nu_0}{4\pi}\right)n_U A_{UL}\phi(\nu). \qquad (7.57)$$

Equation 7.50 can be used to eliminate the stimulated emission coefficient B_{UL} in Equation 7.55 and yield

$$\kappa = \left(\frac{h\nu_0}{c}\right)n_L B_{LU}\left(1 - \frac{n_U}{n_L}\frac{g_L}{g_U}\right)\phi(\nu). \qquad (7.58)$$

The ratio of the emission coefficient to the net absorption coefficient is

$$\frac{j_\nu}{\kappa} = \frac{c n_U A_{UL}}{4\pi n_L B_{LU}}\left(1 - \frac{n_U}{n_L}\frac{g_L}{g_U}\right)^{-1}. \qquad (7.59)$$

Equation 7.50 can be used to eliminate A_{UL} from this ratio also:

$$\frac{j_\nu}{\kappa} = \frac{n_U(8\pi h\nu_0^3/c^2)B_{UL}}{4\pi n_L B_{LU}}\left(1 - \frac{n_U}{n_L}\frac{g_L}{g_U}\right)^{-1} = \frac{2h\nu_0^3}{c^2}\frac{B_{UL}}{B_{LU}}\left(\frac{n_L}{n_U} - \frac{g_L}{g_U}\right)^{-1}. \qquad (7.60)$$

Finally, Equation 7.51 can be used to eliminate both B_{UL} and B_{LU}:

$$\frac{j_\nu}{\kappa} = \frac{2h\nu_0^3}{c^2}\left(\frac{g_U}{g_L}\frac{n_L}{n_U} - 1\right)^{-1}. \qquad (7.61)$$

In LTE, Kirchhoff's law independently implies

$$\frac{j_\nu}{\kappa} = B_\nu(T) = \frac{2h\nu^3}{c^2}\left[\exp\left(\frac{h\nu}{kT}\right) - 1\right]^{-1}, \qquad (7.62)$$

so

$$\frac{g_U}{g_L}\frac{n_L}{n_U} = \exp\left(\frac{h\nu_0}{kT}\right), \qquad (7.63)$$

recovering the Boltzmann equation (Equation 7.43) for LTE (not just for full TE):

$$\frac{n_U}{n_L} = \frac{g_U}{g_L}\exp\left(-\frac{h\nu_0}{kT}\right). \qquad (7.64)$$

Equation 7.58 and the assumption of LTE allows the substitution of

$$B_{LU} = \frac{g_U}{g_L}B_{UL} = \frac{g_U}{g_L}\frac{A_{UL}c^3}{8\pi h\nu_0^3} \qquad (7.65)$$

and

$$\frac{n_U}{n_L}g_L g_U = \exp\left(-\frac{h\nu_0}{kT}\right) \qquad (7.66)$$

to yield the net **line opacity coefficient** in LTE:

$$\kappa(\nu) = \frac{c^2}{8\pi \nu_0^2} \frac{g_U}{g_L} n_L A_{UL} \left[1 - \exp\left(-\frac{h\nu_0}{kT}\right) \right] \phi(\nu) \qquad (7.67)$$

in terms of the spontaneous emission rate A_{UL} only; the stimulated emission coefficient B_{UL} and absorption coefficient B_{LU} have been eliminated.

The quantity

$$\left[1 - \exp\left(-\frac{h\nu_0}{kT}\right) \right] \qquad (7.68)$$

in Equation 7.67 is the sum of two terms, the first representing ordinary absorption and the second accounting for the negative absorption of stimulated emission. In the Rayleigh–Jeans limit $h\nu_0 \ll kT$,

$$\left[1 - \exp\left(-\frac{h\nu_0}{kT}\right) \right] \approx \frac{h\nu_0}{kT} \ll 1. \qquad (7.69)$$

Thus stimulated emission nearly cancels pure absorption and significantly reduces the net line opacity at radio frequencies. Because $\kappa \propto T^{-1}$ and $B_\nu \propto T$, the product κB_ν is independent of temperature. *The brightness of an optically thin ($\tau \ll 1$) radio emission line is proportional to the column density of emitting gas but can be nearly independent of the gas temperature.* Thus the HI line flux (Jy km s^{-1}) of an optically thin galaxy is proportional to the total mass of neutral hydrogen in the galaxy but says nothing about its temperature.

7.4 EXCITATION TEMPERATURE

Even if a macroscopic two-level system is not in LTE, its **excitation temperature** T_x can be *defined* by

$$\frac{n_U}{n_L} \equiv \frac{g_U}{g_L} \exp\left(-\frac{h\nu_0}{kT_x}\right). \qquad (7.70)$$

The excitation temperature is not a real temperature; it only measures the ratio of n_U to n_L. In a two-level system, the excitation temperature is determined by a balance between radiative and collisional excitations and de-excitations. If collisions cause $n_L C_{LU}$ excitations per unit volume per unit time from the lower level to the upper level and $n_U C_{UL}$ de-excitations per unit volume per unit time from the upper level to the lower level, then Equation 7.42 becomes

$$n_U(A_{UL} + B_{UL}\bar{u} + C_{UL}) = n_L(B_{LU}\bar{u} + C_{LU}) \qquad (7.71)$$

and detailed balance requires

$$n_L C_{LU} = n_U C_{UL}. \tag{7.72}$$

Thus

$$\frac{n_U}{n_L} = \frac{B_{LU} \bar{u} + C_{LU}}{A_{UL} + B_{UL} \bar{u} + C_{UL}}. \tag{7.73}$$

Equations 7.47 and 7.51 can be combined to eliminate

$$B_{UL} \bar{u} = \frac{c^3 A_{UL}}{8\pi h \nu_0^3} \frac{8\pi h \nu_0^3}{c^3} \left[\exp\left(\frac{h\nu_0}{kT_b}\right) - 1 \right]^{-1} = A_{UL} \left[\exp\left(\frac{h\nu_0}{kT_b}\right) - 1 \right]^{-1}, \tag{7.74}$$

where T_b is the ambient radiation brightness temperature, in favor of A_{UL}. Equation 7.50 eliminates

$$B_{LU} \bar{u} = B_{UL} \bar{u} \frac{g_U}{g_L} = A_{UL} \frac{g_U}{g_L} \left[\exp\left(\frac{h\nu_0}{kT_b}\right) - 1 \right]^{-1}. \tag{7.75}$$

Finally, Equations 7.64 and 7.72 allow the replacement

$$C_{LU} = \frac{n_U}{n_L} C_{UL} = \frac{g_U}{g_L} \exp\left(-\frac{h\nu_0}{kT_k}\right) C_{UL}, \tag{7.76}$$

where T_k is the kinetic temperature of the gas. The numerator of Equation 7.73 becomes

$$A_{UL} \frac{g_U}{g_L} \left[\exp\left(\frac{h\nu_0}{kT_b}\right) - 1 \right]^{-1} + \frac{g_U}{g_L} \exp\left(-\frac{h\nu_0}{kT_k}\right) C_{UL} \tag{7.77}$$

and the denominator is

$$A_{UL} + A_{UL} \left[\exp\left(\frac{h\nu_0}{kT_b}\right) - 1 \right]^{-1} + C_{UL}. \tag{7.78}$$

Thus

$$\frac{n_U g_L}{n_L g_U} = \frac{A_{UL} + C_{UL} \exp\left(-\dfrac{h\nu_0}{kT_k}\right) \left[\exp\left(\dfrac{h\nu_0}{kT_b}\right) - 1 \right]}{A_{UL} \exp\left(\dfrac{h\nu_0}{kT_b}\right) + C_{UL} \left[\exp\left(\dfrac{h\nu_0}{kT_b}\right) - 1 \right]}, \tag{7.79}$$

$$\exp\left(-\frac{h\nu_0}{kT_x}\right) = \exp\left(-\frac{h\nu_0}{kT_b}\right) \frac{A_{UL} + C_{UL} \exp\left(-\dfrac{h\nu_0}{kT_k}\right) \left[\exp\left(\dfrac{h\nu_0}{kT_b}\right) - 1 \right]}{A_{UL} + C_{UL} \left[1 - \exp\left(-\dfrac{h\nu_0}{kT_b}\right) \right]}. \tag{7.80}$$

If the spontaneous emission rate is much larger than the collision rate, Equation 7.80 yields $T_x \to T_b$; if the collision rate is much higher than the spontaneous emission rate, $T_x \to T_k$. For any A_{UL} and C_{UL}, T_x lies between T_k and T_b.

7.5 MASERS

If the upper energy level is *overpopulated*, that is,

$$\frac{n_U}{n_L} > \frac{g_U}{g_L},$$ (7.81)

then T_x is actually negative,

$$\left[1 - \exp\left(-\frac{h\nu_0}{kT_x}\right)\right]$$ (7.82)

is negative, and Equation 7.67 gives a negative net opacity coefficient κ. Negative net opacity implies brightness gain instead of loss; the intensity of a background source at frequency ν_0 will be amplified. At radio wavelengths this phenomenon is called **maser** (an acronym for microwave amplification by stimulated emission of radiation) amplification. Astronomical masers are common at radio frequencies because $h\nu \ll kT$ and hence $n_U/n_L \approx g_U/g_L$ even in TE. These sources can have line brightness temperatures as high as 10^{15} K, which is much higher than the kinetic temperature of the masing gas. For a clear presentation covering the basics of astronomical masers, written by Reid & Moran, see Verschuur and Kellermann [110, Chapter 6].

Our model for an astronomical maser starts with the radiative transfer equation for a two-level system. Assume for simplicity that $g_U = g_L$ so that Equation 7.50 implies $B_{LU} = B_{UL} \equiv B$ and Equation 7.51 implies $A_{UL} = 8\pi h\nu_0^3 B/c^3 \equiv A$. Then Equation 7.57 simplifies to

$$\frac{dI_\nu}{ds} = -\left(\frac{h\nu_0}{c}\right)(n_L - n_U)B\phi(\nu)I_\nu + \left(\frac{h\nu_0}{4\pi}\right)n_U A\phi(\nu).$$ (7.83)

Next assume that the line profile $\phi(\nu)$ is Gaussian with FWHM $\Delta\nu$ (Figure 7.4) so that Equation 7.37 applies and make the numerical approximation

$$\phi(\nu_0) = \left(\frac{\ln 2}{\pi}\right)^{1/2}\frac{2}{\Delta\nu} = \frac{0.939\ldots}{\Delta\nu} \approx \frac{1}{\Delta\nu}.$$ (7.84)

Then at the line center frequency ν_0,

$$\frac{dI_\nu}{ds} = -\frac{h\nu_0(n_L - n_U)BI_\nu}{c\Delta\nu} + \frac{h\nu_0 n_U A}{4\pi\Delta\nu}.$$ (7.85)

The maser optical depth

$$\tau = \int \kappa \, ds = \int \frac{dI_\nu}{I_\nu} = \frac{h\nu_0 B}{c\Delta\nu}\int (n_U - n_L)ds$$ (7.86)

is called the **maser gain** over the path of integration, and the maser amplifies the intensity of background radiation by the factor $\exp(|\tau|)$. In a laboratory maser, radiation is trapped in a high-Q resonant cavity to create effective path lengths up to 10^9 times the cavity length. In an astronomical maser there is no cavity, so the radiation makes only a single pass and the physical path must be much longer ($s > 10^{13}$ cm) for significant gain to occur.

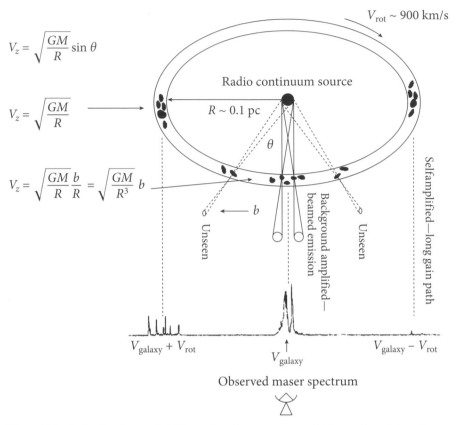

$$V_z = \sqrt{\frac{GM}{R}} \sin\theta$$

$$V_z = \sqrt{\frac{GM}{R}}$$

$$V_z = \sqrt{\frac{GM}{R}}\frac{b}{R} = \sqrt{\frac{GM}{R^3}}\, b$$

$V_{\rm rot} \sim 900$ km/s

Radio continuum source

$R \sim 0.1$ pc

θ

Unseen

$\longleftarrow b$

Unseen

Background amplified— beamed emission

Self amplified—long gain path

$V_{\rm galaxy} + V_{\rm rot}$

$V_{\rm galaxy}$

$V_{\rm galaxy} - V_{\rm rot}$

Observed maser spectrum

Figure 7.6. A simple ring model illustrating the geometry and kinematics of the edge-on 22 GHz water-maser disk orbiting the nucleus of NGC 4258 [73].

Maser emission quickly depopulates the upper energy level, so masers have to be "pumped" to emit continuously. Typically one or more higher energy levels absorb radiation from a pump source (e.g., infrared continuum from a star or an AGN), and radiative decays preferentially repopulate the upper energy level. This radiative pumping process produces no more than one maser photon per pump photon, so the pump energy required is proportional to the frequency $v = E/h$ of the pump photon. If the maser photon emission rate is limited by the pump luminosity, the maser is described as being **saturated**; if the pump power is more than adequate, the maser is **unsaturated**.

Strong, compact maser sources are powerful tools for high-resolution imaging and precision astrometry. They are being used to measure accurate trigonometric distances to individual stars in our Galaxy, the size and structure of our Galaxy, black-hole masses in AGNs, and distances to galaxies up to ~ 150 Mpc [89].

The most spectacular example is the circumnuclear 22 GHz H_2O megamaser disk surrounding the Seyfert nucleus of **NGC 4258** [73] and illustrated in Figure 7.6. The observed maser spectrum has hundreds of narrow lines in three groups clustered around the systemic recession velocity $v \approx 450$ km s^{-1} of NGC 4258 and at velocities

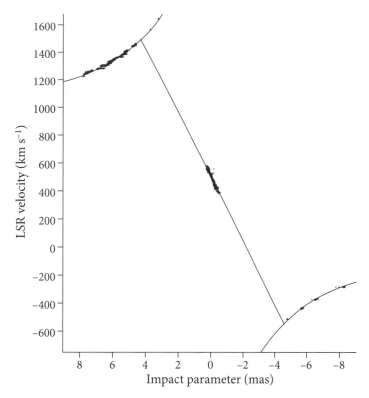

Figure 7.7. The NGC 4258 rotation curve is perfectly Keplerian, indicating that the mass interior to the maser disk is dominated by a single compact object whose density is too high ($> 5 \times 10^{12} M_\odot \, \mathrm{pc}^{-3}$) for it to be a compact star cluster; it must be an SMBH [73].

further redshifted and blueshifted by the $v_{\mathrm{rot}} \sim 900 \, \mathrm{km \, s^{-1}}$ rotation velocities in the disk.

The nearly edge-on maser disk was imaged with high spectral and angular resolution by the VLBA, and its rotation curve is shown in Figure 7.7. The systemic lines are concentrated within a region $< 1 \, \mathrm{mas} \approx 0.03 \, \mathrm{pc}$ wide and centered on the line of sight to the supermassive black hole (SMBH) at the center of the disk. Photons from the central radio continuum source are amplified and beamed in our direction, so systemic maser clouds are visible only when they are within a small angle $|\theta| < 0.07 \, \mathrm{rad}$ from the line of sight. For a constant v_{rot}, their $v_z = (GM/R^3)^{1/2} b$ yields the sloped straight line in Figure 7.7. This slope corresponds to the gravitational acceleration $a_z = GM/R^2$ of the SMBH, and long-term monitoring the velocities of individual systemic-maser lines independently yields $a \approx 8 \, \mathrm{km \, s^{-1} \, yr^{-1}}$. The redshifted and blueshifted maser clouds are visible only near the tangent points of the disk near $\theta = \pm \pi/2 \, \mathrm{rad}$, which have the longest gain paths at nearly constant velocities $v_z = (GM/R)^{1/2}$ that follow the perfectly Keplerian curves in Figure 7.7. Combining the acceleration, velocity, and angular-size measurements yields a unique solution for both the SMBH mass $M \approx 3.8 \times 10^7 M_\odot$ and the distance $D = 7.2 \pm 0.4 \, \mathrm{Mpc}$ to NGC 4258.

7.6 RECOMBINATION LINE SOURCES

Astronomical sources of radio recombination lines are often in local thermodynamic equilibrium (LTE). The spontaneous emission rate from atomic physics and the laws of radiative transfer for spectral lines can be combined to model sources in LTE. However, LTE does not apply to all recombination lines, so departures from LTE must be recognized and treated differently.

7.6.1 Radiative Transfer in LTE

Equation 7.67 gives the absorption coefficient at the center frequency ν_0 of the $n \rightarrow n + 1$ electronic transition of hydrogen in an HII region in local thermodynamic equilibrium (LTE) at electron temperature T_e:

$$\kappa(\nu) = \frac{c^2}{8\pi \nu_0^2} \frac{g_{n+1}}{g_n} n_n A_{n+1,n} \left[1 - \exp\left(-\frac{h\nu_0}{kT_e} \right) \right] \phi(\nu), \quad (7.87)$$

where

$$\nu_0 = \nu_{n,n+1} \approx \frac{2Rc}{n^3} = \frac{4\pi^2 m_e e^4}{h^3 n^3}, \quad (7.88)$$

$$g_n = 2n^2, \quad (7.89)$$

and n_n is the number density of atoms in the nth electronic energy level. At radio frequencies, it can be assumed that $n \gg 1$, $h\nu_0 \ll kT_e$, and $g_{n+1} \approx g_n$. Equation 7.23 gives the spontaneous emission rate

$$A_{n+1,n} \approx \frac{64\pi^6 m_e e^{10}}{3c^3 h^6 n^5} \quad (7.90)$$

and Equation 7.37 parameterizes the normalized line profile

$$\phi(\nu_0) \approx \left(\frac{\ln 2}{\pi} \right)^{1/2} \frac{2}{\Delta \nu}. \quad (7.91)$$

The number density n_n of atoms in the nth electronic energy level is given by the **Saha equation**, a generalization of the Boltzmann equation (for a derivation of the Saha equation, see Rybicki and Lightman [97, Equation 9.47]):

$$n_n = n^2 \left(\frac{h^2}{2\pi m_e kT_e} \right)^{3/2} n_p n_e \exp\left(\frac{\chi_n}{kT_e} \right) \quad (7.92)$$

where $\chi_n < 0$ is the ionization potential of the nth energy level. For large n, $|\chi_n| \ll kT_e$ and the exponential factor $\exp[\chi_n/(kT_e)] \approx 1$ can be ignored. Combining the results from Equations 7.87 through 7.92 yields the opacity coefficient at the line center frequency ν_0:

$$\kappa(\nu_0) \approx \frac{c^2 n^2}{8\pi \nu_0^2} \left(\frac{h^2}{2\pi m_e kT_e} \right)^{3/2} n_e^2 \left(\frac{64\pi^6 m_e e^{10}}{3c^3 h^6 n^5} \right) \frac{h\nu_0}{kT_e} \left[\left(\frac{\ln 2}{\pi} \right)^{1/2} \frac{2}{\Delta \nu} \right]. \quad (7.93)$$

Some algebra reduces this to

$$\kappa(\nu_0) \approx \left(\frac{n_e^2}{T_e^{5/2}\Delta\nu}\right)\left(\frac{4\pi e^6 h}{3m_e^{3/2}k^{5/2}c}\right)\left(\frac{\ln 2}{2}\right)^{1/2}. \tag{7.94}$$

Notice that the electronic energy level n has dropped out; Equation 7.94 is valid for all radio recombination lines with $n \gg 1$. The optical depth $\tau_L = \int \kappa \, ds$ at the line center frequency ν_0 can be expressed in terms of the emission measure defined by Equation 4.57:

$$\frac{EM}{pc\,cm^{-6}} \equiv \int_{los} \left(\frac{n_e}{cm^{-3}}\right)^2 d\left(\frac{s}{pc}\right). \tag{7.95}$$

In astronomically convenient units the line center opacity is

$$\tau_L \approx 1.92 \times 10^3 \left(\frac{T_e}{K}\right)^{-5/2}\left(\frac{EM}{pc\,cm^{-6}}\right)\left(\frac{\Delta\nu}{kHz}\right)^{-1}. \tag{7.96}$$

Because $\tau_L \ll 1$ in all known H II regions, the brightness temperature contributed by a recombination *emission* line at its center frequency ν_0 is

$$T_L \approx T_e\tau_L \approx 1.92 \times 10^3 \left(\frac{T_e}{K}\right)^{-3/2}\left(\frac{EM}{pc\,cm^{-6}}\right)\left(\frac{\Delta\nu}{kHz}\right)^{-1}. \tag{7.97}$$

At frequencies high enough that the free–free continuum is also optically thin, the peak line-to-continuum ratio (which occurs at frequency ν_0) in LTE is

$$\frac{T_L}{T_C} \approx 7.0 \times 10^3 \left(\frac{\Delta\nu}{km\,s^{-1}}\right)^{-1}\left(\frac{\nu}{GHz}\right)^{1.1}\left(\frac{T_e}{K}\right)^{-1.15}\left[1 + \frac{N(He^+)}{N(H^+)}\right]^{-1}, \tag{7.98}$$

where $\Delta\nu$ is the line FWHM expressed as a velocity and the typical He^+/H^+ ion ratio is $N(He^+)/N(H^+) \approx 0.08$. The term in square brackets is necessary because He^+ contributes to the free–free continuum emission but not to the hydrogen recombination line. The line-to-continuum ratio yields an estimate of the electron temperature T_e that is independent of the emission measure so long as the frequency is high enough that the continuum optical depth is small.

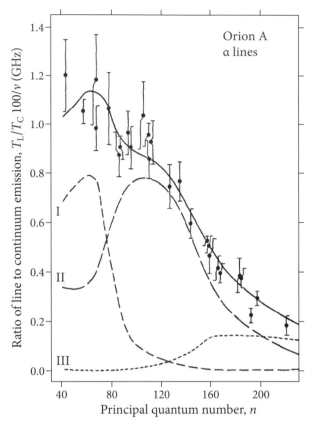

Figure 7.8. A temperature-distribution model for the Orion Nebula H II region based on the line-to-continuum ratios of hydrogen recombination lines [66].

7.6.2 Astronomical Applications

Recombination lines can be used to find the electron temperatures of H II regions in LTE. Solving Equation 7.98 explicitly for T_e gives the useful formula

$$\left(\frac{T_e}{K}\right) \approx \left[7.0 \times 10^3 \left(\frac{\nu}{\text{GHz}}\right)^{1.1} 1.08^{-1} \left(\frac{\Delta v}{\text{km s}^{-1}}\right)^{-1} \left(\frac{T_C}{T_L}\right)\right]^{0.87}. \tag{7.99}$$

By mapping the recombination line-to-continuum ratios T_L/T_C in a number of H$n\alpha$ transitions, Lockman and Brown determined the temperature distribution in the Orion Nebula (Figure 7.8), a nearby H II region.

Differences between the rest and observed frequencies of radio recombination lines are attributed to Doppler shifts from nonzero radial velocities. With a simple rotational model for the disk of our Galaxy, astronomers can convert radial velocities to distances, albeit with some ambiguities, and map the approximate spatial distribution of H II regions in our Galaxy (Figure 7.9). They roughly outline the major spiral arms.

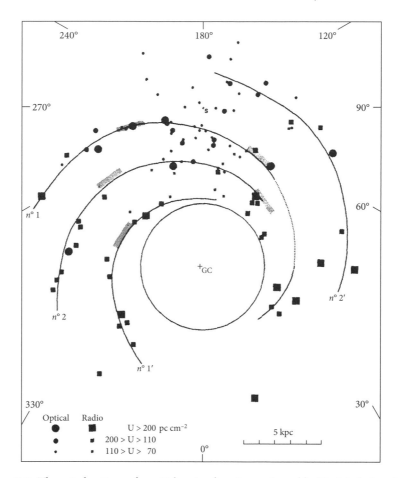

Figure 7.9. The spiral pattern of our Galaxy in plan view, as traced by Hα (circles) and radio recombination lines (squares) [40].

A plot showing the observed electron temperatures of Galactic HII regions (Figure 7.10) reveals that temperature increases with distance from the Galactic center.

The explanation for this trend is the observed decrease in *metallicity* (relative abundance of elements heavier than helium) with galactocentric distance. Power radiated by emission lines of "metals" is the principal cause of HII region cooling.

Radio recombination line strengths are much less affected by dust extinction than optical lines (e.g., the Hα and Hβ lines) are, so radio recombination lines are useful quantitative indicators of the ionization rates and hence star-formation rates in dusty starburst galaxies such as M82 (Figure 7.11).

7.7 MOLECULAR LINE SPECTRA

7.7.1 Molecular Line Frequencies

A molecule is called **polar** if its permanent electric dipole moment (Equation 7.120) is not zero. Symmetric molecules (e.g., the diatomic hydrogen molecule H_2) have no

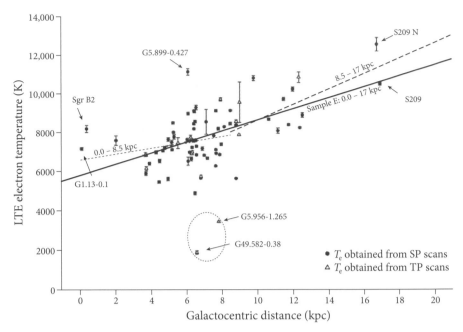

Figure 7.10. Recombination-line observations show that the electron temperatures T_e of H II regions increase with distance from the Galactic center at the rate of $287 \pm 46\,\mathrm{K\,kpc^{-1}}$, probably because metallicity decreases [85].

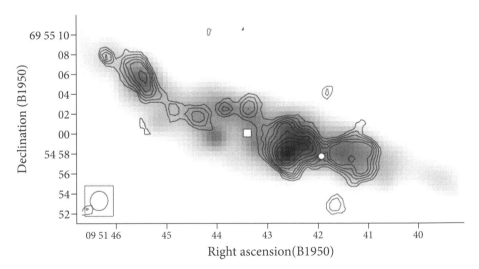

Figure 7.11. M82 imaged in H92α (contours) and 8.3 GHz continuum (gray scale) [93].

permanent electric dipole moment, but most asymmetric molecules (e.g., the carbon monoxide molecule CO) do have asymmetric charge distributions and are polar. The electric dipole moments of polar molecules rotating with constant angular velocity ω appear to vary sinusoidally with that angular frequency, so polar molecules radiate at their rotation frequencies. The intensity of this radiation can be derived from the

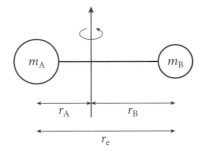

Figure 7.12. A diatomic molecule rotating about its center of mass.

Larmor formula expressed in terms of dipole moments instead of charges and charge separations.

The permitted rotation rates and resulting line frequencies are determined by the quantization of angular momentum. The quantization rule for the permitted electron orbital radii

$$a_n = \frac{n\hbar}{m_e v}$$

in the Bohr atom quantizes the orbital angular momentum $L = m_e a_n v$ in multiples of $\hbar \equiv h/(2\pi)$:

$$\boxed{L = n\hbar.}$$ (7.100)

The rule that angular momentum is an integer multiple of $\hbar$ is universal and applies to the angular momentum of a rotating molecule as well.

Consider a rigid **diatomic molecule** (Figure 7.12) whose two atoms have masses m_A and m_B and whose centers are separated by the equilibrium distance r_e. The individual atomic distances r_A and r_B from the center of mass must obey

$$r_e = r_A + r_B \quad \text{and} \quad r_A m_A = r_B m_B.$$ (7.101)

In the inertial center-of-mass frame,

$$L = I\omega,$$ (7.102)

where I is the moment of inertia and ω is the angular velocity of the rotation. Nearly all of the mass is in the two compact (much smaller than r_e) nuclei, so $I = (m_A r_A^2 + m_B r_B^2)$ and $L = (m_A r_A^2 + m_B r_B^2)\omega$. It is convenient to rewrite this as

$$L = \left(\frac{m_A m_B}{m_A + m_B}\right) r_e^2 \omega$$ (7.103)

or

$$\boxed{L = m r_e^2 \omega,}$$ (7.104)

where

$$m \equiv \left(\frac{m_A m_B}{m_A + m_B} \right) \tag{7.105}$$

is the **reduced mass** of the molecule.

The rotational kinetic energy associated with this angular momentum is

$$E_{\text{rot}} = \frac{I\omega^2}{2} = \frac{L^2}{2I}. \tag{7.106}$$

The quantization of angular momentum to integer multiples of $\hbar$ implies that rotational energy is also quantized. The corresponding energy eigenvalues of the Schrödinger equation are

$$E_{\text{rot}} = \left(\frac{\hbar^2}{2I} \right) J(J + 1), \qquad J = 0, 1, 2, \ldots . \tag{7.107}$$

Note the inverse relation between permitted rotational energies and the moment of inertia I. If the upper-level rotational energy is much higher than kT, few molecules will be collisionally excited to that level and the line emission from molecules in that level will be very weak. For example, the minimum rotational energy of the small and light H_2 molecule is equivalent to a temperature $T = E_{\min}/k \approx 500$ K, which is much higher than the actual temperature of most interstellar H_2. Only relatively massive molecules are likely to be detectable radio emitters in very cold (tens of K) molecular clouds.

Quantization of rotational energy implies that *changes* in rotational energy are quantized. The energy change of permitted transitions is further restricted by the quantum-mechanical selection rule

$$\Delta J = \pm 1. \tag{7.108}$$

Going from J to $J - 1$ releases energy

$$\Delta E_{\text{rot}} = [J(J + 1) - (J - 1)J] \frac{\hbar^2}{2I} = \frac{\hbar^2 J}{I}. \tag{7.109}$$

The frequency of the photon emitted during this rotational transition is

$$\nu = \frac{\Delta E_{\text{rot}}}{h} = \frac{\hbar J}{2\pi I}, \qquad J = 1, 2, \ldots , \tag{7.110}$$

where J is the angular-momentum quantum number corresponding to the upper energy level. In terms of the molecular reduced mass m and equilibrium nuclear

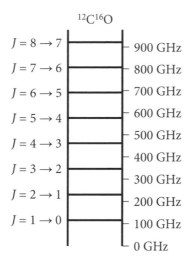

Figure 7.13. The rotational spectrum of $^{12}\text{C}^{16}\text{O}$ looks like a ladder whose rungs indicate the J levels and line frequencies. The $^{12}\text{C}^{16}\text{O}$ molecule has a relatively small moment of inertia, so the lowest rung of this ladder is at $\nu \approx 115\,\text{GHz}$ ($\lambda \approx 2.6\,\text{cm}$); it has no cm-wavelength lines.

separation r_e,

$$\nu = \frac{hJ}{4\pi^2 mr_\text{e}^2}, \qquad J = 1, 2, \ldots . \tag{7.111}$$

Thus a plot of the radio spectrum of a particular molecular species in an interstellar cloud will look like a **ladder** (Figure 7.13) whose steps are all harmonics of the fundamental frequency that is determined solely by the moment of inertia $I = mr_\text{e}^2$ of that species. The relative intensities of lines in the ladder depend on the temperature of the cloud. Since $\nu \propto m^{-1}r_\text{e}^{-2}$, the lowest frequency of line emission depends on the mass and size of the molecule. Large, heavy molecules in cold clouds may be seen at centimeter wavelengths, but smaller and lighter molecules emit only at millimeter wavelengths.

For example, the laboratory spectrum of the $^{12}\text{C}^{16}\text{O}$ carbon-monoxide molecule shows that the fundamental $J = 1 \rightarrow 0$ transition emits a photon at $\nu = 115.27120\,\text{GHz}$. (See the online spectral-line catalog called Splatalogue[1] for accurate frequencies of radio spectral lines.) The distance r_e between the C and O nuclei can be estimated from

$$r_\text{e} = \frac{1}{2\pi}\left(\frac{hJ}{m\nu}\right)^{1/2}, \tag{7.112}$$

where the reduced mass is

$$m = \frac{m_\text{C}m_\text{O}}{m_\text{C} + m_\text{O}} \approx m_\text{H}\left(\frac{12 \cdot 16}{12 + 16}\right) \approx 1.67 \times 10^{-24}\,\text{g} \cdot 6.86 \approx 1.15 \times 10^{-23}\,\text{g}.$$

[1] http://www.splatalogue.net/.

Thus the equilibrium distance between the C and O nuclei is

$$r_e = \frac{1}{2\pi} \left(\frac{6.63 \times 10^{-27} \text{ erg s} \cdot 1}{115.27120 \times 10^9 \text{ Hz} \cdot 1.15 \times 10^{-23} \text{ g}} \right)^{1/2} \approx 1.13 \times 10^{-8} \text{ cm}.$$

The centrifugal forces acting on the nuclei increase as the molecule spins more rapidly, so a nonrigid bond will stretch and r_e will increase slightly with J. Spectral lines emitted by more rapidly rotating $^{12}C^{16}O$ molecules will have frequencies slightly lower than the harmonics $2\nu_{1-0}, 3\nu_{1-0}, \ldots$ of the $J = 1 \rightarrow 0$ line: $2\nu_{1-0} = 230.542416$ GHz and the actual $J = 2 \rightarrow 1$ frequency is $\nu_{2-1} = 230.538000$ GHz. Chemists use these line frequencies to determine r_e, and the difference between $2\nu_{1-0}$ and ν_{2-1} is a measure of the stiffness of the carbon–oxygen chemical bond. Since the actual frequency is only slightly less than the harmonic frequency, the stiffness of the "spring" connecting the atoms is quite high. Consequently the fundamental vibrational frequency of the CO molecule is much higher than the fundamental rotational frequency, and CO emits vibrational lines at mid-infrared wavelengths $\lambda \sim 5\,\mu$m.

Equation 7.111 can be used to calculate frequencies for molecules containing rare isotopes (e.g., $^{13}C^{16}O$) that might be more difficult to measure in the lab:

$$\frac{m(^{13}C^{16}O)}{m(^{12}C^{16}O)} = \frac{13 \cdot 16/(13 + 16)}{12 \cdot 16/(12 + 16)} \approx 1.0460, \tag{7.113}$$

so we expect

$$\frac{\nu_{1-0}(^{13}C^{16}O)}{\nu_{1-0}(^{12}C^{16}O)} \approx \left[\frac{m(^{13}C^{16}O)}{m(^{12}C^{16}O)} \right]^{-1}, \tag{7.114}$$

$$\nu_{1-0}(^{13}C^{16}O) \approx 115.27120/1.0460 \approx 110.20 \text{ GHz}. \tag{7.115}$$

The actual $^{13}C^{16}O$ $J = 1 \rightarrow 2$ frequency is 110.201354 GHz.

Polar diatomic molecules emit a harmonic series of radio spectral lines at millimeter wavelengths. Bigger and heavier linear polyatomic molecules have ladders of lines starting at somewhat lower frequencies. Nonlinear molecules such as the symmetric-top ammonia (NH_3) with two distinct rotational axes have more complex spectra consisting of many parallel ladders (Figure 7.14).

7.7.2 Molecular Excitation

Molecules are excited into $E_{\text{rot}} > 0$ states by ambient radiation and by collisions in a dense gas. The minimum gas temperature T_{min} needed for significant collisional excitation is

$$T_{\text{min}} \sim \frac{E_{\text{rot}}}{k}. \tag{7.116}$$

From Equations 7.107 and 7.111,

$$E_{\text{rot}} = \frac{J(J+1)\hbar^2}{2I} \quad \text{and} \quad \nu = \frac{hJ}{4\pi^2 I}, \tag{7.117}$$

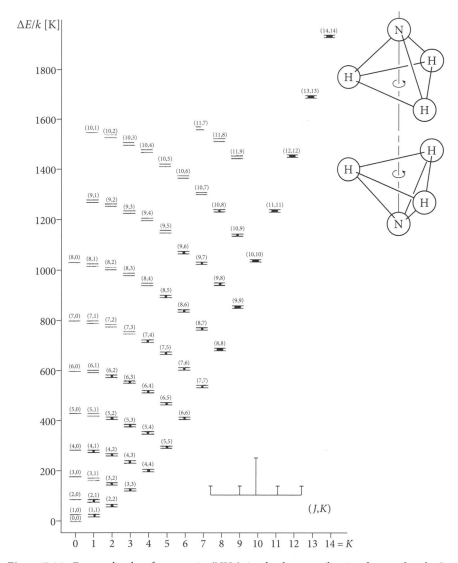

Figure 7.14. Energy levels of ammonia (NH₃) in the lowest vibrational state [115]. On the abscissa, K is the quantum number corresponding to the z-component of the angular momentum. Transitions between the two spin states of the nitrogen atom cause the line splitting shown and yield emission at frequencies near 24 GHz. NH₃ is a very useful thermometer for molecular clouds [52].

so

$$T_{\min} \sim \frac{J(J+1)h^2}{2 \cdot 4\pi^2 I k} = \frac{hJ}{4\pi^2 I} \frac{h(J+1)}{2k} = \nu E_U, \qquad (7.118)$$

where E_U is the rotational energy of the *upper* energy level for the transition. Thus a minimum gas kinetic temperature

$$\boxed{T_{\min} \approx \frac{\nu h(J+1)}{2k}} \qquad (7.119)$$

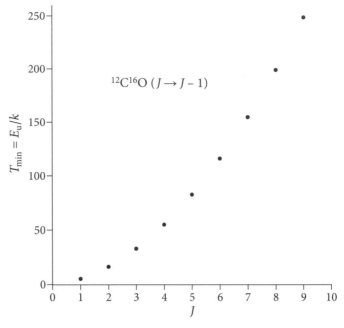

Figure 7.15. The upper-level energies E_U for $^{12}C^{16}O$ $J \to J - 1$ transitions are proportional to $J(J + 1)$. The corresponding minimum temperatures $T_{min} = E_U/k$ required for collisions to excite the molecules are also proportional to $J(J + 1)$, so high-J lines will be weak in cold molecular clouds.

is required to excite the $J \to J - 1$ transition at frequency ν. For example, the minimum gas temperature needed for significant excitation of the $^{12}C^{16}O$ $J = 2$–1 line at $\nu \approx 230.5$ GHz (Figure 7.15) is

$$T_{min} \approx \frac{\nu h (J + 1)}{2k} \approx \frac{230.5 \times 10^9 \, \text{Hz} \cdot 6.63 \times 10^{-27} \, \text{erg s} \, (2 + 1)}{2 \cdot 1.38 \times 10^{-16} \, \text{erg K}^{-1}} \approx 16.6 \, \text{K}.$$

The $T_{min} = E_U/k$ values for many molecular lines may be found in the online spectral-line catalog Splatalogue. If $T_{min} \gg 2.7$ K, then **radiative excitation** by the cosmic microwave background is ineffective.

7.7.3 Molecular Line Strengths

Larmor's formula for a time-varying dipole can be applied to estimate the average power radiated by a rotating polar molecule. The **electric dipole moment** $\vec{p}$ of any charge distribution $\rho(\vec{x})$ is defined as the integral

$$\vec{p} \equiv \int \vec{x} \rho(v) dv, \tag{7.120}$$

over the volume v containing the charges. In the case of two point charges $+q$ and $-q$ with separation r_e,

$$|\vec{p}| = q r_e. \tag{7.121}$$

When a molecule rotates with angular velocity ω, the projection of the dipole moment perpendicular to the line of sight varies with time as $qr_e \exp(-i\omega t)$. Equation 7.101 states that

$$r_A m_A = r_B m_B, \tag{7.122}$$

so

$$\dot{v}_A = \ddot{r}_A = \omega^2 r_A \qquad \text{and} \qquad \dot{v}_B = \ddot{r}_B = \omega^2 r_B. \tag{7.123}$$

Equation 2.133 from the derivation of Larmor's formula states that each charge contributes

$$E_\perp = \frac{q\dot{v}\sin\theta}{rc^2} \tag{7.124}$$

to the radiated electric field at distance r from the source. The fields from both charges add in phase because $r_e \ll \lambda$, so the total radiated field $E_\perp$ is

$$E_\perp = \frac{q(\omega^2 r_A + \omega^2 r_B)\sin\theta}{rc^2} \exp(-i\omega t). \tag{7.125}$$

Thus the instantaneous power emitted is

$$P = \frac{2q^2}{3c^3}\omega^4 |r_e \exp(-i\omega t)|^2 \tag{7.126}$$

and the time-averaged power is

$$\langle P \rangle = \frac{2q^2}{3c^3}(2\pi\nu)^4\frac{r_e^2}{2} = \frac{64\pi^4}{3c^3}\nu^4\left(\frac{qr_e}{2}\right)^2. \tag{7.127}$$

This can be expressed as

$$\langle P \rangle = \frac{64\pi^4}{3c^3}|\mu|^2\nu^4, \tag{7.128}$$

where

$$|\mu|^2 \equiv \left(\frac{qr_e}{2}\right)^2 \tag{7.129}$$

defines the **mean electric dipole moment**. For the radiative transition between upper and lower energy levels U and L, the spontaneous emission coefficient is

$$A_{UL} = \frac{\langle P \rangle}{h\nu}, \tag{7.130}$$

$$\boxed{A_{UL} = \frac{64\pi^4}{3hc^3}|\mu_{UL}|^2\nu^3.} \tag{7.131}$$

The value of μ_{UL} for the $J \rightarrow J - 1$ transition of a linear rotating molecule with dipole moment μ is

$$\boxed{|\mu_{J\rightarrow J-1}|^2 = \frac{\mu^2 J}{2J + 1}.}$$ (7.132)

(This equation reflects the complex angular wave functions involved, and we won't derive it here.)

Dipole moments are often expressed in **debye** units defined by $1\,\text{debye} = 1\,\text{D} \equiv 10^{-18}$ statcoul cm $= 10^{-10}$ statcoul $\times\, 10^{-8}$ cm, where 10^{-10} statcoul $\approx 0.2\,e$ is a typical charge imbalance for a polar molecule and 10^{-8} cm is a typical value for the interatomic separation r_e. For example, the CO molecule has dipole moment $\mu \sim 0.11 \times 10^{-18}$ statcoul cm $= 0.11$ D.

Combining Equations 7.131 and 7.132 yields, in convenient units,

$$\left(\frac{A_{J\rightarrow J-1}}{\mathrm{s}^{-1}}\right) \approx 1.165 \times 10^{-11} \left|\frac{\mu}{\mathrm{D}}\right|^2 \left(\frac{J}{2J+1}\right) \left(\frac{\nu}{\mathrm{GHz}}\right)^3.$$ (7.133)

For example, the spontaneous emission coefficient A_{10} for the CO $J = 1 \rightarrow 0$ line at $\nu \approx 115$ GHz is

$$A_{10} \approx 1.165 \times 10^{-11} \cdot 0.11^2 \cdot \left(\frac{1}{3}\right) \cdot 115^3 \approx 7.1 \times 10^{-8}\,\mathrm{s}^{-1}.$$ (7.134)

This is close to the more accurate Splatalogue value, $A_{10} \approx 7.202 \times 10^{-8}\,\mathrm{s}^{-1}$.

The typical time $A_{10}^{-1} \approx 10^7$ s for a CO molecule to emit a photon spontaneously may be longer than the average time between molecular collisions in an interstellar molecular cloud, so CO can approach LTE with the excitation temperature of the $J = 1 \rightarrow 0$ line being nearly equal to the kinetic temperature T of the molecular cloud. For any molecular transition, there is a **critical density** defined by

$$\boxed{n^* \approx \frac{A_{\mathrm{UL}}}{\sigma v}}$$ (7.135)

at which the radiating molecule suffers collisions at the rate $n^* \sigma v \approx n(\mathrm{H_2})\,\sigma v$ equal to the spontaneous emission rate A_{UL}. Typical collision cross sections are $\sigma \sim 10^{-15}$ cm^2, and the average velocity of the abundant $\mathrm{H_2}$ molecules is $v \approx (3kT/m)^{1/2} \sim 5 \times 10^4$ cm s^{-1} if $T \sim 20$ K. Thus the critical density of the CO $J = 1 \rightarrow 0$ transition is

$$n^* \approx \frac{A_{\mathrm{UL}}}{\sigma v} \approx \frac{7 \times 10^{-8}\,\mathrm{s}^{-1}}{10^{-15}\,\mathrm{cm}^{-2} \cdot 5 \times 10^4\,\mathrm{cm\,s}^{-1}} \approx 1.4 \times 10^3\,\mathrm{cm}^{-3}.$$

Many Galactic molecular clouds have higher densities than this, so Galactic CO $J = 1 \rightarrow 0$ emission is strong and widespread. Also, photons from a particular transition may be repeatedly absorbed and reemitted within the molecular cloud. Such **line trapping** lowers the effective emission rate A_{UL} and reduces the effective value of n^* needed for LTE.

Whether or not a molecular cloud is in LTE, Equation 7.67 with the temperature T replaced by excitation temperature T_x can be used to calculate the line opacity coefficient:

$$\kappa(\nu) = \frac{c^2}{8\pi \nu_0^2} \frac{g_U}{g_L} n_L A_{UL} \left[1 - \exp\left(-\frac{h\nu_0}{kT_x} \right) \right] \phi(\nu). \qquad (7.136)$$

At the line center frequency ν_0, Equation 7.37 gives

$$\phi(\nu_0) = \left(\frac{\ln 2}{\pi} \right)^{1/2} \frac{2}{\Delta \nu} = \left(\frac{\ln 2}{\pi} \right)^{1/2} \frac{2c}{\nu_0 \Delta v},$$

where $\Delta \nu$ is the FWHM line width expressed as a frequency and Δv is the line width in velocity units (e.g., km s^{-1}). The line-center optical depth is $\tau_0 = \int \kappa_0 \, ds$ along the line of sight, and the column density is $N_L = \int n_L \, ds$ along the line of sight, so

$$\tau_0 = \frac{(\ln 2)^{1/2}}{4\pi^{3/2}} \frac{c^3}{\nu_0^3} \frac{g_U}{g_L} \frac{A_{UL}}{\Delta v} N_L \left[1 - \exp\left(\frac{h\nu}{kT_x} \right) \right]. \qquad (7.137)$$

In the Rayleigh–Jeans approximation $h\nu \ll kT_x$, the brightness-temperature difference between the line center and the nearby off-line continuum is

$$\Delta T_b = (T_x - T_c)[1 - \exp(-\tau_0)], \qquad (7.138)$$

where T_c is the brightness temperature of any background continuum emission (e.g., $T_c \approx 2.73$ K from the CMB). In the limit of low line optical depth, the line brightness

$$\Delta T_b \approx (T_x - T_c)\tau_0 = \left(\frac{T_x - T_c}{T_x} \right) N_L \frac{(\ln 2)^{1/2}}{4\pi^{3/2}} \frac{hc^3}{k\nu_0^2} \frac{g_U}{g_L} \frac{A_{UL}}{\Delta v} \qquad (7.139)$$

is proportional to the column density N_L. The spontaneous emission coefficient A_{UL} is proportional to ν^3 (Equation 7.131), so spectral lines tend to have higher brightness temperatures and become more prominent at higher radio frequencies (Figure 7.16).

The hydrogen molecule H_2 is by far the most abundant molecule in interstellar space. Unfortunately, it is symmetric so its dipole moment is zero. Observable but comparatively rare polar molecules such as CO are only tracers, and total column densities of molecular gas must be estimated indirectly by the use of a fairly uncertain **CO to H_2 conversion factor** X_{CO} relating H_2 column density in cm^{-2} to CO velocity-integrated line brightness in K km s^{-1}. The best current value for our Galaxy is

$$\boxed{X_{CO} = (2 \pm 0.6) \times 10^{20} \text{ cm}^{-2} \text{ (K km s}^{-1})^{-1}.} \qquad (7.140)$$

X_{CO} is probably higher in galaxies with low metallicity and lower in starburst galaxies [14]. Color Plate 12 shows the distribution of CO emission tracing molecular gas and obscured star formation in the interacting starburst galaxies NGC 4038/9.

Isotopologues are molecules that differ only in isotopic composition; that is, only in the numbers of neutrons in their component atoms. Lines of the most abundant isotopologue of carbon monoxide, $^{12}C^{16}O$, are often optically thick, so the $^{12}C^{16}O$ line brightness temperature approaches the molecular gas kinetic temperature and is nearly independent of column density. Lines of rarer isotopologues such

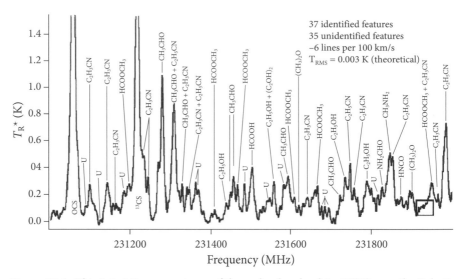

Figure 7.16. This $\lambda \approx 1.3$ mm spectrum of the molecular cloud SgrB2(N) near the Galactic center is completely dominated by molecular lines from known and unknown (U) species [119]. More than 140 different molecules containing up to 13 atoms (HC$_{11}$N) have been identified in space.

as ^{13}C^{16}O or ^{12}C^{18}O are usually optically thin and can be used to measure the column densities needed to estimate the total mass of molecular gas in a source. Intensity ratios of optically thin lines from different J levels can be used to measure excitation temperature, which is close to the kinetic temperature in LTE.

Transitions with high emission coefficients (e.g., the HCN (hydrogen cyanide) $J = 1 \rightarrow 0$ line at $\nu \approx 88.63$ GHz has $A_{\mathrm{UL}} \approx 2.0 \times 10^{-5}$ s^{-1}) are collisionally excited only at very high densities ($n^* \approx 10^5$ cm^{-3} for HCN $J = 1 \rightarrow 0$). They are valuable for highlighting only the very dense gas directly associated with the formation of individual stars.

The discovery of ammonia (NH$_3$) in the direction of the Galactic center by Cheung et al. [22] immediately led to the realization that the interstellar medium must contain regions much denser than previously expected because the critical density needed to excite the NH$_3$ line is $n^* \approx 10^3$ cm^{-3}.

7.8 THE HI 21-CM LINE

Hydrogen is the most abundant element in the interstellar medium (ISM), but the symmetric H$_2$ molecule has no permanent dipole moment and does not emit detectable spectral lines at radio frequencies. Neutral hydrogen (HI) atoms are abundant and ubiquitous in low-density regions of the ISM. They are detectable in the $\lambda \approx 21$ cm ($\nu_{10} = 1420.405751\ldots$ MHz) **hyperfine line**. Two energy levels result from the magnetic interaction between the quantized electron and proton spins. When the relative spins change from parallel to antiparallel, a photon is emitted.

The H I line center frequency is

$$\boxed{\nu_{10} = \frac{8}{3} g_{\mathrm{I}} \left(\frac{m_{\mathrm{e}}}{m_{\mathrm{p}}} \right) \alpha^2 (R_M c) \approx 1420.405751 \ \mathrm{MHz},} \qquad (7.141)$$

where $g_{\mathrm{I}} \approx 5.58569$ is the **nuclear g-factor** for a proton, $\alpha \equiv e^2/(\hbar c) \approx 1/137.036$ is the dimensionless **fine-structure constant**, and $R_M c$ is the hydrogen Rydberg frequency (Equation 7.12).

By analogy with the emission coefficient of radiation by an electric dipole

$$A_{\mathrm{UL}} \approx \frac{64\pi^4}{3hc^3} \nu_{\mathrm{UL}}^3 |\mu_{\mathrm{UL}}|^2, \qquad (7.142)$$

the emission coefficient of this magnetic dipole is

$$A_{\mathrm{UL}} \approx \frac{64\pi^4}{3hc^3} \nu_{\mathrm{UL}}^3 |\mu_{\mathrm{B}}|^2, \qquad (7.143)$$

where μ_{B} is the mean magnetic dipole moment for H I in the ground electronic state ($n = 1$). The magnitude $|\mu_{\mathrm{B}}|$ is called the **Bohr magneton**, and its value is

$$|\mu_{\mathrm{B}}| = \frac{e\hbar}{2m_{\mathrm{e}}c} \approx 9.27401 \times 10^{-21} \ \mathrm{erg \ gauss}^{-1}. \qquad (7.144)$$

Thus the emission coefficient of the 21-cm line is only

$$A_{10} \approx \frac{64\pi^4 (1.42 \times 10^9 \ \mathrm{Hz})^3 (9.27 \times 10^{-21} \ \mathrm{erg \ gauss}^{-1})^2}{3 \cdot 6.63 \times 10^{-27} \ \mathrm{erg \ s} \ (3 \times 10^{10} \ \mathrm{cm \ s}^{-1})^3}, \qquad (7.145)$$

$$\boxed{A_{10} \approx 2.85 \times 10^{-15} \ \mathrm{s}^{-1},} \qquad (7.146)$$

so the radiative half-life of this transition is very long:

$$\tau_{1/2} = A_{10}^{-1} \approx 3.5 \times 10^{14} \ \mathrm{s} \approx 11 \ \mathrm{million \ years}. \qquad (7.147)$$

Such a low emission coefficient implies an extremely low critical density (Equation 7.135) $n^* \ll 1 \ \mathrm{cm}^{-3}$, so collisions can easily maintain this transition in LTE, even in the outermost regions of a normal spiral galaxy and in tidal tails of interacting galaxies.

Regardless of whether the H I is in LTE or not, we can define the H I **spin temperature** T_{s} (the H I analog of the molecular excitation temperature T_{x} defined by Equation 7.70) by

$$\boxed{\frac{n_1}{n_0} \equiv \frac{g_1}{g_0} \exp\left(-\frac{h\nu_{10}}{kT_{\mathrm{s}}} \right),} \qquad (7.148)$$

where the statistical weights of the upper and lower spin states are $g_1 = 3$ and $g_0 = 1$, respectively. Note that

$$\frac{h\nu_{10}}{kT_{\mathrm{s}}} \approx \frac{6.63 \times 10^{-27} \ \mathrm{erg \ s} \cdot 1.42 \times 10^9 \ \mathrm{Hz}}{1.38 \times 10^{-16} \ \mathrm{erg \ K}^{-1} \cdot 150 \ \mathrm{K}} \approx 5 \times 10^{-4} \ll 1 \qquad (7.149)$$

is very small for gas in LTE at $T \approx T_s \approx 150$ K, so in the ISM

$$\frac{n_1}{n_0} \approx \frac{g_1}{g_0} = 3 \quad \text{and} \quad n_H = n_0 + n_1 \approx 4n_0. \tag{7.150}$$

Inserting these weights into Equation 7.67 gives the opacity coefficient of the $\lambda = 21$ cm line:

$$\kappa(\nu) = \frac{c^2}{8\pi \nu_{10}^2} \frac{g_1}{g_0} n_0 A_{10} \left[1 - \exp\left(-\frac{h\nu_{10}}{kT_s} \right) \right] \phi(\nu) \tag{7.151}$$

$$\approx \frac{c^2}{8\pi \nu_{10}^2} \cdot 3 \cdot \frac{n_H}{4} \cdot A_{10} \left(\frac{h\nu_{10}}{kT_s} \right) \phi(\nu), \tag{7.152}$$

$$\kappa(\nu) \approx \frac{3c^2}{32\pi} \frac{A_{10} n_H}{\nu_{10}} \frac{h}{kT_s} \phi(\nu), \tag{7.153}$$

where n_H is the number of neutral hydrogen atoms per cm^3. The neutral hydrogen **column density** along any line of sight is defined as

$$\eta_H \equiv \int_{los} n_H(s)\,ds. \tag{7.154}$$

The total opacity τ of isothermal HI is proportional to the column density. If $\tau \ll 1$, then the integrated HI emission-line brightness T_b is proportional to the column density of HI and is independent of the spin temperature T_s because $T_b \approx T_s \tau$ and $\tau \propto T_s^{-1}$ in the radio limit $h\nu_{10}/(kT_s) \ll 1$. Thus η_H can be determined directly from the integrated line brightness when $\tau \ll 1$. In astronomically convenient units it can be written as

$$\left(\frac{\eta_H}{\text{cm}^{-2}} \right) \approx 1.82 \times 10^{18} \int \left[\frac{T_b(\nu)}{K} \right] d\left(\frac{\nu}{\text{km s}^{-1}} \right), \tag{7.155}$$

where T_b is the observed 21-cm-line brightness temperature at radial velocity ν and the velocity integration extends over the entire 21-cm-line profile. Note that *absorption* by HI in front of a continuum source with continuum brightness temperature $> T_s$, on the other hand, is weighted in favor of colder gas (Figure 7.17).

The equilibrium temperature of cool interstellar HI is determined by the balance of heating and cooling. The primary heat sources are cosmic rays and ionizing photons from hot stars. The main coolant in the cool atomic ISM is radiation from the fine-structure line of singly ionized carbon, CII, at $\lambda = 157.7\,\mu$m. This line is strong only when the temperature is at least

$$kT \approx h\nu = \frac{hc}{\lambda}, \tag{7.156}$$

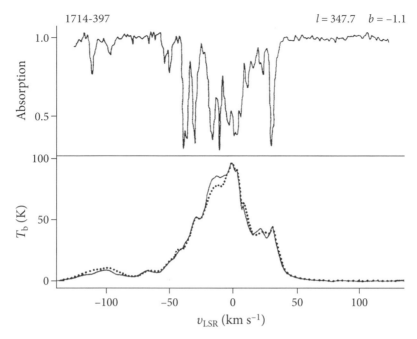

Figure 7.17. The H I absorption and emission spectra toward the source 1714-397 [35].

so the cooling rate increases exponentially above

$$T \approx \frac{hc}{k\lambda} \approx \frac{6.63 \times 10^{-27} \text{ erg s} \cdot 3 \times 10^{10} \text{ cm s}^{-1}}{1.38 \times 10^{-16} \text{ erg K}^{-1} \cdot 157.7 \times 10^{-4} \text{ cm}} \approx 91 \, K. \qquad (7.157)$$

The actual kinetic temperature of H I in our Galaxy can be estimated from the H I line brightness temperatures in directions where the line is optically thick ($\tau \gg 1$) and the brightness temperature approaches the excitation temperature, which is close to the kinetic temperature in LTE. Many lines of sight near the Galactic plane have brightness temperatures as high as 100–150 K, values consistent with the temperature-dependent cooling rate.

7.8.1 Galactic H I

Neutral hydrogen gas in the disk of our Galaxy moves in nearly circular orbits around the Galactic center. Radial velocities v_r measured from the Doppler shifts of H I $\lambda = 21$ cm emission lines encode information about the **kinematic distances** d of H I clouds, and the spectra of H I absorption in front of continuum sources can be used to constrain their distances also. H I is optically thin except in a few regions near the Galactic plane, so the distribution of hydrogen maps out the large-scale structure of the whole Galaxy, most of which is hidden by dust at visible wavelengths.

Figure 7.18 shows a plan view of the Galactic disk. The Sun ($\odot$) lies in the disk and moves in a circular orbit around the Galactic center. The distance to the Galactic center $R_\odot = 8.0 \pm 0.5$ kpc and the Sun's orbital speed $\omega_\odot R_\odot \approx 220$ km s^{-1} have been measured by a variety of means [88]. All H I clouds at galactocentric distance R are assumed to be in circular orbits with angular velocity $\omega(R)$, where $\omega(R)$ is a

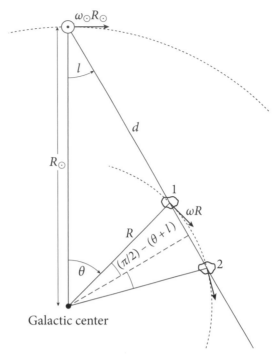

Figure 7.18. In the simplest realistic model for the Galactic disk, the Sun and all H I clouds are in circular orbits about the Galactic center, and the angular orbital velocity ω is a monotonically decreasing function of the orbital radius R. The distance of the Sun from the Galactic center is $R_\odot = 8.0 \pm 0.5$ kpc, and the Sun's orbital speed is $\omega_\odot R_\odot \approx 220$ km s^{-1}. The angle l defines the **Galactic longitude**. For $|l| < \pi/2$, two H I clouds (1 and 2) can have the same radial velocity but be at different distances from the Sun.

monotonically decreasing function of R. For cloud 1 at galactocentric azimuth θ on the line of sight at Galactic longitude l, the observed radial velocity v_r relative to the Sun is given by

$$v_r = \omega R \cos[\pi/2 - (l + \theta)] - \omega_\odot R_\odot \cos(\pi/2 - l). \qquad (7.158)$$

Using the trigonometric identities $\cos[\pi/2 - (l + \theta)] = \sin(l + \theta)$ and $\sin(l + \theta) = \sin\theta \cos l + \cos\theta \sin l$ we obtain

$$v_r = \omega R(\sin\theta \cos l + \cos\theta \sin l) - \omega_\odot R_\odot \sin l \qquad (7.159)$$

$$= R_\odot(\omega - \omega_\odot) \sin l. \qquad (7.160)$$

To apply this equation, we need to determine the **rotation curve** $R\omega(R)$. The maximum radial velocity on the line of sight at longitude l is called the "terminal velocity" v_T. Since ω decreases with R, this velocity occurs at the minimum $R = R_{\min} = R_\odot \sin l$ where the orbit is tangent to the line of sight:

$$v_T = R_\odot[\omega(R_{\min}) - \omega_\odot] \sin l. \qquad (7.161)$$

We can determine the rotation curve from measurements of v_T spanning a wide range of l and thus of $R_{\min}$.

Example. At Galactic longitude $l = 30°$, the terminal velocity is observed to be $v_T \approx$ 130 km s^{-1}. What is R_{min} and the orbital speed $R_{min}\omega(R_{min})$?

$$R_{min} = R_\odot \sin l = 8.0 \text{ kpc} \cdot 0.5 = 4.0 \text{ kpc},$$

$$v_T = R_\odot [\omega(R_{min}) - \omega_\odot] \sin l$$

$$= R_{min}\omega(R_{min}) - R_\odot\omega_\odot \sin l,$$

$$R_{min}\omega(R_{min}) = v_T + R_\odot\omega_\odot \sin l$$

$$= 130 \text{ km s}^{-1} + 220 \text{ km s}^{-1} \cdot 0.5 = 240 \text{ km s}^{-1}.$$

Beware that, for $|l| < \pi/2$, there is a distance ambiguity: clouds 1 and 2 have the same radial velocity but different distances d. There is no distance ambiguity for $|l| > \pi/2$.

7.8.2 H I in External Galaxies

The 1420 MHz H I line is an extremely useful tool for studying gas in the ISM of external galaxies and tracing the large-scale distribution of galaxies in the universe because H I is detectable in most spiral galaxies and in some elliptical galaxies.

Because $\lambda = 21$ cm is such a long wavelength, many galaxies are unresolved by single-dish radio telescopes. For example, the half-power beamwidth of the 100-m GBT is about 9 arcmin at $\lambda = 21$ cm. Thus a single pointing is sufficient to obtain a spectral line representing all of the H I in any but the nearest galaxies.

The observed center frequency of the H I line can be used to measure the radial velocity v_r of a galaxy. The radial velocity of a galaxy is the sum of the recession velocity caused by the uniform Hubble expansion of the universe and the "peculiar" velocity of the galaxy. The radial component of the peculiar velocity reflects motions caused by gravitational interactions with nearby galaxies and is typically ~ 200 km s^{-1} in magnitude. The Hubble velocity is proportional to distance from the Earth, and the **Hubble constant** of proportionality has been measured as $H_0 = 67.8 \pm 0.9$ km s^{-1} Mpc^{-1} [82]. If the radial velocity is significantly larger than the radial component of the peculiar velocity, the observed H I frequency can be used to estimate the distance $D \approx v_r/H_0$ to a galaxy.

Beware that astronomers still use inconsistent radial velocity conventions that were established when most observed radial velocities were much less than the speed of light. The approximation

$$\frac{v_r}{c} \approx \frac{\nu_e - \nu_o}{\nu_e} \qquad (v_r \ll c), \tag{7.162}$$

where ν_e is the line frequency in the source frame and ν_o is the observed frequency, was used to *define* the **radio velocity** for *any* v_r (radio) as

$$v_r(\text{radio}) \equiv c\left(\frac{\nu_e - \nu_o}{\nu_e}\right) \tag{7.163}$$

because radio astronomers measure frequencies, not wavelengths. Optical astronomers measure wavelengths, not frequencies, so the nonrelativistic

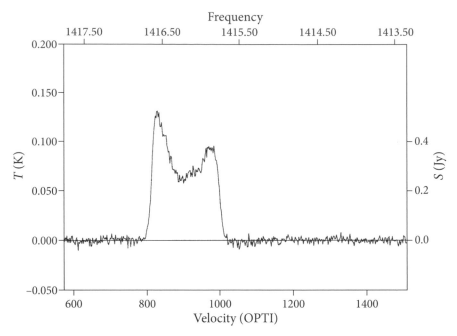

Figure 7.19. This integrated H I spectrum of UGC 11707 was obtained by Haynes et al. [47] with the 140-foot telescope (beamwidth $\approx$ 20 arcmin) and shows the typical two-horned profile of a spiral galaxy. The velocity axis is clearly labeled as showing the "optical" velocity.

approximation

$$\frac{v_r}{c} \approx \frac{\lambda_o - \lambda_e}{\lambda_e} \qquad (v_r \ll c) \tag{7.164}$$

is the basis for the **optical velocity** defined for any v_r(optical) by

$$v_r(\text{optical}) \equiv c\left(\frac{\lambda_o - \lambda_e}{\lambda_e}\right) = cz, \tag{7.165}$$

where z is the redshift defined by Equation 2.127. The optical and radio velocity conventions are not exactly the same, and neither agrees with the relativistically correct radial velocity calculated from Equation 5.142. Occasionally an observer confuses velocity conventions, fails to center the observing passband on the correct frequency, and ends up with only part of the H I spectrum of a galaxy. Outside the local universe ($z \ll 1$) the concept of distance itself becomes more complicated. To calculate distances to astronomical objects with larger redshifts, see Hogg [53, "Distance measures in cosmology"].

For example, the H I emission-line profile (Figure 7.19) of the galaxy UGC 11707 can be used to estimate its distance $D \approx v_r / H_0$. The observed line center frequency is $\nu_o \approx 1416.2$ MHz, so the "radio" and "optical" velocities are

$$v_r(\text{radio}) \approx c\left(1 - \frac{\nu_o}{\nu_e}\right) \approx 3 \times 10^5 \,\text{km s}^{-1}\left(1 - \frac{1416.2\,\text{MHz}}{1420.4\,\text{MHz}}\right) \approx 890\,\text{km s}^{-1},$$

$$v_r(\text{optical}) \approx c\left(\frac{\nu_e}{\nu_o} - 1\right) \approx 3 \times 10^5 \,\text{km s}^{-1}\left(\frac{1420.4\,\text{MHz}}{1416.2\,\text{MHz}} - 1\right) \approx 889\,\text{km s}^{-1}.$$

Using the "optical" velocity gives

$$D \approx \frac{v_r(\text{optical})}{H_0} \approx \frac{889\,\text{km s}^{-1}}{67.8\,\text{km s}^{-1}\,\text{Mpc}^{-1}} \approx 13.1\,\text{Mpc}.$$

If the HI emission from a galaxy is optically thin, then the integrated line flux is proportional to the mass of HI in the galaxy, independent of the unknown HI temperature. It is a straightforward exercise to derive from Equation 7.155 the relation

$$\boxed{\left(\frac{M_H}{M_\odot}\right) \approx 2.36 \times 10^5 \left(\frac{D}{\text{Mpc}}\right)^2 \int \left[\frac{S(v)}{\text{Jy}}\right]\left(\frac{dv}{\text{km s}^{-1}}\right)} \qquad (7.166)$$

for the total **HI mass** M_H of a galaxy. The integral $\int S(v)dv$ over the line is called the **line flux** and is usually expressed in units of Jy km s^{-1}. For example, to estimate the HI mass of UGC 11707, assume $\tau \ll 1$. The single-dish HI line profile of UGC 11707 (Figure 7.19) indicates a line flux

$$\int S(v)dv \approx 0.35\,\text{Jy} \cdot 200\,\text{km s}^{-1} \approx 70\,\text{Jy km s}^{-1},$$

so

$$\left(\frac{M_H}{M_\odot}\right) \approx 2.36 \times 10^5 \cdot (12.4)^2 \cdot 70 \approx 2.5 \times 10^9.$$

Small statistical corrections for nonzero τ can be made from knowledge about the expected opacity as a function of disk inclination, galaxy mass, morphological type, etc.

A well-resolved HI image of a galaxy yields the **total mass** $M(r)$ enclosed within radius r of the center if the gas orbits in circular orbits.

If the mass distribution of every galaxy were spherically symmetric, the gravitational force at radius r would equal the gravitational force of the enclosed mass $M(r)$. This is not a bad approximation, even for disk galaxies. Thus for gas in a circular orbit with orbital velocity v_{rot},

$$\frac{GM}{r^2} = \frac{v_{\text{rot}}^2}{r}, \qquad (7.167)$$

where M is the mass enclosed within the sphere of radius r and v is the orbital velocity at radius r, so

$$v_{\text{rot}}^2 = \frac{GM}{r}. \qquad (7.168)$$

Note that the velocity v_{rot} is the full rotational velocity, not just its radial component $v_{rot} \sin i$, where i is the inclination angle between the galaxy disk and the line of sight. The inclination angle of a thin circular disk can be estimated from the axial ratio

$$\cos i = \frac{\theta_m}{\theta_M}, \tag{7.169}$$

where θ_m and θ_M are the minor- and major-axis angular diameters, respectively. Converting from CGS to astronomically convenient units yields

$$\left[\left(\frac{v_{rot}}{cm\,s^{-1}} \right) \left(\frac{10^5\,cm\,s^{-1}}{km\,s^{-1}} \right) \right]^2 = \left[6.67 \times 10^{-8}\,dyne\,cm^2\,g^{-2} \cdot \left(\frac{M}{g} \right) \left(\frac{2 \times 10^{33}\,g}{M_\odot} \right) \right]$$

$$\times \left[\left(\frac{r}{cm} \right) \left(\frac{3.09 \times 10^{21}\,cm}{kpc} \right) \right]^{-1}, \tag{7.170}$$

$$10^{10} \left(\frac{v_{rot}}{km\,s^{-1}} \right)^2 = \left[6.67 \times 10^{-8} \cdot 2 \times 10^{33} \left(\frac{M}{M_\odot} \right) \right]$$

$$\times \left[3.09 \times 10^{21} \left(\frac{r}{kpc} \right) \right]^{-1}, \tag{7.171}$$

and we obtain the total galaxy mass inside radius r in units of the solar mass:

$$\boxed{ \left(\frac{M}{M_\odot} \right) \approx 2.3 \times 10^5 \left(\frac{v_{rot}}{km\,s^{-1}} \right)^2 \left(\frac{r}{kpc} \right). } \tag{7.172}$$

Thus the total mass of UGC 11707 (Figure 7.20) can be estimated from

$$v_{rot} \sin i \approx \frac{\Delta v_{rot} \sin i}{2} \approx \frac{(1000\,km\,s^{-1} - 800\,km\,s^{-1})}{2} \approx 100\,km\,s^{-1},$$

$$\cos i \approx \frac{minor\,axis}{major\,axis} \approx \frac{0.73 \times 10^{-3}\,rad}{2.0 \times 10^{-3}\,rad} \approx 0.365, \qquad so \qquad \sin i \approx 0.93,$$

$$r \approx \theta_{1/2} D \approx 10^{-3}\,rad \cdot 12.4\,Mpc \approx 12.4\,kpc,$$

so

$$\left(\frac{M}{M_\odot} \right) \approx 2.3 \times 10^5 \cdot (100/0.93)^2 \cdot 12.4 = 3.3 \times 10^{10}.$$

UGC 11707 is a relatively low-mass spiral galaxy.

This "total" mass is really only the mass inside the radius sampled by detectable HI. Even though HI extends beyond most other tracers such as molecular gas or stars, it is clear from plots of HI rotation velocities versus radius that not all of the mass is being sampled, because we don't see the Keplerian relation $v_{rot} \propto r^{-1/2}$ which indicates that all of the mass is enclosed within radius r. Most **rotation curves**, one-dimensional position-velocity diagrams along the major axis, are *flat* at large r,

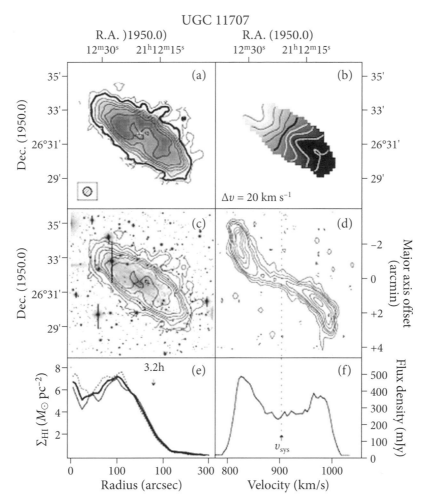

Figure 7.20. H I images of UGC 11707 [103]. The hatched circle in the lower left corner of panel (a) shows the image resolution. The contours in panels (a) and (c) outline the integrated H I brightness distribution. Panel (b) shows contours of constant velocity separated by 20 km s^{-1} and the darker shading indicates approaching gas. Panel (d) is a position-velocity diagram, panel (e) is the radial H I column-density profile, and panel (f) displays the integrated H I spectrum.

suggesting that the enclosed mass $M \propto r$ as far as we can see H I. The large total masses implied by H I rotation curves provided some of the earliest evidence for the existence of cold dark matter in galaxies.

Because detectable H I is so extensive, H I is an exceptionally sensitive tracer of tidal interactions between galaxies. Long streamers and tails of H I trace the interaction histories of pairs and groups of galaxies. See Color Plate 11 showing H I in the M81 group of galaxies and Color Plate 12 revealing long tidal tails of H I from the "Antennae" galaxy pair NGC 4038/9.

Another application of the H I spectra of galaxies is determining departures from smooth Hubble expansion in the local universe via the **Tully–Fisher relation**.

Most galaxies obey the empirical luminosity–velocity relation [108]:

$$L \propto v_{\mathrm{m}}^4, \tag{7.173}$$

where v_{m} is the maximum rotation speed. Arguments based on the virial theorem can explain the Tully–Fisher relation if all galaxies have the same central mass density and density profile, differing only in scale length, and also have the same mass-to-light ratio. Thus a measurement of v_{m} yields an estimate of L that is independent of the Hubble distance D_{H}. The Tully–Fisher distance D_{TF} can be calculated from this "standard candle" L and the apparent luminosity. Apparent luminosities in the near infrared ($\lambda \sim 2\,\mu$m) are favored because the near-infrared mass-to-light ratio of stars is nearly constant and independent of the star-formation history, and because extinction by dust is much less than at optical wavelengths. Differences between D_{TF} and D_{H} are ascribed to the **peculiar velocities** of galaxies caused by intergalactic gravitational interactions. The magnitudes and scale lengths of the peculiar velocity distributions are indications of the average density and clumpiness of mass on megaparsec scales.

7.8.3 Dark Ages and the Epoch of Reionization (EOR)

Most of the baryonic matter in the early universe was fully ionized hydrogen and helium gas, plus trace amounts of heavier elements. This smoothly distributed gas cooled as the universe expanded, and the free protons and electrons recombined to form neutral hydrogen at a redshift $z \approx 1091$ when the age of the universe was about 3.8×10^5 years. The hydrogen remained neutral during the **dark ages** prior to the formation of the first ionizing astronomical sources—massive ($M > 100 M_\odot$) stars, galaxies, quasars, and clusters of galaxies—by gravitational collapse of overdense regions. These astronomical sources gradually started reionizing the universe when it was several hundred million years old ($z \sim 10$) and completely reionized the universe by the time it was about 10^9 years old ($z \sim 6$). This era is called the **epoch of reionization**.

The highly redshifted H$\textsc{i}$ signal was nearly uniform during the dark ages, and it developed structure on angular scales up to several arcmin when the first astronomical sources created bubbles of ionized hydrogen around them. As the bubbles grew and merged, the H$\textsc{i}$ signal developed frequency structure corresponding to the redshifted H$\textsc{i}$ line frequency. The characteristic size of the larger bubbles reached about 10 Mpc at $z \sim 6$ and produced H$\textsc{i}$ signals having angular scales of several arcmin and covering frequency ranges of several MHz. These H$\textsc{i}$ signals encode unique information about the formation of the earliest astronomical sources.

The H$\textsc{i}$ signals produced by the EOR will be very difficult to detect because they are weak (tens of mK), relatively broad in frequency, redshifted to low frequencies (~ 100 MHz) plagued by radio-frequency interference and ionospheric refraction, and lie behind a much brighter (tens of K) foreground of extragalactic continuum radio sources. Nonetheless, the potential scientific payoff is so great that several groups around the world are developing instruments to detect the H$\textsc{i}$ signature of the EOR. Two such instruments are PAPER (Precision Array to Probe the Epoch of Reionization) [78] and the Murchison Widefield Array [107] shown in Color Plate 7.

A Fourier Transforms

A.1 THE FOURIER TRANSFORM

The continuous **Fourier transform** is important in mathematics, engineering, and the physical sciences. Its counterpart for discretely sampled functions is the **discrete Fourier transform (DFT)**, which is normally computed using the so-called fast Fourier transform (FFT). The DFT has revolutionized modern society, as it is ubiquitous in digital electronics and signal processing. Radio astronomers are particularly avid users of Fourier transforms because Fourier transforms are key components in data processing (e.g., periodicity searches) and instruments (e.g., antennas, receivers, spectrometers), and they are the cornerstones of interferometry and aperture synthesis. Useful references include Bracewell [15] (which is on the shelves of most radio astronomers) and the Wikipedia[1] and Mathworld[2] entries for the Fourier transform.

The Fourier transform is a reversible, linear transform with many important properties. Any complex function $f(x)$ of the real variable x that is both integrable $\left(\int_{\infty}^{\infty} |f(x)| dx < \infty \right)$ and contains only finite discontinuities has a complex Fourier transform $F(s)$ of the real variable s, where the product of x and s is dimensionless and unity. For example, the Fourier transform of the waveform $f(t)$ expressed as a *time-domain* signal is the spectrum $F(\nu)$ expressed as a *frequency-domain* signal. If t is given in seconds of time, ν is in s^{-1} = Hz.

The **Fourier transform** of $f(x)$ is defined by

$$F(s) \equiv \int_{-\infty}^{\infty} f(x)\, e^{-2\pi i s x}\, dx, \tag{A.1}$$

which is usually known as the **forward transform**, and

$$f(x) \equiv \int_{-\infty}^{\infty} F(s)\, e^{2\pi i s x}\, ds, \tag{A.2}$$

---

[1] http://en.wikipedia.org/wiki/Fourier_transform.
[2] http://mathworld.wolfram.com/FourierTransform.html.

which is the **inverse transform**. In both cases, $i \equiv \sqrt{-1}$. Alternative definitions of the Fourier transform are based on angular frequency $\omega \equiv 2\pi\nu$, have different normalizations, or the opposite sign convention in the complex exponential. Successive forward and reverse transforms return the original function, so the Fourier transform is cyclic and reversible. The symmetric symbol $\Leftrightarrow$ is often used to mean "is the Fourier transform of"; e.g., $F(s) \Leftrightarrow f(x)$.

The complex exponential (Appendix B.3) is the heart of the transform. A complex exponential is simply a complex number where both the real and imaginary parts are sinusoids. The exact relation is called **Euler's formula**

$$e^{i\phi} = \cos\phi + i\sin\phi, \tag{A.3}$$

which leads to the famous (and beautiful) identity $e^{i\pi} + 1 = 0$ that relates five of the most important numbers in mathematics. Complex exponentials are much easier to manipulate than trigonometric functions, and they provide a compact notation for dealing with sinusoids of arbitrary phase, which form the basis of the Fourier transform.

Complex exponentials (or sines and cosines) are periodic functions, and the set of complex exponentials is complete and orthogonal. Thus the Fourier transform can represent any piecewise continuous function and minimizes the least-square error between the function and its representation. There exist other complete and orthogonal sets of periodic functions; for example, Walsh functions[3] (square waves) are useful for digital electronics. Why do we always encounter complex exponentials when solving physical problems? Why are monochromatic waves sinusoidal, and not periodic trains of square waves or triangular waves? The reason is that the derivatives of complex exponentials are just rescaled complex exponentials. In other words, the complex exponentials are the eigenfunctions of the differential operator. Most physical systems obey linear differential equations. Thus an analog electronic filter will convert a sine wave into another sine wave having the same frequency (but not necessarily the same amplitude and phase), while a filtered square wave will not be a square wave. This property of complex exponentials makes the Fourier transform uniquely useful in fields ranging from radio propagation to quantum mechanics.

A.2 THE DISCRETE FOURIER TRANSFORM

The continuous Fourier transform converts a time-domain signal of infinite duration into a continuous spectrum composed of an infinite number of sinusoids. In astronomical observations we deal with signals that are discretely sampled, usually at constant intervals, and of finite duration or periodic. For such data, only a finite number of sinusoids is needed and the **discrete Fourier transform** (DFT) is appropriate. For almost every Fourier transform theorem or property, there is a related theorem or property for the DFT. The DFT of N data points x_j (where

---

[3]http://en.wikipedia.org/wiki/Walsh_function.

TABLE A.1.
Symmetries between time- and frequency-domain signals.

Time domain	Frequency domain
real	hermitian (real=even, imag=odd)
imaginary	anti-hermitian (real=odd, imag=even)
even	even
odd	odd
real and even	real and even (i.e., cosine transform)
real and odd	imaginary and odd (i.e., sine transform)
imaginary and even	imaginary and even
imaginary and odd	real and odd

$j = 0, \ldots, N - 1$) sampled at uniform intervals and its inverse are defined by

$$X_k \equiv \sum_{j=0}^{N-1} x_j \, e^{-2\pi i j k / N} \tag{A.4}$$

and

$$x_j \equiv \frac{1}{N} \sum_{k=0}^{N-1} X_k \, e^{2\pi i j k / N}. \tag{A.5}$$

Once again, sign and normalization conventions may vary, but this definition is the most common. The continuous variable s has been replaced by the discrete variable (usually an integer) k.

The DFT of an N-point input time series is an N-point frequency spectrum, with Fourier frequencies k ranging from $-(N/2 - 1)$, through the 0-frequency or so-called DC component, and up to the highest Fourier frequency $N/2$. Each bin number represents the integer number of sinusoidal periods present in the time series. The amplitudes and phases represent the amplitudes A_k and phases ϕ_k of those sinusoids. In summary, each bin can be described by $X_k = A_k \, e^{i\phi_k}$.

For real-valued input data, the resulting DFT is **hermitian**—the real part of the spectrum is an even function and the imaginary part is odd, such that $X_{-k} = \overline{X_k}$, where the bar represents complex conjugation. This means that all of the "negative" Fourier frequencies provide no new information. Both the $k = 0$ and $k = N/2$ bins are real valued, and there is a total of $N/2 + 1$ Fourier bins, so the total number of independent pieces of information (i.e., real and complex parts) is N, just as for the input time series. No information is created or destroyed by the DFT.

Other symmetries existing between time- and frequency-domain signals are shown in Table A.1.

DFT versus FFT Example. Estimate the speed increases obtained by computing the FFTs instead of DFTs for transforms of length 10^3, 10^6, and 10^9 points:

$$\text{speed improvement}\,(N = 10^3) \propto \frac{N^2}{N\log_2(N)} = \frac{N}{\log_2(N)} \sim \frac{10^3}{10} \sim 100,$$

$$\text{speed improvement}\,(N = 10^6) \propto \frac{N}{\log_2(N)} \sim \frac{10^6}{20} \sim 5 \times 10^4$$

$$\text{speed improvement}\,(N = 10^9) \propto \frac{N}{\log_2(N)} \sim \frac{10^9}{30} \sim 3 \times 10^7.$$

These speed improvements can be *huge*, and they are one reason why the FFT has become ubiquitous in modern society.

Usually the DFT is computed by a very clever (and truly revolutionary) algorithm known as the **fast Fourier transform (FFT)**. The FFT was discovered by Gauss in 1805 and re-discovered many times since, but most people attribute its modern incarnation to James W. Cooley and John W. Tukey [32] in 1965. The key advantage of the FFT over the DFT is that the operational complexity decreases from $O(N^2)$ for a DFT to $O(N\log_2(N))$ for the FFT. Modern implementations of the FFT[4] allow $O(N\log_2(N))$ complexity for *any* value of N, not just those that are powers of two or the products of only small primes.

A.3 THE SAMPLING THEOREM

Much of modern radio astronomy is now based on **digital signal processing (DSP)**, which relies on continuous radio waves being accurately represented by a series of discrete digital samples of those waves. An amazing theorem which underpins DSP and has strong implications for information theory is known as the **Nyquist– Shannon theorem** or the **sampling theorem**. It states that any bandwidth-limited (or **band-limited**) continuous function confined within the frequency range Δv may be reconstructed *exactly* from uniformly spaced samples separated in time by $\leq (2\Delta v)^{-1}$. The critical sampling rate $(\Delta t)^{-1} = 2\Delta v$ is known as the **Nyquist rate**, and the spacing between samples must satisfy $\Delta t \leq 1/(2\Delta v)$ seconds. A Nyquist-sampled time series contains *all* the information of the original continuous signal, and because the DFT is a reversible linear transform, the DFT of that time series contains all of the information as well.

---

[4]http://www.fftw.org.

Sampling Theorem Examples.
The (young and undamaged) human ear can hear sounds with frequency components up to ~ 20 kHz. Therefore, nearly perfect audio recording systems must sample audio signals at Nyquist frequencies $\nu_{N/2} \geq 40$ kHz. Audio CDs are sampled at 44.1 kHz to give imperfect lowpass audio filters a 2 kHz buffer to remove higher frequencies which would otherwise be aliased into the audible band.

A visual example of an aliased signal is seen in movies where the 24 frame-per-second rate of the movie camera performs "stroboscopic" sampling of a rotating wagon wheel with n uniformly spaced spokes. When the rotation frequency of the wheel is below the Nyquist rate ($12/n$ Hz), the wheel appears to be turning at the correct rate and in the correct direction. When it is rotating faster than $12/n$ Hz but slower than $24/n$ Hz, it appears to be rotating backward and at a slower rate. As the rotation rate approaches $24/n$ Hz, the wheel apparently slows down and then stops when the rotation rate equals twice the Nyquist rate.

The frequency corresponding to the sampled bandwidth, which is also the maximum frequency in a DFT of the Nyquist-sampled signal of length N, is known as the **Nyquist frequency**,

$$\nu_{N/2} = 1/(2\,\Delta t).$$
(A.6)

The Nyquist frequency describes the high-frequency cut-off of the system doing the sampling, and is therefore a property of that system. Any frequencies present in the original signal at higher frequencies than the Nyquist frequency, meaning that the signal was either not properly Nyquist sampled or band limited, will be **aliased** to other lower frequencies in the sampled band as described below. No aliasing occurs for band-limited signals sampled at the Nyquist rate or higher.

In a DFT, where there are N samples spanning a total time $T = N\,\Delta t$, the **frequency resolution** is $1/T$. Each Fourier bin number k represents exactly k sinusoidal oscillations in the original data x_j, and therefore a frequency $\nu = k/T$ in Hz. The Nyquist frequency corresponds to bin $k = \nu_{N/2}T = T/(2\,\Delta T) = NT/(2T) = N/2$. If the signal is not bandwidth limited and higher-frequency components (with $k > N/2$ or $\nu > N/(2T)$ Hz) exist, those frequencies will show up in the DFT shifted to lower frequencies $\nu_a = N/T - \nu$, assuming that $N/(2T) < \nu < N/T$. Such aliasing can be avoided by filtering the input data to ensure that it is properly band limited.

Note that the Sampling theorem does not demand that the original continuous signal be a **baseband** signal, one whose band begins at zero frequency and continues to frequency $\Delta\nu$. The signal can be in any frequency range ν_{min} to ν_{max} such that $\Delta\nu \geq \nu_{max} - \nu_{min}$. Most radio receivers and instruments have a finite bandwidth centered at some high frequency such that the bottom of the band is not at zero frequency. They will either use the technique of heterodyning (Section 3.6.4) to mix the high-frequency band to baseband where it can then be Nyquist sampled or, alternatively, aliasing can be used as part of the sampling scheme. For example, a 1–2 GHz filtered band from a receiver could be mixed to baseband and sampled at 2 GHz, the Nyquist rate for that bandwidth; or the original signal could be sampled at 2 GHz and the 1 GHz bandwidth will be properly Nyquist sampled, but the band will be flipped in its frequency direction by aliasing. This is often called sampling in the "second Nyquist zone." Higher Nyquist zones can be sampled as well, and the band direction flips with each successive zone (e.g., the band direction is normal in odd-numbered Nyquist zones and flipped in even-numbered zones) by aliasing.

A.4 THE POWER SPECTRUM

A useful quantity in astronomy is the **power spectrum** $\overline{F(s)}F(s) = |F(s)|^2$. The power spectrum preserves no phase information from the original function. **Rayleigh's theorem** (sometimes called Plancherel's theorem and related to Parseval's theorem for Fourier series) shows that the integral of the power spectrum equals the integral of the squared modulus of the function (e.g., signal energies are equal in the frequency and time domains):

$$\int_{-\infty}^{\infty} |f(x)|^2\, dx = \int_{-\infty}^{\infty} |F(s)|^2\, ds. \tag{A.7}$$

A.5 BASIC TRANSFORMS

Figure A.1 shows some basic Fourier transform pairs. These can be combined using the Fourier transform theorems below to generate the Fourier transforms of many different functions. There is a nice Java applet[5] on the web that lets you experiment with various simple DFTs.

A.6 BASIC FOURIER THEOREMS

Addition Theorem. The Fourier transform of the sum of two functions $f(x)$ and $g(x)$ is the sum of their Fourier transforms $F(s)$ and $G(s)$. This basic theorem follows from the linearity of the Fourier transform:

$$f(x) + g(x) \Leftrightarrow F(s) + G(s). \tag{A.8}$$

Likewise from linearity, if a is a constant, then

$$af(x) \Leftrightarrow aF(s). \tag{A.9}$$

Shift Theorem. A function $f(x)$ shifted along the x-axis by a to become $f(x - a)$ has the Fourier transform $e^{-2\pi i a s} F(s)$. The magnitude of the transform is the same, only the phases change:

$$f(x - a) \Leftrightarrow e^{-2\pi i a s} F(s). \tag{A.10}$$

[5]http://webphysics.davidson.edu/Applets/mathapps/mathapps_fft.html.

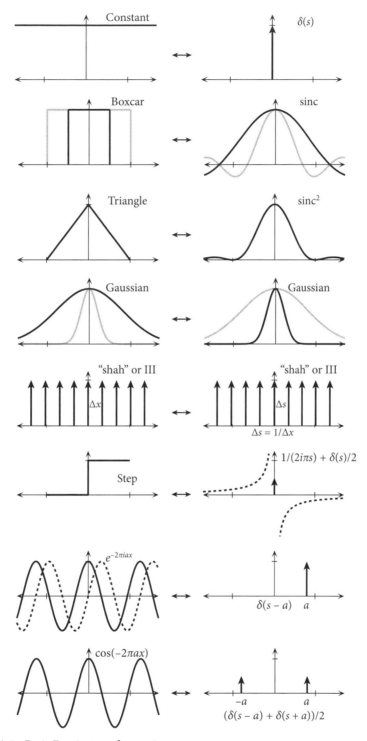

Figure A.1. Basic Fourier transform pairs.

Similarity Theorem. For a function $f(x)$ with a Fourier transform $F(s)$, if the x-axis is scaled by a constant a so that we have $f(ax)$, the Fourier transform becomes $|a|^{-1} F(s/a)$. In other words, a short, wide function in the time domain transforms to a tall, narrow function in the frequency domain, always conserving the area under the transform. This is the basis of the uncertainty principle in quantum mechanics and the diffraction limits of radio telescopes:

$$f(ax) \Leftrightarrow \frac{F(s/a)}{|a|}. \tag{A.11}$$

Modulation Theorem. The Fourier transform of the product $f(x)\cos(2\pi vx)$ is $\frac{1}{2}F(s-v) + \frac{1}{2}F(s+v)$. This theorem is very important in radio astronomy as it describes how signals can be "mixed" to different intermediate frequencies (IFs):

$$f(x)\cos(2\pi vx) \Leftrightarrow \frac{1}{2}F(s-v) + \frac{1}{2}F(s+v). \tag{A.12}$$

Derivative Theorem. The Fourier transform of the derivative of a function $f(x)$, df/dx, is $i2\pi s F(s)$:

$$\frac{df}{dx} \Leftrightarrow i2\pi s F(s). \tag{A.13}$$

Differentiation in the time domain boosts high-frequency spectral components, attenuates low-frequency components, and eliminates the DC component altogether.

A.7 CONVOLUTION AND CROSS-CORRELATION

Convolution shows up in many aspects of astronomy, most notably in the point-source response of an imaging system and in interpolation. Convolution, which we will represent by $*$ (the symbol $\otimes$ is also frequently used for convolution), multiplies one function f by the time-reversed **kernel** function g, shifts g by some amount u, and integrates u from $-\infty$ to $+\infty$. The **convolution** $h(x)$ of the functions f and g is a linear functional defined by

$$h(x) = f * g \equiv \int_{-\infty}^{\infty} f(u)g(x - u)\, du. \tag{A.14}$$

In Figure A.2, notice how the delta-function portion of the function produces an image of the kernel in the convolution. For a time series, that kernel defines the impulse response of the system. For an antenna or imaging system, the kernel is variously called the beam, the point-source response, or the point-spread function.

A very nice applet showing how convolution works is available online[6] (there is also a discrete version[7]); a different visualization tool is also available.[8]

---

[6]http://www.jhu.edu/~signals/convolve/index.html.

[7]http://www.jhu.edu/~signals/discreteconv2/index.html.

[8]https://maxwell.ict.griffith.edu.au/spl/Excalibar/Jtg/Conv.html.

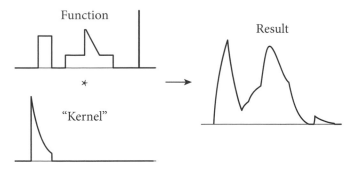

Figure A.2. Convolution example.

The **convolution theorem** is extremely powerful and states that the Fourier transform of the convolution of two functions is the product of their individual Fourier transforms:

$$f * g \Leftrightarrow F \cdot G. \tag{A.15}$$

Cross-correlation is a very similar operation to convolution, except that the kernel is not time-reversed. Cross-correlation is used extensively in interferometry and aperture synthesis imaging, and is also used to perform optimal "matched filtering" of data to detect weak signals in noise. Cross-correlation is represented by the pentagram symbol $\star$ and defined by

$$f \star g \equiv \int_{-\infty}^{\infty} f(u)g(u - x)\, du. \tag{A.16}$$

In this case, unlike for convolution, $f(x) \star g(x) \neq g(x) \star f(x)$.

Closely related to the convolution theorem, the **cross-correlation theorem** states that the Fourier transform of the cross-correlation of two functions is equal to the product of the individual Fourier transforms, where one of them has been complex conjugated:

$$f \star g \Leftrightarrow \overline{F} \cdot G. \tag{A.17}$$

Autocorrelation is a special case of cross-correlation with $f \star f$. The related autocorrelation theorem is also known as the **Wiener–Khinchin theorem** and states

$$f \star f \Leftrightarrow \overline{F} \cdot F = |F|^2. \tag{A.18}$$

In words, *the Fourier transform of an autocorrelation function is the power spectrum, or equivalently, the autocorrelation is the inverse Fourier transform of the power spectrum*. Many radio-astronomy instruments compute power spectra using autocorrelations and this theorem. The following diagram summarizes the relations between a function, its Fourier transform, its autocorrelation, and its power spectrum:

$$
\begin{array}{ccc}
x_j & \Leftrightarrow & X_k \\
\text{(function)} & \text{DFT} & \text{(transform)} \\
\Downarrow & & \Downarrow \\
x_j \star x_j & \Leftrightarrow & |X_k|^2 \\
\text{(autocorrelation)} & \text{DFT} & \text{(power spectrum).}
\end{array}
$$

One important thing to remember about convolution and correlation using DFTs is that they are *cyclic* with a period corresponding to the length of the longest component of the convolution or correlation. Unless this periodicity is taken into account (usually by zero-padding one of the input functions), the convolution or correlation will wrap around the ends and possibly "contaminate" the resulting function.

A.8 OTHER FOURIER TRANSFORM LINKS

Of interest on the web, other Fourier-transform-related links include a Fourier series applet,[9] some tones, harmonics, filtering, and sounds,[10] and a nice online book on the mathematics of the DFT.[11]

---

[9]http://www.jhu.edu/~signals/fourier2/index.html.
[10]http://www.jhu.edu/~signals/listen-new/listen-newindex.htm.
[11]http://ccrma.stanford.edu/~jos/mdft/mdft.html.

B   Mathematical Derivations

B.1 EVALUATION OF PLANCK'S SUM

Planck's sum (Equation 2.83) for the average energy per mode of blackbody radiation is

$$\langle E \rangle = \frac{\sum\limits_{n=0}^{\infty} nh\nu \exp\left(-\frac{nh\nu}{kT}\right)}{\sum\limits_{n=0}^{\infty} \exp\left(-\frac{nh\nu}{kT}\right)}.$$

It is convenient to introduce the variable $\alpha \equiv 1/(kT)$, so

$$\langle E \rangle = \frac{\sum\limits_{n=0}^{\infty} nh\nu \exp(-\alpha nh\nu)}{\sum\limits_{n=0}^{\infty} \exp(-\alpha nh\nu)}.$$

Next consider the quantity

$$-\frac{d}{d\alpha}\left[\ln \sum_{n=0}^{\infty} \exp(-\alpha nh\nu)\right].$$

Using the chain rule to take the derivative yields

$$-\frac{d}{d\alpha}\left[\ln \sum_{n=0}^{\infty} \exp(-\alpha nh\nu)\right] = -\left[\sum_{n=0}^{\infty} \exp(-\alpha nh\nu)\right]^{-1} \frac{d}{d\alpha}\left[\sum_{n=0}^{\infty} \exp(-\alpha nh\nu)\right]$$

$$= \frac{\sum\limits_{n=0}^{\infty} nh\nu \exp(-\alpha nh\nu)}{\sum\limits_{n=0}^{\infty} \exp(-\alpha nh\nu)}.$$

Thus

$$\langle E \rangle = -\frac{d}{d\alpha}\left[\ln \sum_{n=0}^{\infty} \exp(-\alpha n h v)\right].$$

Then,

$$\sum_{n=0}^{\infty} \exp(-\alpha n h v) = 1 + [\exp(-\alpha h v)]^1 + [\exp(-\alpha h v)]^2 + \cdots$$

has the form $1 + x + x^2 + \cdots = (1 - x)^{-1}$, so

$$\sum_{n=0}^{\infty} \exp(-\alpha n h v) = [1 - \exp(-\alpha h v)]^{-1}$$

and

$$\langle E \rangle = -\frac{d \ln[1 - \exp(\alpha h v)]^{-1}}{d\alpha}$$

$$= -[1 - \exp(-\alpha h v)](-1)[1 - \exp(-\alpha h v)]^{-2} h v \exp(-\alpha h v)$$

$$= \frac{h v \exp(-\alpha h v)}{1 - \exp(-\alpha h v)} = \frac{h v}{\exp(\alpha h v) - 1} = \frac{h v}{\exp\left(\dfrac{h v}{kT}\right) - 1}.$$

B.2 DERIVATION OF THE STEFAN–BOLTZMANN LAW

The **Stefan–Boltzmann law** for the **integrated brightness** of blackbody radiation at temperature T (Equation 2.89) is

$$B(T) = \int_0^{\infty} B_v(T)dv = \frac{\sigma T^4}{\pi},$$

where

$$B_v(T) = \frac{2h v^3}{c^2} \frac{1}{\exp\left(\dfrac{h v}{kT}\right) - 1}$$

is Planck's law and σ is the **Stefan–Boltzmann constant**. Although the Stefan–Boltzmann law and constant were first determined experimentally, both can be derived mathematically from Planck's law. For simplicity, define

$$x \equiv \frac{h v}{kT},$$

so

$$B(T) = \int_0^{\infty} \frac{2h}{c^2}\left(\frac{kTx}{h}\right)^3 \left(\frac{1}{e^x - 1}\right)\left(\frac{kT}{h}\right)dx = \frac{2k^4 T^4}{c^2 h^3}\int_0^{\infty} \frac{x^3 dx}{e^x - 1}.$$

The quantity

$$\frac{1}{e^x - 1} = \frac{e^{-x}}{1 - e^{-x}} = e^{-x}\left(\frac{1}{1 - e^{-x}}\right)$$

can be expanded in terms of the infinite series

$$\sum_{m=0}^{\infty} z^m = 1 + z + z^2 + z^3 + \cdots$$

$$= 1 + z(1 + z + z^2 + z^3 + \cdots)$$

$$= 1 + z\sum_{m=0}^{\infty} z^m,$$

$$\sum_{m=0}^{\infty} z^m = \frac{1}{1 - z}.$$

Thus

$$\frac{1}{e^x - 1} = e^{-x}\sum_{m=0}^{\infty} e^{-mx} = e^{-x} + e^{-2x} + e^{-3x} + \cdots$$

and the integral becomes

$$\int_0^{\infty} \frac{x^3 dx}{e^x - 1} = \int_0^{\infty} x^3 \left(\sum_{m=1}^{\infty} e^{-mx}\right) dx.$$

Each integral in this series can be integrated by parts three times:

$$\int_0^{\infty} x^3 e^{-mx} dx = \frac{x^3 e^{-mx}}{-m}\Big|_0^{\infty} - \int_0^{\infty} \frac{3x^2 e^{-mx}}{-m} dx = \frac{3}{m}\int_0^{\infty} x^2 e^{-mx} dx,$$

$$\int_0^{\infty} x^2 e^{-mx} dx = \frac{x^2 e^{-mx}}{-m}\Big|_0^{\infty} - \int_0^{\infty} \frac{2x e^{-mx}}{-m} dx = \frac{2}{m}\int_0^{\infty} x e^{-mx} dx,$$

$$\int_0^{\infty} x e^{-mx} dx = \frac{x e^{-mx}}{-m}\Big|_0^{\infty} - \int_0^{\infty} \frac{x e^{-mx}}{-m} dx = \frac{1}{m}\int_0^{\infty} e^{-mx} dx = \frac{1}{m^2},$$

to give

$$\int_0^{\infty} x^3 e^{-mx} dx = \frac{6}{m^4}$$

and

$$\int_0^{\infty} \frac{x^3 dx}{e^x - 1} = \int_0^{\infty} x^3 \left(\sum_{m=1}^{\infty} e^{-mx}\right) dx = 6\sum_{m=1}^{\infty} \frac{1}{m^4}.$$

The sum

$$\sum_{m=1}^{\infty} \frac{1}{m^4} = \frac{1}{1^4} + \frac{1}{2^4} + \frac{1}{3^4} + \frac{1}{4^4} + \cdots = 1 + \frac{1}{16} + \frac{1}{81} + \frac{1}{256} + \cdots \approx 1.082$$

converges quickly and is the value of the **Riemann zeta function** $\zeta(4) = \pi^4/90 \approx 1.082$. Thus

$$\boxed{\int_0^\infty \frac{x^3 dx}{e^x - 1} = \frac{\pi^4}{15}.}$$

(B.1)

Finally, the integrated brightness of blackbody radiation is

$$B(T) = \frac{2k^4 T^4}{c^2 h^3} \int_0^\infty \frac{x^3 dx}{e^x - 1} = \frac{2k^4 T^4}{c^2 h^3} \left(\frac{\pi^4}{15} \right) = \frac{2\pi^4 k^4}{15 c^2 h^3} T^4 = \frac{\sigma T^4}{\pi},$$

so

$$\sigma = \frac{2\pi^5 k^4}{15 c^2 h^3} \approx 5.67 \times 10^{-5} \frac{\text{erg}}{\text{cm}^2 \text{ s K}^4 \text{ (sr)}}$$

is the value of the Stefan–Boltzmann constant.

Similarly, the integral

$$\int_0^\infty \frac{x^2 dx}{e^x - 1}$$

is needed to evaluate the number density n_γ of blackbody photons:

$$n_\gamma = \frac{8\pi}{c^3} \int_0^\infty \frac{v^2 dv}{\exp\left(\frac{hv}{kT}\right) - 1} = \frac{8\pi}{c^3} \left(\frac{kT}{h} \right)^3 \int_0^\infty \frac{x^2 dx}{e^x - 1}.$$

Following the derivation above,

$$\int_0^\infty \frac{x^2 dx}{e^x - 1} = \int_0^\infty x^2 \left(\sum_{m=1}^{\infty} e^{-mx} \right) dx$$

and

$$\int_0^\infty x^2 e^{-mx} dx = \frac{2}{m} \int_0^\infty x e^{-mx} dx = \frac{2}{m} \left(\frac{1}{m^2} \right) = \frac{2}{m^3},$$

so

$$\boxed{\int_0^\infty \frac{x^2 dx}{e^x - 1} = 2 \sum_{m=1}^{\infty} \frac{1}{m^3} = 2 \left(\frac{1}{1^3} + \frac{1}{2^3} + \frac{1}{3^3} + \cdots \right) \approx 2.404.}$$

(B.2)

B.3 COMPLEX EXPONENTIALS

A complex exponential $e^{i\phi}$, where $i^2 = -1$ and ϕ is any *dimensionless* real variable, is a complex number in which the real and imaginary parts are sines and cosines given by **Euler's formula**

$$\boxed{e^{i\phi} = \cos\phi + i\sin\phi.}$$ (B.3)

Euler's formula can be derived from the Taylor series

$$\cos\phi = 1 - \frac{\phi^2}{2!} + \frac{\phi^4}{4!} - \frac{\phi^6}{6!} + \cdots,$$

$$\sin\phi = \phi - \frac{\phi^3}{3!} + \frac{\phi^5}{5!} - \frac{\phi^7}{7!} + \cdots,$$

$$e^\phi = 1 + \phi + \frac{\phi^2}{2!} + \frac{\phi^3}{3!} + \frac{\phi^4}{4!} + \cdots.$$

Thus

$$e^{i\phi} = 1 + i\phi - \frac{\phi^2}{2!} - \frac{i\phi^3}{3!} + \frac{\phi^4}{4!} + \frac{i\phi^5}{5!} - \frac{i\phi^6}{6!} - \frac{i\phi^7}{7!} + \cdots$$

$$= \left(1 - \frac{\phi^2}{2!} + \frac{\phi^4}{4!} - \frac{\phi^6}{6!} + \cdots\right) + i\left(\phi - \frac{\phi^3}{3!} + \frac{\phi^5}{5!} - \frac{\phi^7}{7!} + \cdots\right)$$

$$= \cos\phi + i\sin\phi.$$

Complex exponentials (or sines and cosines) are widely used to represent periodic functions in physics for the following reasons:

1. They comprise a complete and orthogonal set of periodic functions. This set of functions can be used to approximate any piecewise continuous function, and they are the basis of Fourier transforms (Appendix A.1).
2. They are eigenfunctions of the differential operator—that is, the derivatives of complex exponentials are themselves complex exponentials:

$$\frac{de^{i\phi}}{d\phi} = ie^{i\phi}, \quad \frac{d^2e^{i\phi}}{d\phi^2} = -e^{i\phi}, \quad \frac{d^3e^{i\phi}}{d\phi^3} = -ie^{i\phi}, \quad \frac{d^4e^{i\phi}}{d\phi^4} = e^{i\phi}, \ldots.$$

Most physical systems obey linear differential equations, a low-pass filter consisting of a resistor and a capacitor, for example. A sinusoidal input signal will yield a sinusoidal output signal of the same frequency (but not necessarily with the same amplitude and phase), while a square-wave input will not yield a square-wave output. The response to a square-wave input can be calculated by treating the input square wave as a sum of sinusoidal waves, and the filter output is the sum of these filtered sinusoids. This is the reason why periodic waves or oscillations are almost always treated as combinations of complex exponentials (or sines and cosines).

Real periodic signals can be expressed as the real parts of complex exponentials:

$$\cos\phi = \mathrm{Re}(e^{i\phi}),$$
$$\sin\phi = \mathrm{Im}(e^{i\phi}).$$

Adding and subtracting the equations

$$e^{i\phi} = \cos\phi + i\sin\phi,$$
$$e^{-i\phi} = \cos\phi - i\sin\phi$$

gives the identities

$$\cos\phi = \frac{e^{i\phi} + e^{-i\phi}}{2} \tag{B.4}$$

and

$$\sin\phi = \frac{e^{i\phi} - e^{-i\phi}}{2i}. \tag{B.5}$$

The advantage of complex exponentials over the equivalent sums of sines and cosines is that they are easier to manipulate mathematically. For example, you can use complex exponentials to calculate the output spectrum of a square-law detector (Section 3.6.2) without having to remember trigonometric identities. A square-law detector is a nonlinear device whose output voltage is the square of its input voltage. If the input voltage is $\cos(\omega t)$, the output voltage is

$$\cos^2(\omega t) = \left(\frac{e^{i\omega t} + e^{-i\omega t}}{2}\right)^2$$

$$= \frac{e^{2i\omega t} + 2 + e^{-2i\omega t}}{4}$$

$$= \frac{2\cos(2\omega t) + 2}{4}$$

$$= \frac{1}{2}[\cos(2\omega t) + 1].$$

The output spectrum has two frequency components: one at twice the input frequency ω and the other at zero frequency (DC).

B.4 THE FOURIER TRANSFORM OF A GAUSSIAN

The normalized Gaussian function is usually written as

$$f(x) = \frac{1}{\sqrt{2\pi}\sigma} \exp\left(-\frac{x^2}{2\sigma^2}\right), \tag{B.6}$$

where σ is its rms width. To calculate its Fourier transform

$$F(s) \equiv \int_{-\infty}^{\infty} f(x) \exp(-i2\pi s x) dx, \tag{B.7}$$

it is easier to use the form $f(x) = \exp(-\pi x^2)$, for which $\sigma^2 = 1/(2\pi)$. Then

$$F(s) = \int_{-\infty}^{\infty} \exp(-\pi x^2) \exp(-i2\pi s x) dx \tag{B.8}$$

$$= \int_{-\infty}^{\infty} \exp[-\pi(x^2 + i2sx + s^2 - s^2)]dx \tag{B.9}$$

$$= \exp(-\pi s^2) \int_{-\infty+is}^{\infty+is} \exp[-\pi(x+is)^2]d(x+is) \tag{B.10}$$

$$= \exp(-\pi s^2) \int_{-\infty}^{\infty} \exp(-\pi x^2) dx. \tag{B.11}$$

To evaluate this one-dimensional integral, break it into the product of two integrals and change one dummy variable from x to y to suggest Cartesian coordinates in a plane:

$$\int_{-\infty}^{\infty} \exp(-\pi x^2) dx = \left[\int_{-\infty}^{\infty} \exp(-\pi x^2) dx \int_{-\infty}^{\infty} \exp(-\pi y^2) dy\right]^{1/2} \tag{B.12}$$

$$= \left[\int_{-\infty}^{\infty} \int_{-\infty}^{\infty} \exp[-\pi(x^2 + y^2)]dx\, dy\right]^{1/2}. \tag{B.13}$$

Next transform to polar coordinates r, θ so $r^2 = x^2 + y^2$ and $dx\, dy = r\, dr\, d\theta$:

$$\int_{-\infty}^{\infty} \exp(-\pi x^2) dx = \left[\int_{r=0}^{\infty} \int_{\theta=0}^{2\pi} \exp(-\pi r^2) r\, dr\, d\theta\right]^{1/2}. \tag{B.14}$$

Finally, substitute $u \equiv \pi r^2$ and $du = 2\pi r\, dr$ to get

$$\int_{-\infty}^{\infty} \exp(-\pi x^2) dx = \left[2\pi \int_{u=0}^{\infty} \exp(-u) \frac{du}{2\pi}\right]^{1/2} = \left[-e^{-u}\Big|_0^{\infty}\right]^{1/2} = 1. \tag{B.15}$$

Thus

$$F(s) = \exp(-\pi s^2). \tag{B.16}$$

The Fourier transform of a Gaussian is a Gaussian.

B.5 THE GAUSSIAN PROBABILITY DISTRIBUTION AND NOISE VOLTAGE

The voltage V of random noise has a Gaussian probability distribution

$$P(V) = \frac{1}{(2\pi)^{1/2}\sigma} \exp\left(\frac{-V^2}{2\sigma^2}\right), \tag{B.17}$$

where $P(V)dV$ is the differential probability that the voltage will be within the infinitesimal range V to $V + dV$ and σ is the root mean square (rms) voltage. The probability of measuring *some* voltage must be unity, so

$$\int_{-\infty}^{\infty} P(V)dV = 1. \tag{B.18}$$

The normalization of $P(V)$ in Equation B.17 can be confirmed by evaluating the integral

$$\int_{-\infty}^{\infty} \frac{1}{(2\pi)^{1/2}\sigma} \exp\left(\frac{-V^2}{2\sigma^2}\right) dV = 2 \int_{0}^{\infty} \frac{1}{(2\pi)^{1/2}\sigma} \exp\left(\frac{-V^2}{2\sigma^2}\right) dV \tag{B.19}$$

$$= \left[\frac{2}{(2\pi)^{1/2}\sigma}\right] \int_{0}^{\infty} \exp\left(\frac{-V^2}{2\sigma^2}\right) dV. \tag{B.20}$$

Equation B.15 immediately yields the definite integral

$$\int_{0}^{\infty} \exp(-a^2 x^2) dx = \frac{\pi^{1/2}}{2a}. \tag{B.21}$$

Substituting $a^2 = (2\sigma^2)^{-1}$ gives the desired result:

$$\int_{-\infty}^{\infty} P(V)dV = \left[\frac{2}{(2\pi)^{1/2}\sigma}\right]\left(\frac{\pi^{1/2}}{2}\right)(2\sigma^2)^{1/2} = 1. \tag{B.22}$$

The rms (root mean square) Σ of a normalized distribution is defined by

$$\Sigma^2 \equiv \langle V^2 \rangle - \langle V \rangle^2. \tag{B.23}$$

For the symmetric Gaussian distribution, $\langle V \rangle = 0$, so

$$\Sigma^2 = \langle V^2 \rangle = \int_{-\infty}^{\infty} V^2 P(V)dV \tag{B.24}$$

$$= 2 \int_{0}^{\infty} V^2 \frac{1}{(2\pi)^{1/2}\sigma} \exp\left(\frac{-V^2}{2\sigma^2}\right) dV \tag{B.25}$$

$$= \left[\frac{2}{(2\pi)^{1/2}\sigma}\right] \int_{0}^{\infty} V^2 \exp\left(\frac{-V^2}{2\sigma^2}\right) dV. \tag{B.26}$$

The definite integral

$$\int_0^\infty x^2 \exp(-a^2 x^2) dx = \frac{\pi^{1/2}}{4a^3} \tag{B.27}$$

can be derived by integrating Equation B.21 by parts. Inserting $a^2 = (2\sigma^2)^{-1}$ yields

$$\Sigma^2 = \langle V^2 \rangle = \left[\frac{2}{(2\pi)^{1/2}\sigma} \right] \left(\frac{\pi^{1/2}}{4} \right) (2\sigma^2)^{3/2} = \sigma^2, \tag{B.28}$$

confirming that σ in Equation B.17 is the rms of the Gaussian distribution.

B.6 THE PROBABILITY DISTRIBUTION OF NOISE POWER

A **square-law detector** multiplies the input voltage V by itself to yield an output voltage $V_o = V^2$ that is proportional to the input power. The input voltage distribution is a Gaussian with rms σ (Equation B.17),

$$P(V) = \frac{1}{(2\pi)^{1/2}\sigma} \exp\left(\frac{-V^2}{2\sigma^2} \right). \tag{B.29}$$

The same value of $V_o = V^2$ is produced by both positive and negative values of V and $P(V) = -P(V)$, so

$$P_o(V_o) dV_o = 2P(V) dV \tag{B.30}$$

for all $V_o \geq 0$. Because $dV_o = 2V dV$,

$$P_o(V_o) = \left[\frac{V_o^{-1/2}}{(2\pi)^{1/2}\sigma} \right] \exp\left(\frac{-V_o}{2\sigma^2} \right) \tag{B.31}$$

for $V_o \geq 0$. The distribution of detector output voltage is sharply peaked near $V_o = 0$ and has a long exponentially decaying tail (Figure 3.33).

The mean detector output voltage follows from Equation B.28: $\langle V_o \rangle = \langle V^2 \rangle = \sigma^2$.

The rms σ_o of the detector output voltage is

$$\sigma_o^2 = \langle V_o^2 \rangle - \langle V_o \rangle^2, \tag{B.32}$$

where

$$\langle V_o^2 \rangle = \int_0^\infty V_o^2 P_o(V_o) dV_o = \int_0^\infty V^4 2P(V) dV \tag{B.33}$$

$$= \left[\frac{2}{(2\pi)^{1/2}\sigma} \right] \int_0^\infty V^4 \exp\left(\frac{-V^2}{2\sigma^2} \right) dV. \tag{B.34}$$

Integrating Equation B.27 by parts yields the definite integral

$$\int_0^\infty x^4 \exp(-a^2 x^2) dx = \frac{3\pi^{1/2}}{8a^5}, \tag{B.35}$$

and substituting $a^2 = (2\sigma^2)^{-1}$ gives

$$\langle V_o^2 \rangle = \left[\frac{2}{(2\pi)^{1/2}\sigma} \right] \left(\frac{3\pi^{1/2}}{8} \right) (2\sigma^2)^{5/2} = 3\sigma^4. \tag{B.36}$$

Thus

$$\sigma_o^2 = \langle V_o^2 \rangle - \langle V_o \rangle^2 = 3\sigma^4 - (\sigma^2)^2 = 2\sigma^4. \tag{B.37}$$

The rms $\sigma_o = 2^{1/2}\sigma^2$ of the detector output voltage is $2^{1/2}$ times the mean output voltage σ^2. The rms uncertainty in each independent sample of the measured noise power is $2^{1/2}$ times the mean noise power. If $N \gg 1$ independent samples are averaged, the fractional rms uncertainty of the averaged power is $(2/N)^{1/2}$. This result is the heart of the ideal radiometer equation (Equation 3.154). According to the central limit theorem, the distribution of these averages approaches a Gaussian as N becomes large.

B.7 EVALUATION OF THE FREE–FREE PULSE ENERGY INTEGRAL

The integral in Equation 4.22 is

$$\int_0^{\pi/2} \cos^4 \psi \, d\psi = \int_0^{\pi/2} \cos^2 \psi (1 - \sin^2 \psi) \, d\psi$$

$$= \int_0^{\pi/2} \cos^2 \psi \, d\psi - \int_0^{\pi/2} \cos^2 \psi \sin^2 \psi \, d\psi. \tag{B.38}$$

We have already found that $\langle \cos^2 \psi \rangle = 1/2$ so $\int_0^{\pi/2} \cos^2 \psi \, d\psi = \pi/4$. Integrate the remaining integral by parts using

$$u \equiv \cos^2 \psi \sin \psi \quad \text{and} \quad v \, dv \equiv \sin \psi \, d\psi.$$

Therefore

$$du = \cos^3 \psi - 2 \sin^2 \psi \cos \psi \quad \text{and} \quad v = -\cos \psi,$$

and

$$\int_0^{\pi/2} \cos^2 \psi \sin^2 \psi \, d\psi = -\cos^2 \psi \sin \psi \cos \psi \Big|_0^{\pi/2}$$

$$- \int_0^{\pi/2} -\cos \psi (\cos^3 \psi - 2 \sin^2 \psi \cos \psi) d\psi$$

$$= \int_0^{\pi/2} \cos^4 \psi \, d\psi - 2 \int_0^{\pi/2} \cos^2 \psi \sin^2 \psi \, d\psi,$$

which has the same integral on both sides, so

$$\int_0^{\pi/2} \cos^2 \psi \sin^2 \psi \, d\psi = \frac{1}{3} \int_0^{\pi/2} \cos^4 \psi \, d\psi.$$

Using Equation B.38 we get

$$\int_0^{\pi/2} \cos^4 \psi \, d\psi = \frac{\pi}{4} - \frac{1}{3} \int_0^{\pi/2} \cos^4 \psi \, d\psi,$$

$$\frac{4}{3} \int_0^{\pi/2} \cos^4 \psi \, d\psi = \frac{\pi}{4},$$

so

$$\boxed{\int_0^{\pi/2} \cos^4 \psi \, d\psi = \frac{3\pi}{16}.} \tag{B.39}$$

B.8 THE NONRELATIVISTIC MAXWELLIAN SPEED DISTRIBUTION

Let $v \equiv |\vec{v}|$ be the *speed* of a particle (e.g., an electron) of mass m in a gas in LTE at temperature T. From thermodynamics, recall that the average kinetic energy is $kT/2$ per degree of freedom (e.g., per spatial coordinate for a single particle), so

$$\frac{m\langle v_x^2 \rangle}{2} = \frac{m\langle v_y^2 \rangle}{2} = \frac{m\langle v_z^2 \rangle}{2} = \frac{kT}{2}, \tag{B.40}$$

$$\langle v^2 \rangle = \langle v_x^2 \rangle + \langle v_y^2 \rangle + \langle v_z^2 \rangle = \frac{3kT}{m}. \tag{B.41}$$

Collisions eventually bring the gas into LTE, leading to identical Gaussian distributions (Appendix B.5) for v_x, v_y, and v_z. Writing out only the x-coordinate distribution $P(v_x)$ yields

$$P(v_x) = \frac{1}{\sqrt{2\pi}\sigma_x} \exp\left(-\frac{v_x^2}{2\sigma_x^2}\right), \tag{B.42}$$

where σ_x is the rms (root mean square) value of v_x. The definition of this rms is

$$\sigma_x^2 \equiv \langle v_x^2 \rangle = \int_{-\infty}^{\infty} v_x^2 P(v_x) dv_x = \int_{-\infty}^{\infty} \frac{v_x^2}{\sqrt{2\pi}\sigma_x} \exp\left(-\frac{v_x^2}{2\sigma_x^2}\right) dv_x \tag{B.43}$$

$$= \frac{1}{\sqrt{2\pi}\sigma_x} \frac{1}{2}\sqrt{\pi}\left(\frac{1}{2\sigma_x^2}\right)^{-3/2} = \frac{kT}{m}, \tag{B.44}$$

so

$$P(v_x) = \frac{1}{\sqrt{2\pi}}\left(\frac{m}{kT}\right)^{1/2} \exp\left(-\frac{mv_x^2}{2kT}\right). \tag{B.45}$$

In three dimensions, by isotropy,

$$P(v_x, v_y, v_z)dv_x\, dv_y\, dv_z = P(v_x)P(v_y)P(v_z)dv_x\, dv_y\, dv_z, \qquad \text{(B.46)}$$

$$P(v_x, v_y, v_z) = \left(\frac{m}{2\pi kT}\right)^{3/2} \exp\left(-\frac{mv^2}{2kT}\right). \qquad \text{(B.47)}$$

All velocities in the spherical shell of radius $v = (v_x^2 + v_y^2 + v_z^2)^{1/2}$ correspond to the *speed* v, so

$$f(v) = 4\pi v^2 P(v_x, v_y, v_z), \qquad \text{(B.48)}$$

$$\boxed{f(v) = \frac{4v^2}{\sqrt{\pi}}\left(\frac{m}{2kT}\right)^{3/2} \exp\left(-\frac{mv^2}{2kT}\right).} \qquad \text{(B.49)}$$

This is the nonrelativistic Maxwellian distribution f of speeds $v \ll c$ for particles of mass m at temperature T. If we normalize the speeds by the rms speed $(3kT/m)^{1/2}$, the Maxwellian speed distribution looks like Figure 4.6.

C Special Relativity

C.1 RELATIVITY

The **relativity principle** states that the laws of physics are the same in all rigid inertial frames, so observers in different inertial frames should be able to compare measurements after suitable coordinate transformations.

Figure C.1 shows the inertial coordinate system S and the "primed" system S' moving with a constant velocity v along their common x-axis. The planes $y = 0$ and $z = 0$ always coincide with the planes $y' = 0$ and $z' = 0$, and clocks in both frames were synchronized by setting $t = t' = 0$ at the instant when $x = x'$. An **event** is an observable confined to one point in space and time, such as a camera flash firing. The event coordinates in S and S' are (x, y, z, t) and (x', y', z', t'), respectively.

According to "everyday" **Galilean relativity**, the event coordinates in S are related to those in S' by the **Galilean transform**

$$x = x' + vt', \qquad y = y', \qquad z = z', \qquad t = t', \tag{C.1}$$

$$x' = x - vt, \qquad y' = y, \qquad z' = z, \qquad t' = t. \tag{C.2}$$

However, everyday experience with slowly moving objects cannot be extrapolated to speeds approaching the vacuum speed of light c. Everyday experience suggests that $t = t'$ in all inertial frames, so Isaac Newton believed in "absolute, true, and mathematical time, in and of itself and of its own nature, without reference to anything external." If $t = t'$ and $x = x' + vt'$, parallel velocities simply add. The speed of a photon emitted in the $+x'$-direction by a flash at rest in the S' system will be seen by an observer in the S frame as

$$c_x = \frac{dx}{dt} = \frac{d(x' + vt')}{dt'} = \frac{dx'}{dt'} + v = c'_x + v. \tag{C.3}$$

Thus Galilean relativity is inconsistent with both observation and Maxwell's equations, which correctly predict that the speed of light in a vacuum is the same for all observers in all inertial frames ($c_x = c'_x$), regardless of their relative velocities v. Apparently time is not absolute, and identical clocks in the S and S' frames do not run at the same rates: $t \neq t'$.

The **Lorentz transform** is the only coordinate transform consistent with both relativity and the existence of some still-unspecified **invariant speed** c (for example,

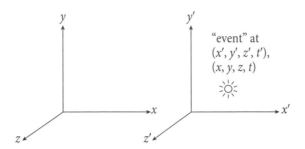

Figure C.1. An "event" is an observable confined to a single point in space and time. Its coordinates are (x, y, z, t) in the "unprimed" inertial rest frame S and (x', y', z', t') in the "primed" frame S' moving to the right with constant speed v. The x- and x'-axes always overlap, and identical clocks at rest in the two frames were synchronized by setting $t = t' = 0$ at the instant when $x = x'$.

the vacuum speed of light, or even ∞). The assumption that there exists *some* invariant speed is actually weaker than assuming $t = t'$ because it turns out that $c = \infty$ reduces the Lorentz transform to the Galilean transform.

The Lorentz transform can be derived with two additional assumptions (see Rindler [90, p. 39] or Rindler [91, p. 12]):

1. space is **homogeneous**, and
2. space is **isotropic**.

These assumptions say that the laws of physics are invariant under translation and rotation of the coordinate frames.

Homogeneity implies that any transformation from one inertial frame to another must be **linear**; that is, $y' = Ay + B$ where A and B are constants. Adding nonlinear terms (e.g., $y' = Ay + B + Cy^2$) would cause the transform itself to vary under coordinate translations, so $C = 0$ in a homogeneous space. Because the coordinate frames were chosen such that $y = 0$ when $y' = 0$, linearity also requires that $B = 0$, leaving $y' = Ay$, where A is a still-unspecified scale factor.

Isotropy implies that the observers in both frames agree on their relative speed $|v|$ because a $180°$ coordinate rotation exchanges the roles of the two frames. That rotation should have no effect if space is isotropic, so $y' = Ay$ implies $y = A'y'$ and only $A = A' = \pm 1$ is consistent with isotropy. The negative solution $A = -1$ can be rejected because it implies $y = -y'$ even when $v = 0$. These arguments can also be applied to z and z', so the Lorentz transform for the y and z coordinates is

$$y = y', \qquad z = z', \tag{C.4}$$

in agreement with the Galilean transform (Equation C.1).

For the x-coordinate, linearity requires

$$x = \gamma'(x' + vt'), \qquad x' = \gamma(x - vt), \tag{C.5}$$

where γ' and γ are still-unspecified constant scale factors. In Galilean relativity, $\gamma = \gamma' = 1$.

Invoking isotropy and reversing the directions of S and S' gives

$$x = \gamma'(x' - vt'), \qquad x' = \gamma(x + vt) \tag{C.6}$$

and reversing the roles of the two frames gives

$$x = \gamma(x' - vt'), \qquad x' = \gamma'(x + vt). \tag{C.7}$$

Equations C.6 and C.7 together imply $\gamma = \gamma'$; that is, the observers in S and S' agree on the value of the coordinate scale factor γ associated with their relative velocity v.

We now make use of the assumption that there is some speed c which is the same in all inertial frames. Maxwell's equations and experiment both show that c is the speed of light in a vacuum, but for this argument, the invariant speed might be any speed, even $c = \infty$, in which limit the Lorentz transform turns out to be identical to the Galilean transform. Then $x = ct$ implies $x' = ct'$ and Equation C.5 implies

$$ct = \gamma t'(c + v), \qquad ct' = \gamma t(c - v). \tag{C.8}$$

The product

$$c^2 tt' = \gamma^2 tt'(c + v)(c - v) \tag{C.9}$$

of these two equations can be solved for the **Lorentz factor**:

$$\boxed{\gamma = \left(1 - \frac{v^2}{c^2}\right)^{-1/2} = (1 - \beta^2)^{-1/2},} \tag{C.10}$$

where the dimensionless velocity β is defined as

$$\boxed{\beta \equiv \frac{v}{c}.} \tag{C.11}$$

Again, the negative solution to Equation C.10 can be rejected as unphysical. Notice that $\gamma \geq 1$ for all possible $\beta \leq 1$. The x-coordinate Lorentz transform becomes

$$x = \gamma(x' + vt'), \qquad x' = \gamma(x - vt). \tag{C.12}$$

Eliminating x or x' from this pair of equations yields the Lorentz time transform

$$t = \gamma(t' + \beta x'/c), \qquad t' = \gamma(t - \beta x/c). \tag{C.13}$$

In summary, the **Lorentz coordinate transform** of special relativity is

$$\boxed{x = \gamma(x' + vt'), \qquad y = y', \qquad z = z', \qquad t = \gamma(t' + \beta x'/c),} \tag{C.14}$$

$$\boxed{x' = \gamma(x - vt), \qquad y' = y, \qquad z' = z, \qquad t' = \gamma(t - \beta x/c).} \tag{C.15}$$

Note that in the limit $\beta \to 0$ ($v \ll c$), the Lorentz transform reduces to the Galilean transform (Equation C.1), as it must to agree with "everyday" observations involving small velocities. Equations C.14 and C.15 also show that the Lorentz transform reduces to the Galilean transform for any finite v in the limit $c \to \infty$, so the assumption that there exists *some* invariant velocity is not restrictive.

If $(\Delta x, \Delta y, \Delta z, \Delta t)$ and $(\Delta x', \Delta y', \Delta z', \Delta t')$ are the coordinate differences between two events, the **differential Lorentz transform** is

$$\Delta x = \gamma(\Delta x' + v\Delta t'), \quad \Delta y = \Delta y', \quad \Delta z = \Delta z', \quad \Delta t = \gamma(\Delta t' + \beta\Delta x'/c),$$

$$\text{(C.16)}$$

$$\Delta x' = \gamma(\Delta x - v\Delta t), \quad \Delta y' = \Delta y, \quad \Delta z' = \Delta z, \quad \Delta t' = \gamma(\Delta t - \beta\Delta x/c).$$

$$\text{(C.17)}$$

The differences Δ can be finite because the Lorentz transform is linear. This makes the differential Lorentz transform easy to apply to physical problems such as determining the lengths of rulers.

C.2 TIME DILATION AND LENGTH CONTRACTION

The phenomenon of relativistic **time dilation** follows from the differential time transform $\Delta t = \gamma(\Delta t' + \beta\Delta x'/c)$. If successive ticks of a clock at rest in the primed frame (so $\Delta x' = 0$) are separated by $\Delta t' = 1$ in time, they will be separated by $\Delta t = \gamma \geq 1$ in the unprimed frame. Likewise a clock at rest in the unprimed frame appears to run slow by the same factor γ when observed in the primed frame.

Relativistic **length contraction** can also be derived from Equation C.16. Suppose the ends of a ruler of unit length in the primed frame ($\Delta x' = 1$) emit two flashes of light at the same time in the unprimed frame ($\Delta t = 0$). The time $\Delta t'$ between the flashes in the primed frame can be calculated from $\Delta t = 0 = \gamma(\Delta t' + v\Delta x'/c^2)$ to be $\Delta t' = -v\Delta x'/c^2$. The ruler length in the unprimed frame is the distance between the flashes. It is shorter by the factor

$$\Delta x = \gamma(\Delta x' + v\Delta t') = \gamma(1 - \beta^2) = 1/\gamma. \qquad \text{(C.18)}$$

A ruler at rest in the unprimed frame also appears to be shorter by the factor $1/\gamma$ when observed in the primed frame.

C.3 VELOCITY ADDITION FORMULAS

Relativistic velocities do not add linearly. Let the velocity of a particle be $\vec{u} = (u_x, u_y, u_z)$ in the unprimed frame and $\vec{u}' = (u'_x, u'_y, u'_z)$ in the primed frame:

$$u_x \equiv \frac{dx}{dt} = \frac{dx}{dt'}\frac{dt'}{dt}. \qquad \text{(C.19)}$$

The differential Lorentz transforms $dx = \gamma(dx' + vdt')$ and $dt = \gamma(dt' + \beta dx'/c)$ yield

$$\frac{dx}{dt'} = \gamma\left(\frac{dx'}{dt'} + v\right) = \gamma(u'_x + v), \tag{C.20}$$

$$\frac{dt}{dt'} = \gamma\left(1 + \frac{\beta}{c}\frac{dx'}{dt'}\right) = \gamma\left(1 + \frac{vu'_x}{c^2}\right), \tag{C.21}$$

so

$$\boxed{u_x = \frac{u'_x + v}{(1 + vu'_x/c^2)}} \tag{C.22}$$

and, by symmetry,

$$\boxed{u'_x = \frac{u_x - v}{(1 - vu_x/c^2)}.} \tag{C.23}$$

For the velocity components perpendicular to v,

$$u_y \equiv \frac{dy}{dt} = \frac{dy}{dt'}\frac{dt'}{dt} = \frac{dy'}{dt'}\frac{dt'}{dt}, \tag{C.24}$$

so

$$\boxed{u_y = \frac{u'_y}{\gamma(1 + vu'_x/c^2)} \quad \text{and} \quad u_z = \frac{u'_z}{\gamma(1 + vu'_x/c^2)},} \tag{C.25}$$

and likewise,

$$\boxed{u'_y = \frac{u_y}{\gamma(1 - vu_x/c^2)} \quad \text{and} \quad u'_z = \frac{u_z}{\gamma(1 - vu_x/c^2)}.} \tag{C.26}$$

C.4 MASS, ENERGY, AND POWER

The ratio of a moving object's **relativistic mass** m to its **rest mass** m_0 follows from a thought experiment in Rindler [90]. Imagine two identical electrons, one at rest in the unprimed frame and the other at rest in the primed frame moving with velocity v. Let one electron be slightly displaced from the other along the y-axis. At the moment the electrons pass by each other, their Coulomb repulsion will accelerate them in the $\pm y$-directions. Because $dy = dy'$, both electrons must experience the same y displacement, but time dilation will make the moving electron take a factor γ longer to do so. Invoking momentum conservation, both observers conclude that the mass of the electron at rest in the moving frame is a factor γ larger:

$$m = \gamma m_0. \tag{C.27}$$

Likewise, its total **energy**

$$E = mc^2 = \gamma m_0 c^2 \tag{C.28}$$

has been multiplied by γ. Applying the chain rule for derivatives to the mass–energy transform,

$$P \equiv \frac{dE}{dt} = \frac{dE}{dt'}\frac{dt'}{dt} = \frac{dE}{dE'}\frac{dE'}{dt'}\frac{dt'}{dt} = \gamma P' \gamma^{-1} = P' \tag{C.29}$$

shows that **power** is a relativistic invariant.

D Wave Propagation in a Plasma

The propagation of radio waves is governed by **Maxwell's equations**:

$$\vec{\nabla} \cdot \vec{D} = 4\pi\rho, \tag{D.1}$$

$$\vec{\nabla} \cdot \vec{B} = 0, \tag{D.2}$$

$$\vec{\nabla} \times \vec{E} = -\frac{1}{c}\frac{\partial \vec{B}}{\partial t}, \tag{D.3}$$

$$\vec{\nabla} \times \vec{H} = \frac{4\pi \vec{J}}{c} + \frac{1}{c}\frac{\partial \vec{D}}{\partial t}. \tag{D.4}$$

D.1 DISPERSION AND REFLECTION IN A LOW-DENSITY PLASMA

The interstellar medium (ISM) of our Galaxy is a nearly perfect vacuum, so $\vec{D} \approx \vec{E}$ and $\vec{B} \approx \vec{H}$. However, the **current density** (charge flow rate per unit area) $\vec{J}$ need not be zero because there are free electrons in the ISM, with mean density $n_e \sim 0.03$ cm^{-3}. The transverse electric field $E = E_0 \exp(ikx - i\omega t)$ of a monochromatic plane wave with angular frequency ω traveling in the x-direction past a free electron at $x = 0$ causes the electron to oscillate with frequency ω and generate an oscillating electric current. The ISM density is low enough that damping by electron collisions can be ignored, so the electron's equation of motion is

$$F = m_e \dot{v} = eE = eE_0 \exp(-i\omega t), \tag{D.5}$$

where $\dot{v}$ is the electron's acceleration. Integrating Equation D.5 yields the electron velocity

$$v = \int_0^t \dot{v}\, dt = -\frac{eE_0}{i\omega m_e}\exp(-i\omega t). \tag{D.6}$$

Since the resulting current density is proportional to the applied electric field strength, the ISM obeys **Ohm's law**

$$\vec{J} = \sigma \vec{E},$$ (D.7)

where the constant of proportionality σ is the called the **conductivity** of the medium. The current density is the rate at which charge flows through a unit area, so $J = e n_e v$ and the conductivity of the ISM is

$$\sigma = \frac{J}{E} = e n_e \left(-\frac{e}{i \omega m_e} \right) = i \left(\frac{e^2 n_e}{\omega m_e} \right).$$ (D.8)

The conductivity of a **collisionless plasma** is purely imaginary. That means that the current density and electric field are 90 degrees out of phase, and radio waves can propagate without suffering resistive power losses.

For a plane wave traveling in the x-direction,

$$\vec{\nabla} \to ik \qquad \text{and} \qquad \frac{\partial}{\partial t} \to -i\omega,$$ (D.9)

and Maxwell's two curl equations become

$$ikE = \frac{i\omega H}{c},$$ (D.10)

$$-ikH = \frac{4\pi\sigma E}{c} - \frac{i\omega E}{c}.$$ (D.11)

Using the first to eliminate H from the second gives

$$-ikH = -ik \left(\frac{ckE}{\omega} \right) = \frac{4\pi\sigma E}{c} - \frac{i\omega E}{c}.$$ (D.12)

Thus

$$k^2 = \left(\frac{\omega}{c} \right)^2 \left(1 + i \frac{4\pi\sigma}{\omega} \right).$$ (D.13)

The ISM conductivity (Equation D.8) can be expressed in terms of the **plasma frequency** defined by

$$\nu_p \equiv \left(\frac{e^2 n_e}{\pi m_e} \right)^{1/2} \approx 8.97 \text{ kHz} \left(\frac{n_e}{\text{cm}^{-3}} \right)^{1/2}.$$ (D.14)

The plasma frequency in the ISM where $n_e \sim 0.03 \text{ cm}^{-3}$ is only $\nu_p \sim 0.3$ kHz.

In terms of ν_p the magnitude of the wave vector k obeys

$$k^2 = \left(\frac{\omega}{c} \right)^2 \left(1 - \frac{\nu_p^2}{\nu^2} \right),$$ (D.15)

where $\nu = \omega/(2\pi)$. The wave vector is imaginary for radio waves whose frequency ν is less than the plasma frequency ν_p. Low-frequency radiation cannot propagate through the plasma, and radio waves with $\nu < \nu_p$ incident on a plasma are reflected. For example, the maximum electron density in the F layer of the Earth's **ionosphere** is $n_e \approx 10^6$ cm^{-3}, so celestial radio radiation at frequencies lower than $\nu \sim 10$ MHz is reflected back into space by the ionosphere and cannot be observed from the ground.

The plasma **index of refraction** is

$$\mu = \frac{ck}{\omega} = \left(1 - \frac{\nu_p^2}{\nu^2}\right)^{1/2}. \tag{D.16}$$

In the limit $\nu_p \ll \nu$ appropriate for radio waves in the ISM,

$$\mu \approx 1 - \frac{\nu_p^2}{2\nu^2}, \tag{D.17}$$

and the **group velocity** ν_g of radio pulses traveling through the ISM depends on frequency:

$$\nu_g \approx \mu c \approx c\left(1 - \frac{\nu_p^2}{2\nu^2}\right). \tag{D.18}$$

D.2 FARADAY ROTATION IN A MAGNETIZED PLASMA

If the ISM contains a magnetic field of strength B, all nonrelativistic electrons orbit around field lines with angular frequency

$$\omega_G = \frac{eB}{m_e c}, \tag{D.19}$$

where $\nu_G = \omega_G/(2\pi)$ is called the **gyro frequency** (Section 5.1.1). In the (non-inertial) coordinate frame rotating with the electron, circularly polarized radiation at frequency ω will drive the electron at frequency $\omega + \omega_G$ or $\omega - \omega_G$ depending on the sense (left handed or right handed) of the circular polarization. Because the index of refraction in a collisionless plasma depends on frequency (Equation D.16), there are two indices of refraction in a magnetized plasma, which is said to be **birefringent**. A linearly polarized wave is the sum of left- and right-handed circularly polarized waves, so the position angle of the linearly polarized wave will rotate as the wave travels parallel to the magnetic field. This rotation is called **Faraday rotation**, and it is a useful tool for measuring the line-of-sight component $B_\parallel$ of the magnetic field in a radio source. Equations describing Faraday rotation are easily derived by extending the derivation in Appendix D.1.

Equation D.5 can be expanded to include the force on an electron from a steady ambient magnetic field of strength B:

$$F = m_e \dot{\nu} = eE_0 \exp(-i\omega t) + \frac{e}{c}\nu B_\parallel, \tag{D.20}$$

where $B_\parallel$ is the component of the magnetic field parallel to the direction of the electromagnetic field $E = E_0 \exp(ikx - i\omega t)$. If the incident wave is circularly polarized:

$$\vec{E} = E_y \hat{y} \pm i E_z \hat{z}, \tag{D.21}$$

then the electron velocity is

$$\vec{v} = v(\hat{y} \pm i\hat{z}) \exp(-i\omega t) \tag{D.22}$$

and

$$v = i \left[\frac{eE}{m_e(\omega \pm \omega_G)} \right]. \tag{D.23}$$

The resulting conductivity of the magnetized plasma is

$$\sigma = i \left[\frac{e^2 n_e}{m_e(\omega \pm \omega_G)} \right], \tag{D.24}$$

and the refractive index takes on two values

$$\mu = \left[1 - \frac{v_p^2}{v(v \pm v_G)} \right]^{1/2}. \tag{D.25}$$

Faraday rotation changes the polarization position angle by

$$\Delta\chi \approx \lambda^2 \left[\frac{e^3}{2\pi(m_e c^2)^2} \int_{los} n_e(l) B_\parallel(l) dl \right] \equiv \lambda^2 \, \mathrm{RM}, \tag{D.26}$$

where the quantity in square brackets defines the **rotation measure** RM. The sign convention is that RM is positive for a magnetic field pointing from the source to the observer. In astronomically convenient units,

$$\left(\frac{\mathrm{RM}}{\mathrm{rad\ m^{-2}}} \right) \approx 8.1 \times 10^5 \int_{los} \left(\frac{n_e}{\mathrm{cm^{-3}}} \right) \left(\frac{B_\parallel}{\mathrm{gauss}} \right) \left(\frac{dl}{\mathrm{pc}} \right). \tag{D.27}$$

The phenomenon of Faraday rotation can also lead to **Faraday depolarization** when the rotation occurs within the emission region.

E Essential Equations

The **specific intensity** I_ν of radiation is defined by

$$I_\nu \equiv \frac{dP}{(\cos\theta \, d\sigma) \, d\nu \, d\Omega},$$ (2.2)

where dP is the power received by a detector with projected area $(\cos\theta \, d\sigma)$ in the solid angle $d\Omega$ and in the frequency range ν to $\nu + d\nu$. Likewise I_λ is the brightness per unit wavelength:

$$I_\lambda \equiv \frac{dP}{(\cos\theta \, d\sigma) \, d\lambda \, d\Omega}.$$ (2.3)

These two quantities are related by

$$\frac{I_\lambda}{I_\nu} = \left| \frac{d\nu}{d\lambda} \right| = \frac{c}{\lambda^2} = \frac{\nu^2}{c}.$$ (2.5)

The **flux density** S_ν of a source is the spectral power received per unit detector area:

$$S_\nu \equiv \int_{\text{source}} I_\nu(\theta, \phi) \cos\theta \, d\Omega.$$ (2.9)

If the source is compact enough that $\cos\theta \approx 1$ then

$$S_\nu \approx \int_{\text{source}} I_\nu(\theta, \phi) d\Omega.$$ (2.10)

The MKS units of flux density are W m^{-2} Hz^{-1}; 1 jansky (Jy) $\equiv 10^{-26}$ W m^{-2} Hz^{-1}.

The **spectral luminosity** L_ν of a source is the total power per unit frequency radiated at frequency ν; its MKS units are W Hz^{-1}. In free space and at distances d much greater than the source size, the **inverse-square law**

$$L_\nu = 4\pi d^2 S_\nu$$

(2.15)

relates the spectral luminosity *of an isotropic source* to its flux density.

The **linear absorption coefficient** at frequency ν of an absorber is defined as the probability $dP(\nu)$ that a photon will be absorbed in a layer of thickness ds:

$$\kappa(\nu) \equiv \frac{dP(\nu)}{ds}.$$

(2.18)

The **opacity** or **optical depth** τ is defined as the sum of those infinitesimal probabilities through the absorber, starting at the source end:

$$\tau \equiv \int_{s_{\text{out}}}^{s_{\text{in}}} -\kappa(s')ds'.$$

(2.23)

The **emission coefficient** at frequency ν is the infinitesimal increase dI_ν in specific intensity per infinitesimal distance ds:

$$j_\nu \equiv \frac{dI_\nu}{ds}.$$

(2.26)

The **equation of radiative transfer** is

$$\frac{dI_\nu}{ds} = -\kappa I_\nu + j_\nu.$$

(2.27)

For any substance in Local Thermodynamic Equilibrium (LTE), **Kirchhoff's law** connects the emission and absorption coefficients via the specific intensity B_ν of blackbody radiation:

$$\frac{j_\nu}{\kappa} = B_\nu(T).$$

(2.30)

The **brightness temperature** of a source with *any* specific intensity I_ν is defined as

$$T_{\text{b}}(\nu) \equiv \frac{I_\nu c^2}{2k\nu^2}.$$

(2.33)

For an **opaque body** in LTE, Kirchhoff's law connects the **emission coefficient** e_ν (the spectral power per unit area emitted by the body divided by the spectral power per unit area emitted by a blackbody) to the **absorption coefficient** a_ν (fraction of radiation absorbed by the body) and the **reflection coefficient** r_ν (fraction of radiation reflected by the body):

$$e_\nu = a_\nu = 1 - r_\nu. \tag{2.47}$$

The **spectral energy density of radiation** is

$$u_\nu = \frac{1}{c} \int I_\nu \, d\Omega. \tag{2.76}$$

The **Rayleigh–Jeans approximation** for the specific intensity of blackbody radiation when $h\nu \ll kT$ is

$$B_\nu = \frac{2kT\nu^2}{c^2} = \frac{2kT}{\lambda^2}. \tag{2.79}$$

The **energy of a photon** is

$$E = h\nu. \tag{2.81}$$

Planck's equation for the specific intensity of blackbody radiation at any frequency is

$$B_\nu = \frac{2h\nu^3}{c^2} \frac{1}{\exp\left(\dfrac{h\nu}{kT}\right) - 1}. \tag{2.86}$$

The **total intensity of blackbody radiation** is

$$B(T) \equiv \int_0^\infty B_\nu(T) \, d\nu = \frac{\sigma T^4}{\pi}, \tag{2.89}$$

where the **Stefan–Boltzmann constant** σ is defined by

$$\sigma \equiv \frac{2\pi^5 k^4}{15c^2 h^3} \approx 5.67 \times 10^{-5} \frac{\text{erg}}{\text{cm}^2 \, \text{s} \, \text{K}^4 \, \text{sr}}. \tag{2.90}$$

The **total energy density of blackbody radiation** is

$$u = \frac{4\sigma T^4}{c} = a T^4, \tag{2.93}$$

where $a \equiv 4\sigma/c \approx 7.56577 \times 10^{-15}$ erg cm^{-3} K^{-4} is the **radiation constant**. The **photon number density of blackbody radiation** is

$$\left(\frac{n_\gamma}{\text{cm}^{-3}}\right) \approx 20.3 \left(\frac{T}{\text{K}}\right)^3. \tag{2.100}$$

The **mean photon energy of blackbody radiation** is

$$\langle E_\gamma \rangle \approx 2.70 \, kT. \tag{2.101}$$

The **frequency of the peak blackbody brightness** *per unit frequency B_ν* is

$$\left(\frac{\nu_{\text{max}}}{\text{GHz}}\right) \approx 59 \left(\frac{T}{\text{K}}\right). \tag{2.104}$$

The wavelength of the peak blackbody brightness *per unit wavelength B_λ* is given by **Wien's displacement law:**

$$\left(\frac{\lambda_{\text{max}}}{\text{cm}}\right) \approx 0.29 \left(\frac{T}{\text{K}}\right)^{-1}. \tag{2.106}$$

The **flux density of isotropic radiation** is

$$S_\nu = \pi I_\nu. \tag{2.109}$$

The **Nyquist approximation** for the spectral power generated by a warm resistor in the limit $h\nu \ll kT$ is

$$P_\nu = kT. \tag{2.117}$$

At any frequency, the exact **Nyquist formula** is

$$P_\nu = \frac{h\nu}{\exp\left(\frac{h\nu}{kT}\right) - 1}. \tag{2.119}$$

The **critical density** needed to close the universe is

$$\rho_c = \frac{3H_0^2}{8\pi G} \approx 8.6 \times 10^{-30} \text{ g cm}^{-3}.$$

(2.126)

Redshift z is defined by

$$z \equiv \frac{\lambda_o - \lambda_e}{\lambda_e} = \frac{\lambda_o}{\lambda_e} - 1 = \frac{\nu_e}{\nu_o} - 1,$$

(2.127)

where λ_e and ν_e are the wavelength and frequency emitted by a source at redshift z, and λ_o and ν_o are the observed wavelength and frequency at $z = 0$.
Redshift z and **expansion scale factor** a are related by

$$(1 + z) = a^{-1}.$$

(2.128)

The **CMB temperature at redshift** z is

$$T = T_0(1 + z).$$

(2.129)

The **radiated electric field** at distance r from a charge q at angle θ from the acceleration $\dot{v}$ is

$$E_\perp = \frac{q\dot{v}\sin\theta}{rc^2}.$$

(2.136)

In a vacuum, the **Poynting flux**, or power per unit area, is

$$|\vec{S}| = \frac{c}{4\pi}E^2.$$

(2.139)

The total power emitted by an accelerated charge is given by **Larmor's formula**

$$P = \frac{2}{3}\frac{q^2\dot{v}^2}{c^3},$$

(2.143)

which is valid only if $v \ll c$.
Exponential notation for trigonometric functions is

$$e^{-i\omega t} = \cos(\omega t) - i\sin(\omega t).$$

(3.2)

Electric current is defined as the time derivative of electric charge:

$$I \equiv \frac{dq}{dt}.$$

(3.4)

The **power pattern** of a **short dipole** antenna is

$$P \propto \sin^2 \theta.$$

(3.14)

The **power emitted** by a short ($l \ll \lambda$) dipole driven by a current $I = I_0 e^{-i\omega t}$ is

$$\langle P \rangle = \frac{\pi^2}{3c} \left(\frac{I_0 l}{\lambda} \right)^2.$$

(3.17)

Radiation resistance is defined by

$$R \equiv \frac{2 \langle P \rangle}{I_0^2}.$$

(3.25)

Energy conservation implies the **average power gain** of any lossless antenna is

$$\langle G \rangle = 1$$

(3.32)

and

$$\int_{\text{sphere}} G \, d\Omega = 4\pi.$$

(3.33)

The **beam solid angle** is defined by

$$\Omega_A \equiv \frac{4\pi}{G_{\text{max}}} = \frac{1}{G_{\text{max}}} \int_{4\pi} G(\theta, \phi) \, d\Omega.$$

(3.34)

The **effective area** of an antenna is defined by

$$A_e \equiv 2P_\nu / S_\nu,$$

(3.35)

where P_ν is the output power density produced by an unpolarized point source of total flux density S_ν.

The **average effective area** of any lossless antenna is

$$\langle A_{\mathrm{e}} \rangle = \frac{\lambda^2}{4\pi}. \tag{3.41}$$

Reciprocity implies

$$G(\theta, \phi) \propto A_{\mathrm{e}}(\theta, \phi). \tag{3.44}$$

Reciprocity and energy conservation imply

$$A_{\mathrm{e}}(\theta, \phi) = \frac{\lambda^2 G(\theta, \phi)}{4\pi}. \tag{3.46}$$

Antenna temperature is defined by

$$T_{\mathrm{A}} \equiv \frac{P_{\nu}}{k}. \tag{3.47}$$

The antenna temperature produced by an **unpolarized point source** of flux density S is

$$T_{\mathrm{A}} = \frac{A_{\mathrm{e}} S}{2k}. \tag{3.48}$$

If $A_{\mathrm{e}} \approx 2761 \ \mathrm{m}^2$, the point-source sensitivity is $1 \ \mathrm{K \ Jy^{-1}}$.

For a uniform compact source of brightness temperature T_{b} covering solid angle Ω_{s},

$$\frac{T_{\mathrm{A}}}{T_{\mathrm{b}}} = \frac{\Omega_{\mathrm{s}}}{\Omega_{\mathrm{A}}}. \tag{3.56}$$

The **main beam solid angle** is defined by the integral over the main beam to the first zero only:

$$\Omega_{\mathrm{MB}} \equiv \frac{1}{G_{\max}} \int_{\mathrm{MB}} G(\theta, \phi) \, d\Omega \tag{3.57}$$

and is used in the definition of **main beam efficiency**:

$$\eta_{\mathrm{B}} \equiv \frac{\Omega_{\mathrm{MB}}}{\Omega_{\mathrm{A}}}. \tag{3.58}$$

The height z at axial distance r above the vertex of a **paraboloidal reflector** of focal length f is

$$z = \frac{r^2}{4f}. \tag{3.60}$$

The **far-field distance** of an aperture of diameter D used at wavelength λ is

$$R_{\text{ff}} \approx \frac{2D^2}{\lambda}. \tag{3.64}$$

In the far field, the **electric field pattern** of an aperture antenna is the Fourier transform of the aperture illumination:

$$l \equiv \sin\theta, \tag{3.69}$$

$$u \equiv \frac{x}{\lambda}, \tag{3.72}$$

$$f(l) = \int_{\text{aperture}} g(u)e^{-i2\pi lu}\,du. \tag{3.74}$$

The **power pattern of a uniformly illuminated linear aperture** is

$$P(\theta) \propto \text{sinc}^2\left(\frac{\theta D}{\lambda}\right), \tag{3.79}$$

where $\text{sinc}(x) \equiv \sin(\pi x)/(\pi x)$, and the **half-power beamwidth** is

$$\theta_{\text{HPBW}} \approx 0.89\frac{\lambda}{D}. \tag{3.82}$$

The **half-power beamwidth** (HPBW) of a a typical radio telescope with tapered illumination is

$$\theta_{\text{HPBW}} \approx 1.2\frac{\lambda}{D}. \tag{3.96}$$

The **two-dimensional aperture field pattern** is

$$f(l, m) \propto \int_{-\infty}^{\infty}\int_{-\infty}^{\infty} g(u, v)e^{-i2\pi(lu+mv)}\,du\,dv, \tag{3.97}$$

where m is the y-axis analog of l on the x-axis, and $v \equiv y/\lambda$. The electric field pattern of a two-dimensional aperture is the two-dimensional Fourier transform of the aperture field illumination.

The **power pattern of a uniformly illuminated rectangular aperture** is

$$G \approx \frac{4\pi D_x D_y}{\lambda^2} \operatorname{sinc}^2\left(\frac{\theta_x D_x}{\lambda}\right) \operatorname{sinc}^2\left(\frac{\theta_y D_y}{\lambda}\right). \tag{3.108}$$

Aperture efficiency is defined by

$$\eta_A \equiv \frac{\max(A_e)}{A_{\text{geom}}}. \tag{3.111}$$

The **beam solid angle of a Gaussian beam** is

$$\Omega_A = \left(\frac{\pi}{4\ln 2}\right) \theta_{\text{HPBW}}^2 \approx 1.133\, \theta_{\text{HPBW}}^2. \tag{3.118}$$

The **surface efficiency η_s of a reflector** whose surface errors ϵ have rms σ is given by the **Ruze equation**:

$$\eta_s = \exp\left[-\left(\frac{4\pi\sigma}{\lambda}\right)^2\right]. \tag{3.129}$$

Noise temperature is defined by

$$T_N \equiv \frac{P_\nu}{k}. \tag{3.149}$$

The **system noise temperature** is the sum of noise contributions from all sources:

$$T_s = T_{\text{cmb}} + T_{\text{rsb}} + \Delta T_{\text{source}} + [1 - \exp(-\tau_A)]T_{\text{atm}} + T_{\text{spill}} + T_r + \cdots. \tag{3.150}$$

The **ideal total-power radiometer equation** is

$$\sigma_T \approx T_s \left[\frac{1}{\Delta\nu\, \tau}\right]^{1/2}. \tag{3.154}$$

The **practical total-power radiometer equation** includes the effects of gain fluctuations:

$$\sigma_T \approx T_{\mathrm{s}}\left[\frac{1}{\Delta \nu\, \tau} + \left(\frac{\Delta G}{G}\right)^2\right]^{1/2}. \tag{3.158}$$

The **Dicke-switching radiometer equation** is

$$\sigma_T \approx 2T_{\mathrm{s}}\left[\frac{1}{\Delta \nu\, \tau}\right]^{1/2}. \tag{3.162}$$

The **rms confusion** caused by unresolved continuum sources in a Gaussian beam with HPBW θ at frequency ν is

$$\left(\frac{\sigma_{\mathrm{c}}}{\mathrm{mJy\ beam}^{-1}}\right) \approx \begin{cases} 0.2\left(\dfrac{\nu}{\mathrm{GHz}}\right)^{-0.7}\left(\dfrac{\theta}{\mathrm{arcmin}}\right)^2 & (\theta > 0.17\ \mathrm{arcmin}), \\[2ex] 2.2\left(\dfrac{\nu}{\mathrm{GHz}}\right)^{-0.7}\left(\dfrac{\theta}{\mathrm{arcmin}}\right)^{10/3} & (\theta < 0.17\ \mathrm{arcmin}). \end{cases}$$

$$\tag{3.163}$$

Individual sources fainter than the **confusion limit** $\approx 5\sigma_{\mathrm{c}}$ cannot be detected reliably.

Radiometer input noise temperature T_{r} can be measured by the Y-**factor** method; it is

$$T_{\mathrm{r}} = \frac{T_{\mathrm{h}} - Y T_{\mathrm{c}}}{Y - 1}. \tag{3.168}$$

The response of a two-element interferometer to a source of brightness distribution $I_\nu(\hat{s})$ is the **complex visibility**

$$\mathcal{V}_\nu = \int I_\nu(\hat{s}) \exp(-i2\pi \vec{b}\cdot\hat{s}/\lambda)\, d\Omega. \tag{3.186}$$

To minimize **bandwidth smearing** in bandwidth $\Delta \nu$, the image angular radius $\Delta\theta$ should satisfy

$$\Delta\theta\, \Delta\nu \ll \theta_{\mathrm{s}}\nu. \tag{3.192}$$

To minimize **time smearing** in an image of angular radius $\Delta\theta$ the averaging time should satisfy

$$\Delta\theta\, \Delta t \ll \frac{\theta_{\mathrm{s}} P}{2\pi} \approx \theta_{\mathrm{s}}\cdot 1.37\times 10^4\ \mathrm{s}. \tag{3.194}$$

The source brightness distribution $I_\nu(l, m)$ and the visibilities $\mathcal{V}_\nu(u, v, w)$ for an interferometer in three dimensions are related by

$$\mathcal{V}_\nu(u, v, w) = \int\int \frac{I_\nu(l, m)}{(1 - l^2 - m^2)^{1/2}} \exp[-i2\pi(ul + vm + wn)]dl\, dm. \qquad (3.197)$$

For a **two-dimensional interferometer** confined to the (u, v) plane, the source brightness distribution $I_\nu(l, m)$ is the Fourier transform of the fringe visibilities $\mathcal{V}_\nu(u, v)$:

$$\frac{I_\nu(l, m)}{(1 - l^2 - m^2)^{1/2}} = \int\int \mathcal{V}_\nu(u, v, 0) \exp[+i2\pi(ul + vm)]du\, dv. \qquad (3.198)$$

The **point-source sensitivity** (or brightness sensitivity in units of flux density per beam solid angle) for an interferometer with N antennas, each with effective area A_e, is

$$\sigma_S = \frac{2kT_s}{A_e[N(N-1)\Delta\nu\,\tau]^{1/2}}. \qquad (3.203)$$

The **brightness sensitivity** (K) corresponding to a point source sensitivity σ_S and a beam solid angle Ω_A is

$$\sigma_T = \left(\frac{\sigma_S}{\Omega_A}\right)\frac{\lambda^2}{2k}, \qquad (3.204)$$

where $\Omega_A = \pi\theta_{\text{HPBW}}^2/(4\ln 2) \approx 1.133\theta_0^2$ for a Gaussian beam of HPBW θ_{HPBW}. The (nonrelativistic) **Maxwellian distribution** of particle speeds v is

$$f(v) = \frac{4v^2}{\sqrt{\pi}}\left(\frac{m}{2kT}\right)^{3/2}\exp\left(-\frac{mv^2}{2kT}\right). \qquad (4.34)$$

The **free–free emission coefficient** is

$$j_\nu = \frac{\pi^2 Z^2 e^6 n_e n_i}{4c^3 m_e^2}\left(\frac{2m_e}{\pi kT}\right)^{1/2}\ln\left(\frac{b_{max}}{b_{min}}\right), \qquad (4.39)$$

where

$$b_{min} \approx \frac{Ze^2}{m_e v^2}. \qquad (4.43)$$

The **free–free absorption coefficient** is

$$\kappa = \frac{1}{\nu^2 T^{3/2}} \left[\frac{Z^2 e^6}{c} n_e n_i \frac{1}{\sqrt{2\pi (m_e k)^3}} \right] \frac{\pi^2}{4} \ln\left(\frac{b_{max}}{b_{min}} \right). \tag{4.52}$$

At frequencies low enough that $\tau \gg 1$, the HII region becomes opaque, its spectrum approaches that of a blackbody with temperature $T \sim 10^4$ K, and the flux density varies as $S \propto \nu^2$. At very high frequencies, $\tau \ll 1$, the HII region is nearly transparent, and

$$S_\nu \propto \frac{2kT\nu^2}{c^2} \tau(\nu) \propto \nu^{-0.1}. \tag{4.54}$$

On a log-log plot, the overall spectrum of a uniform HII region has a break near the frequency at which $\tau \approx 1$.

The **emission measure** of a plasma is defined by

$$\frac{EM}{pc\ cm^{-6}} \equiv \int_{los} \left(\frac{n_e}{cm^{-3}} \right)^2 d\left(\frac{s}{pc} \right). \tag{4.57}$$

The **free–free optical depth** of a plasma is

$$\tau \approx 3.28 \times 10^{-7} \left(\frac{T}{10^4\ K} \right)^{-1.35} \left(\frac{\nu}{GHz} \right)^{-2.1} \left(\frac{EM}{pc\ cm^{-6}} \right). \tag{4.60}$$

The **ionization rate** Q_H of Lyman continuum photons produced per second required to maintain an HII region is

$$\left(\frac{Q_H}{s^{-1}} \right) \approx 6.3 \times 10^{52} \left(\frac{T}{10^4\ K} \right)^{-0.45} \left(\frac{\nu}{GHz} \right)^{0.1} \left(\frac{L_\nu}{10^{20}\ W\ Hz^{-1}} \right), \tag{4.62}$$

where L_ν is the free–free luminosity at any frequency ν high enough that $\tau(\nu) \ll 1$.

The **magnetic force** on a moving charge is

$$\vec{F} = \frac{q(\vec{v} \times \vec{B})}{c}. \tag{5.1}$$

The **gyro frequency** is defined by

$$\omega_G \equiv \frac{qB}{mc}. \tag{5.4}$$

The (nonrelativistic) electron gyro frequency in MHz is

$$\left(\frac{\nu_{\mathrm{G}}}{\mathrm{MHz}}\right) = 2.8\left(\frac{B}{\mathrm{gauss}}\right). \tag{5.7}$$

The **Lorentz transform** is

$$x = \gamma(x' + vt'), \qquad y = y', \qquad z = z', \qquad t = \gamma(t' + \beta x'/c), \tag{5.12}$$

$$x' = \gamma(x - vt), \qquad y' = y, \qquad z' = z, \qquad t' = \gamma(t - \beta x/c), \tag{5.13}$$

where

$$\beta \equiv v/c \tag{5.14}$$

and

$$\gamma \equiv (1 - \beta^2)^{-1/2} \tag{5.15}$$

is called the **Lorentz factor**. If $(\Delta x', \Delta y', \Delta z', \Delta t')$ and $(\Delta x, \Delta y, \Delta z, \Delta t)$ are the coordinate differences between two events, the **differential form** of the (linear) Lorentz transform is

$$\Delta x = \gamma(\Delta x' + v\Delta t'), \quad \Delta y = \Delta y', \quad \Delta z = \Delta z', \quad \Delta t = \gamma(\Delta t' + \beta\Delta x'/c),$$

$$\tag{5.16}$$

$$\Delta x' = \gamma(\Delta x - v\Delta t), \quad \Delta y' = \Delta y, \quad \Delta z' = \Delta z, \quad \Delta t' = \gamma(\Delta t - \beta\Delta x/c).$$

$$\tag{5.17}$$

The **Thomson cross section** of an electron is defined by

$$\sigma_{\mathrm{T}} \equiv \frac{8\pi}{3}\left(\frac{e^2}{m_e c^2}\right)^2. \tag{5.33}$$

Magnetic energy density is given by

$$U_B = \frac{B^2}{8\pi}. \tag{5.35}$$

The **synchrotron power of one electron** is

$$P = 2\sigma_{\rm T}\beta^2\gamma^2 c\, U_B \sin^2\alpha. \qquad (5.37)$$

Synchrotron power averaged over all pitch angles α is

$$\langle P \rangle = \frac{4}{3}\sigma_{\rm T}\beta^2\gamma^2 c U_B. \qquad (5.42)$$

The **synchrotron spectrum of a single electron** is

$$P(\nu) = \frac{\sqrt{3}e^3 B\sin\alpha}{m_e c^2}\left(\frac{\nu}{\nu_{\rm c}}\right)\int_{\nu/\nu_{\rm c}}^{\infty} K_{5/3}(\eta)d\eta, \qquad (5.66)$$

where $K_{5/3}$ is a modified Bessel function and the **critical frequency** is

$$\nu_{\rm c} = \frac{3}{2}\gamma^2\nu_{\rm G}\sin\alpha \approx \gamma^2\nu_{\rm G} \propto E^2 B_\perp. \qquad (5.67)$$

The observed **energy distribution of cosmic-ray electrons** in our Galaxy is roughly a power law:

$$n(E)dE \approx K E^{-\delta}dE, \qquad (5.70)$$

where $n(E)dE$ is the number of electrons per unit volume with energies E to $E + dE$ and $\delta \approx 5/2$. The corresponding **synchrotron emission coefficient** is

$$j_\nu \propto B^{(\delta+1)/2}\nu^{(1-\delta)/2}. \qquad (5.78)$$

The (negative sign convention) **spectral index** of both synchrotron radiation and inverse-Compton radiation is

$$\alpha = \frac{\delta - 1}{2}. \qquad (5.79)$$

The **effective temperature of a relativistic electron** emitting at frequency ν in magnetic field B is

$$\left(\frac{T_e}{\rm K}\right) \approx 1.18 \times 10^6 \left(\frac{\nu}{\rm Hz}\right)^{1/2}\left(\frac{B}{\rm gauss}\right)^{-1/2}. \qquad (5.85)$$

At a sufficiently low frequency ν,

$$S_\nu \propto \nu^{-5/2} \tag{5.89}$$

and

$$\left(\frac{B}{\text{gauss}}\right) \approx 1.4 \times 10^{12} \left(\frac{\nu}{\text{Hz}}\right)\left(\frac{T_b}{\text{K}}\right)^{-2}. \tag{5.91}$$

For a given synchrotron luminosity, the **electron energy density** is

$$U_e \propto B^{-3/2}. \tag{5.98}$$

The **total energy density** of both cosmic rays and magnetic fields is

$$U = (1 + \eta)U_e + U_B, \tag{5.100}$$

where η is the ion/electron energy ratio.

At **minimum total energy**, the ratio of particle to field energy is ~ 1 (**equipartition**):

$$\frac{\text{particle energy}}{\text{field energy}} = \frac{(1+\eta)U_e}{U_B} = \frac{4}{3}. \tag{5.107}$$

The **minimum-energy magnetic field** is

$$B_{\min} = [4.5(1 + \eta)c_{12}L]^{2/7} R^{-6/7} \text{ gauss} \tag{5.109}$$

and the corresponding **total energy** is

$$E_{\min}(\text{total}) = c_{13}[(1 + \eta)L]^{4/7} R^{9/7} \text{ ergs}. \tag{5.110}$$

The **synchrotron lifetime** is approximately

$$\tau \approx c_{12} B_\perp^{-3/2}, \tag{5.112}$$

where the functions c_{12} and c_{13} in Gaussian CGS units are plotted in Figures 5.10 and 5.11. Frequency limits $\nu_{\min} = 10^7$ Hz and $\nu_{\max} = 10^{11}$ Hz are commonly used.

The **Eddington limit** for luminosity is

$$\left(\frac{L_E}{L_\odot}\right) \approx 3.3 \times 10^4 \left(\frac{M}{M_\odot}\right). \tag{5.117}$$

The **nonrelativistic Thomson-scattering power** is

$$P = \sigma_{\mathrm{T}} c U_{\mathrm{rad}}. \tag{5.132}$$

The **relativistic Doppler equation** is

$$\nu' = \nu[\gamma(1 + \beta\cos\theta)]. \tag{5.142}$$

The net **inverse-Compton power** emitted is

$$P_{\mathrm{IC}} = \frac{4}{3}\sigma_{\mathrm{T}} c\beta^2\gamma^2 U_{\mathrm{rad}}. \tag{5.152}$$

The **IC/synchrotron power ratio** is

$$\frac{P_{\mathrm{IC}}}{P_{\mathrm{syn}}} = \frac{U_{\mathrm{rad}}}{U_B}. \tag{5.154}$$

The **average frequency** $\langle\nu\rangle$ of upscattered photons having initial frequency ν_0 is

$$\frac{\langle\nu\rangle}{\nu_0} = \frac{4}{3}\gamma^2. \tag{5.160}$$

The **maximum rest-frame brightness temperature** of an incoherent synchrotron source is limited by inverse-Compton scattering to

$$T_{\max} \sim 10^{12}\ \mathrm{K}. \tag{5.163}$$

The **apparent transverse velocity** of a moving source component is

$$\beta_\perp(\text{apparent}) = \frac{\beta\sin\theta}{1 - \beta\cos\theta}. \tag{5.167}$$

For any β the angle θ_{m} that maximizes $\beta_\perp(\text{apparent})$ satisfies

$$\cos\theta_{\mathrm{m}} = \beta \tag{5.170}$$

and

$$\sin\theta_{\mathrm{m}} = \gamma^{-1}. \tag{5.171}$$

The largest apparent transverse speed is

$$\max[\beta_\perp(\text{apparent})] = \beta\gamma.$$ (5.172)

The **transverse Doppler shift** (at $\theta = \pi/2$) is

$$\frac{\nu}{\nu'} = \gamma^{-1}.$$ (5.180)

The **Doppler boosting** for **Doppler factor** $\delta \equiv \nu/\nu'$ is in the range

$$\delta^{2+\alpha} < \frac{S}{S_0} < \delta^{3+\alpha}.$$ (5.183)

Thermal and nonthermal **radio luminosities of star-forming galaxies** are

$$\left(\frac{L_{\mathrm{T}}}{\mathrm{W\,Hz^{-1}}}\right) \approx 5.5 \times 10^{20} \left(\frac{\nu}{\mathrm{GHz}}\right)^{-0.1} \left[\frac{\mathrm{SFR}(M > 5M_\odot)}{M_\odot\,\mathrm{yr^{-1}}}\right]$$ (5.184)

and

$$\left(\frac{L_{\mathrm{NT}}}{\mathrm{W\,Hz^{-1}}}\right) \approx 5.3 \times 10^{21} \left(\frac{\nu}{\mathrm{GHz}}\right)^{-0.8} \left[\frac{\mathrm{SFR}(M > 5M_\odot)}{M_\odot\,\mathrm{yr^{-1}}}\right].$$ (5.185)

The **minimum mean density** of a pulsar with period P is

$$\rho > \frac{3\pi}{G P^2}.$$ (6.5)

A rotating **magnetic dipole** radiates power

$$P_{\mathrm{rad}} = \frac{2}{3}\frac{(\ddot{m}_\perp)^2}{c^3}.$$ (6.10)

The **spin-down luminosity** of a pulsar is

$$-\dot{E} \equiv -\frac{dE_{\mathrm{rot}}}{dt} = \frac{-4\pi^2 I \dot{P}}{P^3}.$$ (6.20)

The **minimum magnetic field strength** of a pulsar is

$$\left(\frac{B}{\text{gauss}}\right) > 3.2 \times 10^{19} \left(\frac{P \dot{P}}{\text{s}}\right)^{1/2}.$$

(6.26)

The **characteristic age** of a pulsar is defined by

$$\tau \equiv \frac{P}{2\dot{P}}.$$

(6.31)

The **braking index** of a pulsar in terms of its observable period P and the first and second time derivatives is

$$n = 2 - \frac{P \ddot{P}}{\dot{P}^2}.$$

(6.37)

At frequency ν the **refractive index** of a cold plasma is

$$\mu = \left[1 - \left(\frac{\nu_{\text{p}}}{\nu}\right)^2\right]^{1/2},$$

(6.39)

where ν_{p} is the **plasma frequency**

$$\nu_{\text{p}} = \left(\frac{e^2 n_{\text{e}}}{\pi m_{\text{e}}}\right)^{1/2} \approx 8.97 \text{ kHz} \left(\frac{n_{\text{e}}}{\text{cm}^{-3}}\right)^{1/2}.$$

(6.40)

The **group velocity** of pulses is

$$v_{\text{g}} \approx c \left(1 - \frac{\nu_{\text{p}}^2}{2\nu^2}\right).$$

(6.42)

The **dispersion delay** of a pulsar is

$$\left(\frac{t}{\text{sec}}\right) \approx 4.149 \times 10^3 \left(\frac{\text{DM}}{\text{pc cm}^{-3}}\right) \left(\frac{\nu}{\text{MHz}}\right)^{-2},$$

(6.45)

where

$$\text{DM} \equiv \int_0^d n_{\text{e}} \, dl$$

(6.46)

in units of pc cm^{-3} is the **dispersion measure** of a pulsar at distance d.
The **Bohr radius** of a hydrogen atom is

$$a_n = \frac{n^2 \hbar^2}{m_e e^2} \approx 0.53 \times 10^{-8} \text{cm} \cdot n^2. \tag{7.6}$$

The frequency of a **recombination line** is

$$\nu = R_M c \left[\frac{1}{n^2} - \frac{1}{(n + \Delta n)^2} \right], \qquad \text{where} \qquad R_M \equiv R_\infty \left(1 + \frac{m_e}{M} \right)^{-1}. \tag{7.12}$$

The approximate recombination **line separation frequency** $\Delta\nu \equiv \nu(n) - \nu(n+1)$
for $n \gg 1$ is

$$\frac{\Delta\nu}{\nu} \approx \frac{3}{n}. \tag{7.15}$$

The **spontaneous emission rate** is

$$A_{n+1,n} \approx \frac{64\pi^6 m_e e^{10}}{3c^3 h^6 n^5} \approx 5.3 \times 10^9 \left(\frac{1}{n^5} \right) \text{s}^{-1}. \tag{7.23}$$

The normalized **Gaussian line profile** is

$$\phi(\nu) = \frac{c}{\nu_0} \left(\frac{M}{2\pi kT} \right)^{1/2} \exp \left[-\frac{Mc^2}{2kT} \frac{(\nu - \nu_0)^2}{\nu_0^2} \right], \tag{7.32}$$

where

$$\Delta\nu = \left(\frac{8\ln 2 \, k}{c^2} \right)^{1/2} \left(\frac{T}{M} \right)^{1/2} \nu_0 \tag{7.35}$$

and

$$\phi(\nu_0) = \left(\frac{\ln 2}{\pi} \right)^{1/2} \frac{2}{\Delta\nu}. \tag{7.37}$$

Rate balance is given by

$$n_U A_{UL} + n_u B_{UL} \bar{u} = n_L B_{LS} \bar{u}. \tag{7.42}$$

The **detailed balance equations** connecting **Einstein coefficients** are

$$\frac{g_L}{g_U}\frac{B_{LU}}{B_{UL}} = 1, \qquad (7.50)$$

$$\frac{A_{UL}}{B_{UL}} = \frac{8\pi h \nu_0^3}{c^3}. \qquad (7.51)$$

The **spectral line radiative transfer equation** is

$$\frac{dI_\nu}{ds} = -\left(\frac{h\nu_0}{c}\right)(n_L B_{LU} - n_U B_{UL})\phi(\nu)I_\nu + \left(\frac{h\nu_0}{4\pi}\right)n_U A_{UL}\phi(\nu). \qquad (7.57)$$

The **Boltzmann equation** for a two-level system is

$$\frac{n_U}{n_L} = \frac{g_U}{g_L}\exp\left(-\frac{h\nu_0}{kT}\right). \qquad (7.64)$$

The **line opacity coefficient** in LTE is

$$\kappa = \frac{c^2}{8\pi \nu_0^2}\frac{g_U}{g_L}n_L A_{UL}\left[1 - \exp\left(-\frac{h\nu_0}{kT}\right)\right]\phi(\nu). \qquad (7.67)$$

The **excitation temperature** T_x is defined by

$$\frac{n_U}{n_L} \equiv \frac{g_U}{g_L}\exp\left(-\frac{h\nu_0}{kT_x}\right). \qquad (7.70)$$

The **recombination-line opacity coefficient** is

$$\kappa(\nu_0) \approx \left(\frac{n_e^2}{T_e^{5/2}\Delta\nu}\right)\left(\frac{4\pi e^6 h}{3m_e^{3/2}k^{5/2}c}\right)\left(\frac{\ln 2}{2}\right)^{1/2} \qquad (7.94)$$

and the **recombination line opacity** is

$$\tau_L \approx 1.92 \times 10^3 \left(\frac{T_e}{K}\right)^{-5/2}\left(\frac{EM}{pc\,cm^{-6}}\right)\left(\frac{\Delta\nu}{kHz}\right)^{-1}. \qquad (7.96)$$

The **recombination line brightness temperature** is given by

$$T_L \approx T_e \tau_L \approx 1.92 \times 10^3 \left(\frac{T_e}{K} \right)^{-3/2} \left(\frac{EM}{pc\, cm^{-6}} \right) \left(\frac{\Delta v}{kHz} \right)^{-1}.$$ (7.97)

The **recombination line/continuum ratio** is

$$\frac{T_L}{T_C} \approx 7.0 \times 10^3 \left(\frac{\Delta v}{km\, s^{-1}} \right)^{-1} \left(\frac{v}{GHz} \right)^{1.1} \left(\frac{T_e}{K} \right)^{-1.15} \left[1 + \frac{N(He^+)}{N(H^+)} \right]^{-1},$$ (7.98)

where $[1 + N(He^+)/N(H^+)] \approx 1.08$.
The **electron temperature** from the line/continuum ratio is

$$\left(\frac{T_e}{K} \right) \approx \left[7.0 \times 10^3 \left(\frac{v}{GHz} \right)^{1.1} 1.08^{-1} \left(\frac{\Delta v}{km\, s^{-1}} \right)^{-1} \left(\frac{T_C}{T_L} \right) \right]^{0.87}.$$ (7.99)

Quantization of angular momentum is given by

$$L = n\hbar.$$ (7.100)

The **angular momentum of a diatomic molecule** is

$$L = mr_e^2 \omega,$$ (7.104)

where

$$m \equiv \left(\frac{m_A m_B}{m_A + m_B} \right)$$ (7.105)

is the **reduced mass** and r_e is the separation of the atoms with masses m_A and m_B.
The **rotational energy levels** of a diatomic molecule with moment of inertia I are

$$E_{rot} = \frac{J(J+1)\hbar^2}{2I}, \qquad J = 0, 1, 2, \ldots.$$ (7.107)

For a transition satisfying the **selection rule**

$$\Delta J = \pm 1,$$ (7.108)

the **line frequency** is

$$\nu = \frac{hJ}{4\pi^2 m r_e^2}. \tag{7.111}$$

The **minimum temperature** needed to excite the $J \rightarrow J - 1$ transition at frequency ν is

$$T_{\min} \approx \frac{\nu h(J+1)}{2k}. \tag{7.119}$$

The **spontaneous emission coefficient** is

$$A_{\mathrm{UL}} = \frac{64\pi^4}{3hc^3} \nu_{\mathrm{UL}}^3 |\mu_{\mathrm{UL}}|^2, \tag{7.131}$$

where

$$|\mu_{J\rightarrow J-1}|^2 = \frac{\mu^2 J}{2J+1} \tag{7.132}$$

and μ is the electric dipole moment of the molecule.
The **critical density** is

$$n^* \approx \frac{A_{\mathrm{UL}}}{\sigma \upsilon}, \tag{7.135}$$

where $\sigma \sim 10^{-15}$ cm^{-2} is the collision cross section and $\upsilon \sim 10^5$ cm s^{-1} is the typical H_2 molecular velocity.
The **CO-to-H_2 conversion factor** X_{CO} in our Galaxy is

$$X_{\mathrm{CO}} = (2 \pm 0.6) \times 10^{20} \text{ cm}^{-2} \text{ (K km s}^{-1})^{-1}. \tag{7.140}$$

The **HI hyperfine line frequency** is

$$\nu_{10} = \frac{8}{3} g_{\mathrm{I}} \left(\frac{m_e}{m_p} \right) \alpha^2 (R_M c) \approx 1420.405751 \text{ MHz}. \tag{7.141}$$

The HI hyperfine line **emission coefficient** is

$$A_{10} \approx 2.85 \times 10^{-15} \text{ s}^{-1}. \tag{7.146}$$

The **HI spin temperature** T_s is defined by

$$\frac{n_1}{n_0} \equiv \frac{g_1}{g_0} \exp\left(-\frac{h\nu_{10}}{kT_s}\right),$$

(7.148)

where $g_1/g_0 = 3$.

The HI line **opacity coefficient** is

$$\kappa(\nu) \approx \frac{3c^2}{32\pi} \frac{A_{10}n_H}{\nu_{10}} \frac{h}{kT_s} \phi(\nu).$$

(7.153)

The hydrogen **column density** η_H is defined as the integral of density along the line of sight:

$$\eta_H \equiv \int_{los} n_H(s)\,ds.$$

(7.154)

If the HI line is optically thin ($\tau \ll 1$) then the **HI column density** is

$$\left(\frac{\eta_H}{cm^{-2}}\right) \approx 1.82 \times 10^{18} \int \left[\frac{T_b(\nu)}{K}\right] d\left(\frac{\nu}{km\,s^{-1}}\right).$$

(7.155)

If $\tau \ll 1$ the **hydrogen mass** of a galaxy is

$$\left(\frac{M_H}{M_\odot}\right) \approx 2.36 \times 10^5 \left(\frac{D}{Mpc}\right)^2 \int \left[\frac{S(\nu)}{Jy}\right] \left(\frac{d\nu}{km\,s^{-1}}\right).$$

(7.166)

The **total mass** of a galaxy is

$$\left(\frac{M}{M_\odot}\right) \approx 2.33 \times 10^5 \left(\frac{\nu_{rot}}{km\,s^{-1}}\right)^2 \left(\frac{r}{kpc}\right).$$

(7.172)

F Constants, Units, and Dimensions

F.1 PHYSICAL CONSTANTS

Symbol	Name	Value	Units
a	radiation constant	7.56577×10^{-15}	$\mathrm{erg\,cm^{-3}\,K^{-4}}$
a_0	Bohr radius	5.29177×10^{-9}	cm
c	speed of light in vacuum	2.99792×10^{10}	$\mathrm{cm\,s^{-1}}$
e	electron charge (magnitude)	4.80325×10^{-10}	statcoulomb (or esu)
eV	electron volt	1.60218×10^{-12}	erg
G	gravitational constant	6.67428×10^{-8}	$\mathrm{dyne\,cm^2\,g^{-2}}$
h	Planck's constant	6.62607×10^{-27}	erg s
k	Boltzmann's constant	1.38065×10^{-16}	$\mathrm{erg\,K^{-1}}$
m_e	electron mass	9.10938×10^{-28}	g
m_p	proton mass	1.67262×10^{-24}	g
μ_B	Bohr magneton	9.27401×10^{-21}	$\mathrm{erg\,gauss^{-1}}$
R_∞	Rydberg constant	1.09737×10^5	$\mathrm{cm^{-1}}$
$R_\infty c$	Rydberg frequency	3.28984×10^{15}	$\mathrm{s^{-1}}$
σ	Stefan–Boltzmann constant	5.67040×10^{-5}	$\mathrm{erg\,s^{-1}\,cm^{-2}\,sr^{-1}\,K^{-4}}$
σ_T	Thomson cross section	6.65245×10^{-25}	$\mathrm{cm^2}$
u	atomic mass unit	1.66054×10^{-24}	g

F.2 ASTRONOMICAL CONSTANTS

Symbol	Name	Value	Units
au	astronomical unit	1.49598×10^{13}	cm
		≈ 500	light seconds
H_0	Hubble constant	67.8	$\mathrm{km\,s^{-1}\,Mpc^{-1}}$
kpc	kiloparsec	10^3	pc
$L_\odot$	solar luminosity	3.826×10^{33}	$\mathrm{erg\,s^{-1}}$
ly	light year	9.4606×10^{17}	cm
$M_\odot$	solar mass	1.989×10^{33}	g
Mpc	megaparsec	10^6	pc
pc	parsec	3.0856×10^{18}	cm
$R_\odot$	solar radius	6.9558×10^{10}	cm
yr	tropical year	3.1557×10^7	s
		$\approx 10^{7.5}$	s

F.3 MKS (SI) AND GAUSSIAN CGS UNITS

The **International System of Units**, or Système Internationale d'Unités (**SI**), was adopted in 1960 to succeed the **MKS** (meter, kilogram, second) system of units dating from 1799 and the **CGS** (centimeter, gram, second) system introduced in 1874. The various CGS units for the electric and magnetic fields are inconveniently large or small and have different dimensions, so the *practical units* **ampere** for electric current, **ohm** for resistance, and **volt** for electromotive force were introduced in 1893. SI is the direct descendant of the MKS system, with base units meter, kilogram, and second, plus ampere for current and kelvin for temperature. The kilogram is defined by the mass of the international prototype of the kilogram. The other base units are defined by laboratory measurements: the second is the duration of 9,192,631,770 periods of the radiation corresponding to the transition between the two hyperfine levels of the ground state of the cesium 133 atom, the meter is the distance traveled by light in vacuum during a time interval of 1/299,792,458 of a second, the ampere is that constant current which, if maintained in two straight parallel conductors of infinite length, of negligible circular cross section, and placed 1 meter apart in vacuum, would produce between these conductors a force equal to 2×10^{-7} newton per meter of length, and the kelvin is the fraction 1/273.16 of the thermodynamic temperature of the triple point of water. Thus the SI speed of light in a vacuum is *exactly* 299,792,458 m s^{-1} *by definition*. For the purposes of this book, the terms MKS and SI are interchangeable.

Type	MKS unit	CGS unit	conversion
length	m	cm	10^2 cm = 1 m
			2.54 cm $\equiv$ 1 inch (exactly)
mass	kg	g	10^3 g = 1 kg
time	s	s	
energy	joule	erg	10^7 erg = 1 joule = 1 kg m^2 s^{-2}
force	newton	dyne	10^5 dyne = 1 newton = 1 kg m s^{-2}
frequency	Hz	Hz	1 Hz = 1 s^{-1}
power	W	erg s^{-1}	10^7 erg s^{-1} = 1 W = 1 kg m^2 s^{-3}
temperature	kelvin	kelvin	
charge	coulomb	statcoulomb	3×10^9 statcoulomb $\leftrightarrow$ 1 coulomb
			(1 statcoulomb = 1 esu)
current	ampere	statampere	3×10^9 statamp $\leftrightarrow$ 1 amp
			(1 amp = 1 coulomb s^{-1})
electric field	volt m^{-1}	statvolt cm^{-1}	$(1/3) \times 10^{-4}$ statvolt cm^{-1} $\leftrightarrow$
			1 volt m^{-1}
magnetic field	tesla	gauss	10^4 gauss $\leftrightarrow$ 1 tesla
resistance	ohm	sec cm^{-1}	$(1/9) \times 10^{-11}$ s cm^{-1} $\leftrightarrow$ 1 ohm
voltage	volt	statvolt	$(1/3) \times 10^{-2}$ statvolt $\leftrightarrow$ 1 volt
			(1 volt = 1 joule coulomb^{-1})

Engineers and most physicists prefer MKS units, so radio astronomers use MKS units to describe their equipment and the results of their observations. Most astrophysicists prefer Gaussian CGS units to describe astrophysical processes and astronomical sources. Thus both systems of units appear in the literature. Be careful when converting between them because astrophysical quantities can be so large or small that conversion mistakes are not intuitively obvious—everyday experience doesn't show that the mass of the Sun is 2×10^{30} kg and not 2×10^{30} g.

The MKS and CGS units of length, mass, time, and temperature have the same *dimensions*, so it is possible to state that 10^2 cm $= 1$ m, for example. However, the MKS and CGS units for electromagnetic quantities such as charge, current, voltage, etc. are trickier because *they do not have the same dimensions*. The Gaussian CGS unit of charge (statcoulomb) is *defined* so that Coulomb's law for the magnitude of the electrostatic force F between two charges q_1 and q_2 separated by distance r is simply

$$F = \frac{q_1 q_2}{r^2}. \tag{F.1}$$

The MKS unit of charge (coulomb) is *defined* in terms of currents and Ampere's law, and Coulomb's law becomes

$$F = \frac{c^2}{10^7} \frac{q_1 q_2}{r^2}, \tag{F.2}$$

where $c = 2.99729458 \times 10^8$ m s^{-1} is the vacuum speed of light in MKS units. The MKS form of Coulomb's law is usually written as

$$F = \frac{1}{4\pi\epsilon_0} \frac{q_1 q_2}{r^2}, \tag{F.3}$$

where ϵ_0 is called the **permittivity of free space** and

$$\epsilon_0 = \frac{10^7}{4\pi c^2} \approx 8.854 \times 10^{-12} \tag{F.4}$$

has units m^{-2} s$^2 =$ C^2 N^{-1} m^{-2}. The corresponding MKS magnetic quantity is the **permeability of free space**

$$\mu_0 \equiv 4\pi \times 10^{-7} \text{ newton ampere}^{-2} = (\epsilon_0 c^2)^{-1}. \tag{F.5}$$

Comparing forces in Equations F.1 and F.2 shows that

$$1 \text{ coulomb} = 10c \text{ statcoulomb} \tag{F.6}$$

Consequently statcoul and coulomb have different dimensions, and the **numerical conversion factor** factor relating them is

$$1 \text{ coulomb} \leftrightarrow 2.99729458 \times 10^9 \text{ statcoulomb}, \tag{F.7}$$

where the symbol $\leftrightarrow$ indicates **numerical conversion** relating *given amounts* of dimensionally incompatible quantities. For example, given that the charge of an electron is $e = -4.80325 \times 10^{-10}$ statcoulomb in CGS units, Equation F.7 can be used to show that the electron charge is $e = -1.60253 \times 10^{-19}$ coulomb in MKS units. The charge conversion factor is simplified to 3×10^9 in the table above, where 3 is the conventional dimensionless shorthand for the exact and dimensionally correct

$2.99729458 \ (\mathrm{m \ s^{-1}}) = 10^{-8}c$; likewise, $(1/3)$ for $1/[2.99729458 \ (\mathrm{m \ s^{-1}})]$ and $1/9$ for $1/[2.99729458 \ (\mathrm{m \ s^{-1}})]^2$.

Entire equations can also be converted between CGS and MKS if the dimensions are carefully preserved. For example, the CGS form of Larmor's formula (Equation 2.143) for the power P radiated by a charge q under acceleration $\dot{v}$ is

$$P = \frac{2}{3} \frac{q^2 \dot{v}^2}{c^3},$$

so

$$\left(\frac{P}{\mathrm{erg \ s^{-1}}} \right) = \frac{2}{3} \left(\frac{q}{\mathrm{statcoul}} \right)^2 \left(\frac{\dot{v}}{\mathrm{cm \ s^{-1}}} \right)^2 \left(\frac{c}{\mathrm{cm \ s^{-1}}} \right)^{-3}.$$

The numerical conversion

$$3 \times 10^9 \ \mathrm{statcoulomb} \leftrightarrow 1 \ \mathrm{coulomb}$$

implies the dimensionally correct equality

$$2.99729458 \ \mathrm{m \ s^{-1}} \times 10^9 \ \mathrm{statcoulomb} = 10c \ \mathrm{statcoulomb} = 1 \ \mathrm{coulomb},$$

so

$$\left(\frac{P}{10^{-7} \ \mathrm{W}} \right) = \frac{2}{3} \left(\frac{q}{(10c)^{-1} \ \mathrm{coul}} \right)^2 \left(\frac{\dot{v}}{10^{-2} \ \mathrm{m \ s^{-1}}} \right)^2 \left(\frac{c}{10^{-2} \ \mathrm{m \ s^{-1}}} \right)^{-3},$$

$$10^7 \left(\frac{P}{\mathrm{W}} \right) = \frac{2}{3} (10c)^2 \left(\frac{q}{\mathrm{coul}} \right)^2 (10^4) \left(\frac{\dot{v}}{\mathrm{m \ s^{-1}}} \right)^2 (10^{-6}) \left(\frac{c}{\mathrm{m \ s^{-1}}} \right)^{-3},$$

$$\left(\frac{P}{\mathrm{W}} \right) = \frac{2}{3} \left(\frac{c^2}{10^7} \right) \left(\frac{q}{\mathrm{coul}} \right)^2 \left(\frac{\dot{v}}{\mathrm{m \ s^{-1}}} \right)^2 \left(\frac{c}{\mathrm{m \ s^{-1}}} \right)^{-3}$$

$$= \frac{2}{3} \left(\frac{1}{4\pi \epsilon_0} \right) \left(\frac{q}{\mathrm{coul}} \right)^2 \left(\frac{\dot{v}}{\mathrm{m \ s^{-1}}} \right)^2 \left(\frac{c}{\mathrm{m \ s^{-1}}} \right)^{-3}.$$

Thus Larmor's equation in MKS units is

$$P = \frac{1}{6\pi \epsilon_0} \frac{q^2 \dot{v}^2}{c^3}.$$

MKS units are used in practical electromagnetics—voltmeters read MKS volts, ammeters read MKS amperes, a 9 volt battery delivers 9 MKS volts, etc. An important advantage of Gaussian CGS units is that the Lorentz force law becomes

$$\vec{F} = q(\vec{E} + \vec{\beta} \times \vec{B}), \tag{F.8}$$

where $\vec{\beta} \equiv \vec{v}/c$ is dimensionless, so $\vec{E}$ and $\vec{B}$ have the same dimensions (base units $\mathrm{cm^{-1/2} \ g^{1/2} \ s^{-1}}$) in Gaussian CGS units, and 1 statvolt $\mathrm{cm^{-1}} = 1$ gauss, emphasizing the relativistic result that electric and magnetic fields are equivalent fields viewed in different reference frames. This equivalence is not apparent in the MKS version of Equation F.8:

$$\vec{F} = q(\vec{E} + \vec{v} \times \vec{B}), \tag{F.9}$$

which drops the speed of light and also shows that $\vec{E}$ and $\vec{B}$ have different dimensions in the MKS system. Thus the equality

$$|\vec{E}| = |\vec{B}| \tag{F.10}$$

makes sense in the CGS system but not in MKS.

For a detailed explanation of the common systems of units, see the appendix in Jackson [54].

F.4 OTHER CONSTANTS AND UNITS

Symbol	Value
AB magnitude	$-2.5\log[S(\mathrm{Jy})]+8.90 = -2.5\log_{10}[S/(3631\ \mathrm{Jy})]$
arcmin	1/60 deg
arcsec	1/60 arcmin
Å	10^{-10} m
D	1 debye $\equiv 10^{-18}$ statcoulomb cm
dB	$10\log_{10}(P_1/P_2)$
deg	$(\pi/180)$ rad
e	$2.71828\ldots$
GHz	10^9 Hz
Jy	10^{-26} W m^{-2} Hz^{-1} = 10^{-23} erg s^{-1} cm^{-2} Hz^{-1}
MHz	10^6 Hz
mJy	10^{-3} Jy
μm	10^{-6} m
μJy	10^{-6} Jy
π	$3.14159\ldots$
radian (rad)	angle subtending unit arc length on a unit circle
	= $180/\pi \approx 57.296$ deg, = $640000/\pi \approx 206265$ arcsec
steradian (sr)	solid angle subtending unit area on a unit sphere
THz	10^{12} Hz

F.5 RADAR AND WAVEGUIDE FREQUENCY BANDS

Band name	Frequency (GHz)	Mnemonic
P	$0.25 < \nu < 0.50$	Previous
L	$1 < \nu < 2$	Long (wavelength)
S	$2 < \nu < 4$	Short (wavelength)
C	$4 < \nu < 8$	Compromise (between S and X)
X	$8 < \nu < 12$	X shape of crosshairs
Ku	$12 < \nu < 18$	Kurz — under
K	$18 < \nu < 26.5$	Kurz (short in German)
Ka	$26.5 < \nu < 40$	Kurz — above
Q	$33 < \nu < 50$	
U	$40 < \nu < 60$	U is before V in alphabet
V	$50 < \nu < 75$	Very absorbed by O_2
W	$75 < \nu < 110$	W is after V in alphabet

F.6 DIMENSIONAL ANALYSIS

The **dimensions** of any physical quantity can be written as a product of powers of the fundamental physical dimensions of length, mass, and time. The dimensions of velocity can be written as (length) × (time)$^{-1}$, for instance. The derived dimensions temperature and charge are used for convenience.

Both sides of every equation should have the same dimensions because valid laws of physics are independent of the units used. **Dimensional analysis** provides a useful check for errors in derivations of new equations. For example, the total flux of a blackbody radiator at temperature T is

$$S(T) = \sigma T^4.$$

The dimensions of flux are power per unit area (e.g., units erg s^{-1} cm^{-2}), so the dimensions of the Stefan–Boltzmann constant σ should be power × area^{-1} × (temperature)$^{-4}$ (e.g., units erg s^{-1} cm^{-2} K^{-4}). However, the total brightness of a blackbody radiator is

$$B(T) = \frac{\sigma T^4}{\pi},$$

and the dimensions of brightness are power × area^{-1} *per unit solid angle* (e.g., units erg s^{-1} cm^{-2} sr^{-1}), indicating that σ has dimensions of power × area^{-1} *per unit solid angle* × (temperature)$^{-4}$ (e.g., dimensions erg s^{-1} cm^{-2} sr^{-1} K^{-4}). This does not violate the rule that both sides of the equation should have the same dimensions because the extra sr^{-1} is itself dimensionless: solid angle has dimensions of (angle)^2 = (length/length)^2. For clarity, the extra sr^{-1} should be written explicitly where appropriate.

Just because angles and solid angles are dimensionless doesn't mean they don't have units. The natural unit for angle is the radian, defined as the angle subtended at the center of a circle by an arc whose length equals the radius of the circle. Only this angle has dimensions (length/length) in which both lengths have the same units. Any other unit of angle, the degree for example, does not have that property and must be called out explicitly in equations. Likewise, the only natural unit for solid angle is the steradian, defined as the solid angle subtended by a unit area on the surface of a sphere of unit radius.

The dimensions of some quantities can be written in more than one way, and the simplest is not always the clearest. For example, the noise power per unit frequency generated by a resistor at temperature T is

$$P_\nu = kT.$$

Both P_ν and kT have dimensions of energy (e.g., erg or joule). However, it is easier to think of P_ν as a power per unit frequency (e.g., erg s^{-1} Hz^{-1} or W Hz^{-1}), even though power per unit frequency also has dimensions of energy because the dimension of Hz^{-1} is time (s) which cancels the s^{-1}.

Finally, the arguments of many functions are dimensionless. Examples include $\sin\theta$, where θ is an angle (dimensions of length/length), $\cos(\omega t)$, where the dimensions of angular frequency ω (inverse time) and t (time) cancel, and $\exp[h\nu/(kT)]$, where the numerator $h\nu$ and denominator (kT) both have dimensions of energy.

G Symbols and Abbreviations

G.1 GREEK SYMBOLS

Symbol	Definition	Reference
α	fine-structure constant	Equation 7.141
	magnetic inclination angle	Section 6.1.4
	pitch angle	Section 5.2.3
	right ascension	Figure 6.1
	spectral index	Equation 4.55
α_0	initial spectral index	Equation 5.127
α_H	hydrogen recombination coefficient	Equation 4.7
β	$\equiv v/c$	Equation C.11
	ecliptic latitude	Section 6.3
γ	Lorentz factor $\equiv (1 - \beta^2)^{-1/2}$	Equation C.10
δ	declination	Section 5.6.4
	Doppler factor	Equation 5.179
	phase difference (polarization)	Equation 2.51
	phase error caused by reflector surface error	Equation 3.119
	power law slope of electron energies	Equation 5.70
δ_0	initial power law slope of electron energies	Equation 5.126
Δx	change in a quantity x	
$\Delta \nu$	bandwidth	
	line profile FWHM	Figure 7.4
ϵ	reflector surface error	Section 3.3.4
ϵ_0	permittivity of free space	Equation F.4
η	antenna efficiency	Section 3.1.3
	baryon/photon ratio	Section 2.6.3
	ion/electron energy ratio	Equation 5.100
η_A	aperture efficiency	Equation 3.111
η_B	main beam efficiency	Equation 3.58
η_H	H I column density	Equation 7.154
η_s	reflector surface efficiency	Equation 3.129

Symbol	Definition	Reference
θ	angle, polar angle	
θ_{HPBW}	half power beam width	Figure 3.15
θ_{m}	angle for maximum superluminal motion	Equations 5.170, 5.171
κ	linear absorption coefficient	Equation 2.18
λ	ecliptic longitude	Equation 6.50
	wavelength	Section 1.1.1
λ	subscript $\lambda \equiv$ per unit wavelength	Equation 2.4
λ_{c}	waveguide cutoff wavelength	Equation 3.141
λ_{D}	Debye length	Equation 4.46
μ	mean electric dipole moment	Equation 7.129
	plasma refractive index	Equation 6.39
μ_0	permeability of free space	Equation F.5
μ_{B}	Bohr magneton	Equation 7.144
μJy	microjansky $\equiv 10^{-6}$ Jy	Appendix F.4
μm	micrometer $=$ micron $\equiv 10^{-6}$ m	Appendix F.4
ν	frequency in Hz $=$ cycles s^{-1} $= \omega/(2\pi)$	Appendix F.3
ν	subscript $\nu \equiv$ per unit frequency	Equation 2.4
ν_0	line center frequency	Figure 7.4
ν_{10}	H I line frequency	Equation 7.141
ν_{c}	critical frequency	Equation 5.67
	waveguide cutoff frequency	Equation 3.142
ν_{e}	frequency in the source frame	
ν_{g}	group velocity	Equation D.18
ν_{G}	electron gyro frequency	Equation 5.7
ν_{o}	frequency in the observer's frame	
ν_{p}	plasma frequency	Equation 6.40
Π	unit rectangle function	Equation 3.74
ρ	mass density	
σ	cross section	
	root mean square (rms)	
	Stefan–Boltzmann constant	Equation 2.90
	surface area element	
σ_{c}	rms confusion	Section 3.6.3.2
σ_S	rms fluctuation in peak flux density	Section 3.7.6
σ_T	rms image brightness fluctuation	Section 3.7.6
	rms noise temperature fluctuation	Equation 3.154
σ_{T}	Thomson cross section	Equation 5.33
τ	duration, lifetime, smoothing time	
	opacity $=$ optical depth	Equation 2.23
	pulsar characteristic age	Equation 6.31
τ_{g}	geometric delay (time)	Equation 3.172
τ_{L}	line opacity at the line center frequency	Equation 7.96
τ_{s}	synchrotron lifetime	Equation 5.111
τ_{z}	atmospheric zenith opacity	Section 1.1.2

Symbol	Definition	Reference
ϕ	angle	
	azimuth angle	Section 1.2
	fringe phase	Equation 3.176
	pulsar pulse phase	Section 6.3
	visibility phase	Equation 3.185
$\phi(\nu)$	normalized line profile	Figure 7.4
χ_n	ionization potential of nth energy level	Equation 7.92
ψ	free–free interaction angle	Figure 4.2
ω	angular frequency in rad s^{-1} $= 2\pi\nu$	Equation 2.50
$\omega_\odot$	Sun's Galactic orbital angular velocity	Section 7.8.1
ω_B	(angular) magnetic orbital frequency	Equation 5.21
ω_G	(angular) gyro frequency (rad s^{-1})	Equation 5.4
Ω	rotation angular velocity	
	solid angle	
	ohm	Appendix F.3
Ω_A	beam solid angle	Equations 3.42, 3.34
Ω_{MB}	main beam solid angle	Equation 3.57

G.2 OTHER SYMBOLS AND ABBREVIATIONS

Symbol	Definition	Reference
$\equiv$	is defined as	
$\propto$	is proportional to	
$\dot{f}$	time derivative of f	
$\ddot{f}$	second time derivative of f	
$\langle x \rangle$	average value of x	
$\lvert x \rvert$	magnitude of x	
$\hat{x}$	unit vector in the x–direction	
$\vec{x}$	vector quantity x	
$\overline{x}$	complex conjugate of x	
$\Leftrightarrow$	is the Fourier transform of	Equation A.1
$*$	convolution operator	Equation A.14
$\star$	cross-correlation operator	Equation A.16
$\leftrightarrow$	numerical conversion between units	Appendix F.3
a	acceleration	
	dimensionless scaling factor	Equation A.11
	dust grain diameter	Section 2.8
	radiation constant	Equation 2.94
	scale factor of the expanding universe	Equation 2.128
$a(\nu)$	absorption coefficient of an opaque body	Section 2.2.4

Symbol	Definition	Reference
a_0	Bohr radius	Appendix F.1
a_n	Bohr radius of nth energy level	Equation 7.6
A	geometric area of an aperture	Section 3.3
	visibility amplitude	Equation 3.184
A_0	antenna peak collecting area	Equation 3.43
A_{10}	H I line emission coefficient	Equation 7.146
A_e	effective collecting area	Equation 3.35
$A_{n,n-1}$	spontaneous emission rate, level n to level $n-1$	Section 7.2.2
A_{UL}	Einstein coefficient of spontaneous emission	Figure 7.5
Å	angstrom $\equiv 10^{-10}$ m	Appendix F.4
AGN	active galactic nucleus	Figure 5.22
ALMA	Atacama Large Millimeter Array	Color Plate 5
ASKAP	Australian SKA Pathfinder	Color Plate 6
au	astronomical unit	Appendix F.2
b	baseline length	Figure 3.41
	free–free scattering impact parameter	Figure 4.2
	Galactic latitude	Section 1.3
$\vec{b}$	baseline vector	Figure 3.41
B_λ	blackbody spectral brightness per unit wavelength	Equation 2.88
B_ν	blackbody spectral brightness per unit frequency	Equation 2.86
B	blackbody total brightness	Equation 2.89
	magnetic field strength	Appendix F.3
B_{LU}	Einstein coefficient of absorption	Figure 7.5
B_{UL}	Einstein coefficient of stimulated emission	Figure 7.5
c	vacuum speed of light	Appendix F.1
c_{12}	minimum magnetic energy function	Equation 5.109
c_{13}	minimum total energy function	Equation 5.110
C_{LU}	collisional excitation rate coefficient	Equation 7.71
C_{UL}	collisional de-excitation rate coefficient	Equation 7.71
C band	$4 \le \nu(\text{GHz}) \le 8$	Appendix F.5
CARMA	Combined Array for Research in Mm λ Astronomy	
cm	centimeter	Appendix F.3
CMB(R)	cosmic microwave background radiation	Section 2.6
d	distance	
$\vec{p}$	electric dipole moment	Equation 7.120
d_H	Hubble distance	Equation 2.123
D	diameter	
	directivity	Section 3.1.3
D	1 debye $\equiv 10^{-18}$ statcoulomb cm	Appendix F.4
dB	decibel	Appendix F.4
DC	Direct Current	
DFT	discrete Fourier transform	Appendix A.2
DM	dispersion measure	Equation 6.46

Symbol	Definition	Reference
e	eccentricity	Section 6.1.9
	electron charge (magnitude)	Appendix F.1
	Euler's number $= 2.71828\ldots$	
$e(\nu)$	emission coefficient of an opaque body	Section 2.2.4
E	electric field strength	Appendix F.3
	energy	
$\dot{E}$	spin-down luminosity	Equation 6.20
$\langle E_\gamma \rangle$	mean photon energy of blackbody radiation	Equation 2.101
EM	emission measure	Equation 4.57
EOR	Epoch of Reionization	Section 7.8.3
eV	electron volt	Appendix F.1
f	electric field on an aperture	Section 3.2.3
	focal length	Figure 3.7
	pulsar pulse frequency	Section 6.3
$f(v)$	nonrelativistic Maxwellian velocity distribution	Equation 4.34
$f(x)$	(inverse) Fourier transform of $F(s)$	Equation A.2
f/D	focal ratio	Section 3.2.1
$\vec{F}$	force	
$F(s)$	(forward) Fourier transform of $f(x)$	Equation A.1
F_ν	spectral power/area from an isotropic radiator	Equation 2.107
$F_{\rm n}$	noise factor	Equation 3.169
FFT	fast Fourier transform	Appendix A.2
FR	Fanaroff–Riley	Section 5.6.1
FR I	edge-dimmed source morphology	Section 5.6.1
FR II	edge-brightened source morphology	Section 5.6.1
FWHM	full width between half-maximum points	
g	electric field strength on an aperture	Section 3.2.3
	statistical weight of an energy state	Section 7.3.1
g	gram	Appendix F.3
$\langle g_{\rm ff} \rangle$	free–free Gaunt factor	Equation 4.59
G	gravitational constant	Appendix F.1
	antenna power gain	Section 3.1.3
	receiver power gain	Section 3.6.3.1
G_0	antenna peak power gain	Section 3.1.3
GBT	Green Bank Telescope	Color Plate 1
GeV	Giga eV $\equiv 10^9$ eV	Appendix F.1
GHz	Gigahertz $\equiv 10^9$ Hz	Appendix F.4
GMRT	Giant Metre wave Radio Telescope	
h	height	
	Planck's constant	Appendix F.1
$\hbar$	reduced Planck's constant $\equiv h/(2\pi)$	
H	atomic hydrogen	
H_0	Hubble constant	Appendix F.2

Symbol	Definition	Reference
H_2	hydrogen diatomic molecule	
HI	neutral atomic hydrogen	
HII	ionized hydrogen H^+	
HPBW	Half Power Beam Width	
Hz	hertz = frequency in cycles per second (s^{-1})	Appendix F.3
i	inclination angle	
	$\sqrt{-1}$	
I	electric current	Equation 3.4
	moment of inertia	Section 6.1.5
	total intensity	Equation 2.6
I_λ	specific intensity(per unit wavelength)	Equation 2.3
I_ν	specific intensity (per unit frequency)	Equation 2.2
I_p	polarized flux density	Equation 2.57
IC	inverse Compton	Section 5.5
IF	intermediate frequency	Section 3.6.4
III	shah function	Equation 5.61
IMF	initial mass function	Section 4.2
IR	infrared	
ISM	interstellar medium	
j_ν	(spectral) emission coefficient	Equation 2.26
J	angular momentum quantum number	Section 7.7.1
J	joule	Appendix F.3
Jy	jansky $\equiv 10^{-26}$ W m^{-2} Hz^{-1}	Appendix F.4
k	Boltzmann's constant	Appendix F.1
$\vec{k}$	wave vector	Equation 2.49
K	kelvins	Appendix F.3
K band	$18 \le \nu(GHz) \le 26.5$	Appendix F.5
Ka band	$26.5 \le \nu(GHz) \le 40$	Appendix F.5
Ku band	$12 \le \nu(GHz) \le 18$	Appendix F.5
kg	kilogram	Appendix F.3
kpc	kiloparsec $\equiv 10^3$ pc	Appendix F.2
l	Galactic longitude	Figure 7.18
(l, m, n)	direction cosines in (u, v, w) coordinates	Figure 3.45
L	angular momentum	
	luminosity	Equation 2.16
L band	$1 \le \nu(GHz) \le 2$	Appendix F.5
L_ν	spectral luminosity	Equation 2.15
$L_\odot$	solar luminosity	Appendix F.2
L_E	Eddington luminosity	Equation 5.117
LO	local oscillator	Section 3.6.4
LOFAR	LOw Frequency Array for Radio astronomy	Color Plate 8

Symbol	Definition	Reference
LTE	local thermodynamic equilibrium	Section 2.2.2
ly	light year	Appendix F.2
m	meter	Appendix F.3
m	mass	
	reduced mass	Equation 7.105
$m_\perp$	perpendicular magnetic moment	Section 6.1.4
m_e	electron mass	Appendix F.1
m_p	proton mass	Appendix F.1
M	subreflector magnification factor	Equation 3.147
$M_\odot$	solar mass	Appendix F.2
M_{Ch}	Chandrasekhar mass limit	Equation 6.6
M_H	H I mass	Equation 7.166
MJy	megajansky $\equiv 10^6$ Jy	
mJy	millijansky $\equiv 10^{-3}$ Jy	
mm	millimeter $\equiv 10^{-3}$ m	Appendix F.3
MHz	megahertz $\equiv 10^6$ Hz	Appendix F.4
Mpc	megaparsec $\equiv 10^6$ pc	Appendix F.2
MWA	Murchison Widefield Array	Color Plate 7
n	electronic quantum number	Figure 7.1
	pulsar braking index	Section 6.1.8
n^*	critical density	Equation 7.135
n_γ	photon number density	Equation 2.95
n_b	baryon number density	Section 2.6.3
n_e	electron number density	
n_H	number density of neutral hydrogen atoms	
$\dot{n}_H$	hydrogen recombination rate	Equation 4.6
n_i	ion number density	
n_L	number density in lower energy state	Equation 7.42
n_n	number density of atoms in nth energy level	Equation 7.92
n_p	proton number density	
n_U	number density in upper energy state	Equation 7.42
N	number	
N_ν	number of modes per unit frequency	Equation 2.72
N_e	electron column density	Section 1.1.4
$\dot{N}_{IC}$	photon scattering rate per electron	Equation 5.158
NF	noise figure	Equation 3.171
NRAO	National Radio Astronomy Observatory	
p	pressure	
	degree of polarization	Equation 2.58
P	power	
	probability	
	pulsar rotation period	Section 6.1.2
P band	$0.25 \leq \nu(\text{GHz}) \leq 0.50$	Appendix F.5

Symbol	Definition	Reference
$\dot{P}$	pulsar period derivative	Equation 6.17
P_ν	power per unit frequency	
$P(x)$	probability distribution of x	
P_{IC}	inverse Compton power	Section 5.5
P_n	normalized antenna power pattern	
P_{syn}	synchrotron power	Section 5.2
pc	parsec	Appendix F.2
PDBI	Plateau de Bure Interferometer	
pwv	precipitable water vapor	Section 1.1.2
q	electric charge	Appendix F.3
Q_H	production rate of Lyα photons	Equation 4.10
Q band	$33 \le \nu(\text{GHz}) \le 50$	Appendix F.5
QSO	quasi-stellar object	
r	radial coordinate or distance	
$r(\nu)$	reflection coefficient of an opaque body	Section 2.2.4
r_e	equilibrium interatomic distance	Figure 7.12
R	resistance	Equation 3.21
	correlator response	Equation 3.175
R_0	radiation resistance of space	Equation 3.29
R_∞	Rydberg constant	Equation 7.10
$R_\infty c$	Rydberg frequency	Equation 7.11
$R_\odot$	distance from the Sun to the Galactic center	Figure 7.18
	solar radius	Appendix F.2
R_{ff}	far-field distance	Equation 3.64
R_M	Rydberg constant for nuclear mass M	Equation 7.12
R_S	Strömgren sphere radius	Figure 4.1
rad	radian	Appendix F.4
RF	radio frequency	
RFI	radio frequency interference	
rms	root mean square	
s	second of time	Appendix F.3
s	coordinate from source to detector	Figure 2.7
$\hat{s}$	unit vector in source direction	Figure 3.41
$\hat{s}_0$	phase reference direction	Figure 3.45
S	flux or flux density	Equation 2.14
S band	$2 \le \nu(\text{GHz}) \le 4$	Appendix F.5
S_ν	flux density	Equation 2.9
$\vec{S}$	Poynting flux	Equation 2.138
sinc(x)	sinc function $\equiv \sin(\pi x)/(\pi x)$	Equation 3.77
SKA	Square Kilometer Array	
SMA	SubMillimeter Array	
SMBH	supermassive black hole	
sr	steradian	

Symbol	Definition	Reference
t	time	
t	dispersion delay	Equation 6.43
$t_\star$	age of universe at recombination	Section 2.6.2
T	temperature	
	total time	
T_A	antenna temperature	Section 3.1.6
T_0	CMB temperature at redshift zero	Section 2.6
	noise factor reference temperature $= 290$ K	Equation 3.169
T_{atm}	kinetic temperature of the atmosphere	
T_b	brightness temperature	Equation 2.33
T_C	continuum brightness temperature	Section 7.6.1
T_{cmb}	cosmic microwave background temperature	$T_{cmb} \approx 2.73$ K
T_e	electron temperature	
T_L	line brightness temperature at the line center	Section 7.6.1
T_{min}	minimum kinetic temperature	Equation 7.119
T_N	noise temperature	Equation 3.149
T_r	radiometer noise temperature	Section 3.6.1
T_s	HI spin temperature	Equation 7.148
	system noise temperature	Equation 3.150
T_x	excitation temperature	Equation 7.70
TE	thermodynamic equilibrium	Section 2.2.2
THz	terahertz $\equiv 10^{12}$ Hz	Appendix F.4
TOA	pulsar pulse Time Of Arrival	Section 6.3
u	atomic mass unit	Appendix F.1
	dimensionless coordinate, e.g., $u \equiv x/\lambda$	Equation 3.72
	total radiation energy density	Equation 2.92
$\bar{u}$	line profile weighted radiation energy density	Equation 7.39
u_ν	spectral energy density of radiation	Equation 2.76
(u, v, w)	dimensionless coordinates for interferometers	Figure 3.45
U band	$40 < \nu(\text{GHz}) < 60$	Appendix F.5
U_B	magnetic energy density	Equation 5.35
U_e	electron energy density	Equation 5.93
U_{rad}	radiation energy density	Section 5.5
UV	ultraviolet	
v	dimensionless coordinate, e.g., $v \equiv y/\lambda$	Figure 3.45
	velocity	
v_g	group velocity	Equation 6.41
	waveguide group velocity	Equation 3.143
v_r	radial velocity	
v_{rot}	rotational velocity	Equation 7.167
V	voltage	Equation 3.20
	volume	
V_o	square law detector output voltage	Section 3.6.2

Symbol	Definition	Reference
$\mathcal{V}$	complex visibility	Equation 3.182
V band	$50 \leq \nu(\text{GHz}) \leq 75$	Appendix F.5
VLA	Very Large Array	Color Plate 4
VLBA	Very Long Baseline Array	Figure 1.5
VLBI	Very Long Baseline Interferometry	
w	dimensionless coordinate, e.g., $w \equiv z/\lambda$	Figure 3.45
W band	$75 \leq \nu(\text{GHz}) \leq 110$	Appendix F.5
W	free–free pulse radiated energy	Equation 4.24
WMAP	Wilkinson Microwave Anisotropy Probe	
WSRT	Westerbork Synthesis Radio Telescope	Color Plate 3
x	Cartesian coordinate	
(x', y', z', t')	coordinates in moving frame	Figure C.1
x_j	(inverse) discrete Fourier transform of X_k	Equation A.5
X_{CO}	CO to H_2 conversion factor	Equation 7.140
X_k	(forward) discrete Fourier transform of x_j	Equation A.4
X band	$8 \leq \nu(\text{GHz}) \leq 12$	Appendix F.5
y	Cartesian coordinate	
yr	year	Appendix F.2
Y	Y factor	Equation 3.167
z	Cartesian coordinate	
	redshift	Equation 2.127
	zenith angle	Figure 2.10
Z	number of electrons removed from an ion	

H References and Links

H.1 REFERENCE BOOKS

Bracewell, R. (2000). *The Fourier Transform and Its Applications*. McGraw-Hill: New York. The classic textbook on Fourier transforms.

Burke, B. F., and Graham-Smith, F. (2002). *An Introduction to Radio Astronomy* (2nd ed.). Cambridge University Press: Cambridge. A very readable descriptive introduction to radio astronomy. Most of the equations are only presented, not derived.

Christiansen, W. N., and Högbom, J. A. (1985). *Radio Telescopes*. Cambridge University Press: Cambridge. Principles of design for a wide range of radio telescopes.

Draine, B. T. (2011). *Physics of the Interstellar and Intergalactic Medium*. Princeton University Press: Princeton. Graduate-level textbook on the interstellar and intergalactic medium with an extensive discussion of dust emission and absorption.

Evans, J. V., and Hagfors, T. (eds) (1968). *Radar Astronomy*. McGraw-Hill: New York. The textbook based on the 1960 MIT summer course on radar astronomy.

Goldsmith, P. F. (ed) (1988). *Instrumentation and Techniques for Radio Astronomy*. IEEE Press: New York. Reprints of classic radio astronomy papers.

Griffiths, D. J. (2012). *Introduction to Electrodynamics* (4th ed.). Addison-Wesley. A very accessible introduction to vector calculus and Maxwell's equations.

Jackson, J. D. (1962). *Classical Electrodynamics*. Wiley: New York. The standard textbook for electromagnetism, with an appendix explaining systems of units.

Kraus, J. D. (1986). *Radio Astronomy*. Cygnus-Quasar Books: Powell, OH. Revised edition of the classic but idiosyncratic general textbook, with an emphasis on radio telescope antennas and receivers from an engineer's viewpoint.

Longair, M. S. (1992). *High Energy Astrophysics* (2nd ed.). Cambridge University Press: Cambridge. Two-volume textbook containing physically insightful derivations of the Larmor equation and formulas for free–free emission, synchrotron radiation, and inverse-Compton scattering.

Lorimer, D. L., and Kramer, M. (2005). *Handbook of Pulsar Astronomy*. Cambridge University Press: Cambridge. Comprehensive review of pulsar observational techniques and results.

Lyne, A. G., and Graham-Smith, F. (1998). *Pulsar Astronomy* (2nd ed.). Cambridge University Press: Cambridge. A very readable book covering most of pulsar astronomy.

Osterbrock, D. E. (1989). *Astrophysics of Gaseous Nebulae and Active Galactic Nuclei* (2nd ed.). University Science Books: Mill Valley, CA. The classic text covering free–free continuum and hydrogen recombination lines at the advanced graduate level.

Pacholczyk, A. B. (1970). *Radio Astrophysics*. Freeman: San Francisco. Textbook with mathematically complete derivations of formulas for free–free emission, synchrotron radiation, and inverse-Compton radiation.

Rybicki, G. B., and Lightman, A. P. (1979). *Radiative Processes in Astrophysics*. Wiley: New York. A very good textbook on radiation fundamentals and astrophysical emission mechanisms.

Stanimirovic, S., Altschuler, D. R., Goldsmith, P. F., and Salter, C. J. (eds) (2002). *Single-Dish Radio Astronomy: Techniques and Applications*. ASP: San Francisco. Everything you wanted to know about single-dish observing, from the 2001 Arecibo summer school.

Sullivan, W. T. (2009). *Cosmic Noise: A History of Early Radio Astronomy*. Cambridge University Press: Cambridge. The standard history of early radio astronomy.

Taylor, G. B., Carilli, C. L., and Perley, R. A. (eds) (1999). *Synthesis Imaging in Radio Astronomy II*. ASP: San Francisco. Everything you wanted to know about interferometry, but were afraid to ask, from the 1998 VLA synthesis-imaging summer school.

Thompson, A. R., Moran, J. M., and Swenson, G. W. (2001). *Interferometry and Synthesis in Radio Astronomy* (2nd ed.). Wiley: New York. The bible of radio interferometry.

Wilson, T. L., Rohlfs, K., and Hüttemeister, S. (2009). *Tools of Radio Astronomy* (5th ed.). Springer: Berlin. For many years, the only complete radio-astronomy textbook in print; successor to Rohlfs and Wilson, *Tools of Radio Astronomy*.

H.2 LINKS

http://xxx.lanl.gov/archive/astro-ph. arXiv astronomy and astrophysics preprints, but *caveat emptor* as they have not all been refereed.

http://adsabs.harvard.edu/. The Astrophysical Data System (ADS) searches the astronomy/astrophysics literature by subject, author, etc.

http://cdsweb.u-strasbg.fr/astroWeb/astroweb.html. AstroWeb links to most everything astronomical on the web.

http://cdsweb.u-strasbg.fr/Cats.html. The CDS (Strasbourg astronomical Data Center) collects and distributes astronomical data catalogues, related to observations of stars and galaxies, and other Galactic and extragalactic objects.

http://www.cv.nrao.edu/course/astr534/ERA.shtml. The online radio astronomy course *Essential Radio Astronomy*.

http://nedwww.ipac.caltech.edu. The NASA/IPAC Extragalactic Database. NED is built around a master list of extragalactic objects for which cross-identifications of names have been established, accurate positions and redshifts entered to the extent possible, and some basic data collected. Bibliographic references relevant to individual objects have been compiled, and abstracts of extragalactic interest are kept online. Detailed and referenced photometry, position, and redshift data have been taken from large compilations and from the literature.

http://www.nrao.edu/. The NRAO (National Radio Astronomy Observatory) site describing the NRAO telescopes, how to propose for observing time, schedule observations, reduce data, etc.

http://www.cv.nrao.edu/nvss/. The NVSS (NRAO VLA Sky Survey) site provides radio continuum views of the entire sky north of $\delta = -40°$, including a catalog of 1.8×10^6 sources stronger than 2.5 mJy at 1.4 GHz and postage-stamp images with $45''$ resolution.

http://simbad.harvard.edu. SIMBAD (Set of Identifications, Measurements, and Bibliography for Astronomical Data) contains data, cross-identifications, observational measurements, and bibliographies for celestial objects outside the Solar System: stars, galaxies, and nonstellar objects within our galaxy, or in external galaxies.

Bibliography

[1] ALMA Partnership (2015, March). First results from high angular resolution ALMA observations toward the HL Tau region. *ArXiv e-prints*.

[2] Alpher, R. A., H. Bethe, and G. Gamow (1948, April). The origin of chemical elements. *Physical Review 73*, 803–804.

[3] Alpher, R. A. and R. Herman (1948, November). Evolution of the universe. Nature *162*, 774–775.

[4] Baade, W. and R. Minkowski (1954, January). Identification of the radio sources in Cassiopeia, Cygnus A, and Puppis A. ApJ *119*, 206.

[5] Baade, W. and F. Zwicky (1934, May). Cosmic rays from super-novae. *Proceedings of the National Academy of Science 20*, 259–263.

[6] Baars, J. W. M., R. Genzel, I. I. K. Pauliny-Toth, and A. Witzel (1977, October). The absolute spectrum of CAS A - An accurate flux density scale and a set of secondary calibrators. A&A *61*, 99–106.

[7] Bahcall, J. N., S. Kirhakos, D. P. Schneider, R. J. Davis, T. W. B. Muxlow, S. T. Garrington, R. G. Conway, and S. C. Unwin (1995, October). Hubble Space Telescope and MERLIN observations of the jet in 3C 273. ApJ *452*, L91.

[8] Balick, B. and R. L. Brown (1974, December). Intense sub-arcsecond structure in the galactic center. ApJ *194*, 265–270.

[9] Barrett, A. H. (1970). Fundamentals of radio astronomy. Lectures prepared for course at MIT taught by the author.

[10] Blain, A. W. and M. S. Longair (1993, September). Submillimetre cosmology. MNRAS *264*, 509.

[11] Blake, C. and J. Wall (2002, March). A velocity dipole in the distribution of radio galaxies. Nature *416*, 150–152.

[12] Blandford, R. D. and R. L. Znajek (1977, May). Electromagnetic extraction of energy from Kerr black holes. MNRAS *179*, 433–456.

[13] Blumenthal, G. R. and R. J. Gould (1970). Bremsstrahlung, synchrotron radiation, and Compton scattering of high-energy electrons traversing dilute gases. *Reviews of Modern Physics 42*, 237–271.

[14] Bolatto, A. D., M. Wolfire, and A. K. Leroy (2013, August). The CO-to-H_2 conversion factor. ARA&A *51*, 207–268.

[15] Bracewell, R. N. (2000). *The Fourier Transform and its Applications*. Boston: McGraw Hill.

[16] Bradt, H. (2008, September). *Astrophysics Processes*. Cambridge University Press.

[17] Breton, R. P., V. M. Kaspi, M. Kramer, M. A. McLaughlin, M. Lyutikov, S. M. Ransom, I. H. Stairs, R. D. Ferdman, F. Camilo, and A. Possenti (2008, July). Relativistic spin precession in the double pulsar. *Science 321*, 104–.

[18] Brussaard, P. J. and H. C. van de Hulst (1962, July). Approximation formulas for nonrelativistic bremsstrahlung and average Gaunt factors for a Maxwellian electron gas. *Reviews of Modern Physics 34*, 507–520.

[19]   Burgay, M., M. Kramer, and M. A. McLaughlin (2014, September). The double pulsar J0737–3039A/B: A decade of surprises. *Bulletin of the Astronomical Society of India 42*, 101–119.

[20]   Burke, B. F. and F. Graham-Smith (1997). *An Introduction to Radio Astronomy*. Cambridge, UK: Cambridge University Press.

[21]   Cane, H. V. (1979, November). Spectra of the non-thermal radio radiation from the galactic polar regions. MNRAS *189*, 465–478.

[22]   Cheung, A. C., D. M. Rank, C. H. Townes, D. D. Thornton, and W. J. Welch (1968, December). Detection of NH_3 molecules in the interstellar medium by their microwave emission. *Physical Review Letters 21*, 1701–1705.

[23]   Churchwell, E. (2002). Ultra-compact HII regions and massive star formation. ARA&A *40*, 27–62.

[24]   Condon, J. J. (1992). Radio emission from normal galaxies. ARA&A *30*, 575–611.

[25]   Condon, J. J. (2002, December). Continuum 1: General aspects. In S. Stanimirovic, D. Altschuler, P. Goldsmith, and C. Salter (Eds.), *Single-Dish Radio Astronomy: Techniques and Applications*, Volume 278 of *Astronomical Society of the Pacific Conference Series*, pp. 155–171.

[26]   Condon, J. J., M. L. Anderson, and G. Helou (1991, July). Correlations between the far-infrared, radio, and blue luminosities of spiral galaxies. ApJ *376*, 95–103.

[27]   Condon, J. J., J. J. Broderick, and G. A. Seielstad (1989, April). A 4.85 GHz sky survey. I - Maps covering delta between 0 and + 75 deg. AJ *97*, 1064–1073.

[28]   Condon, J. J., W. D. Cotton, and J. J. Broderick (2002, August). Radio sources and star formation in the local universe. AJ *124*, 675–689.

[29]   Condon, J. J., W. D. Cotton, E. B. Fomalont, K. I. Kellermann, N. Miller, R. A. Perley, D. Scott, T. Vernstrom, and J. V. Wall (2012, October). Resolving the radio source background: Deeper understanding through confusion. ApJ *758*, 23.

[30]   Condon, J. J., W. D. Cotton, E. W. Greisen, Q. F. Yin, R. A. Perley, G. B. Taylor, and J. J. Broderick (1998, May). The NRAO VLA Sky Survey. AJ *115*, 1693–1716.

[31]   Condon, J. J., Z.-P. Huang, Q. F. Yin, and T. X. Thuan (1991, September). Compact starbursts in ultraluminous infrared galaxies. ApJ *378*, 65–76.

[32]   Cooley, J. W. and J. W. Tukey (1965, April). An algorithm for the machine calculation of complex Fourier series. *Mathematics of Computation 19*, 297–301.

[33]   Cordes, J. M. and T. J. W. Lazio (2002, July). NE2001. I. a new model for the galactic distribution of free electrons and its fluctuations. *ArXiv Astrophysics e-prints*.

[34]   Demorest, P. B., T. Pennucci, S. M. Ransom, M. S. E. Roberts, and J. W. T. Hessels (2010, October). A two-solar-mass neutron star measured using Shapiro delay. Nature *467*, 1081–1083.

[35]   Dickey, J. M., S. R. Kulkarni, C. E. Heiles, and J. H. van Gorkom (1983, November). A survey of H I absorption at low latitudes. ApJS *53*, 591–621.

[36]   Evans, J. V. and T. Hagfors (1968). *Radar Astronomy*. New York: McGraw Hill.

[37]   Fabian, A. C. (2012, September). Observational evidence of active galactic nuclei feedback. ARA&A *50*, 455–489.

[38]   Fanaroff, B. L. and J. M. Riley (1974, May). The morphology of extragalactic radio sources of high and low luminosity. MNRAS *167*, 31P–36P.

[39]   Fishman, G. J., F. R. Harnden, Jr., W. N. Johnson, III, and R. C. Haymes (1969, October). The period and hard-X spectrum of NP 0532 in 1967. ApJ *158*, L61.

[40]   Georgelin, Y. M. and Y. P. Georgelin (1976, May). The spiral structure of our Galaxy determined from H II regions. A&A *49*, 57–79.

[41] Gruber, D. E., W. A. Heindl, R. E. Rothschild, W. Coburn, R. Staubert, I. Kreykenbohm, and J. Wilms (2001, November). Stability of the cyclotron resonance scattering feature in Hercules X-1 with RXTE. ApJ *562*, 499–507.

[42] Hasinger, G. (1996, December). The extragalactic X-ray and gamma-ray background. A&AS *120*, C607.

[43] Haslam, C. G. T., C. J. Salter, H. Stoffel, and W. E. Wilson (1982a, January). A 408 MHz all-sky continuum survey. II - The atlas of contour maps. A&AS *47*, 1+.

[44] Haslam, C. G. T., C. J. Salter, H. Stoffel, and W. E. Wilson (1982b, January). A 408 MHz all-sky continuum survey. II - The atlas of contour maps. A&AS *47*, 1.

[45] Hauser, M. G., R. G. Arendt, T. Kelsall, E. Dwek, N. Odegard, J. L. Weiland, H. T. Freudenreich, W. T. Reach, R. F. Silverberg, S. H. Moseley, Y. C. Pei, P. Lubin, J. C. Mather, R. A. Shafer, G. F. Smoot, R. Weiss, D. T. Wilkinson, and E. L. Wright (1998, November). The COBE diffuse infrared background experiment search for the cosmic infrared background. I. Limits and detections. ApJ *508*, 25–43.

[46] Hauser, M. G. and E. Dwek (2001). The cosmic infrared background: Measurements and implications. ARA&A *39*, 249–307.

[47] Haynes, M. P., L. van Zee, D. E. Hogg, M. S. Roberts, and R. J. Maddalena (1998, January). Asymmetry in high-precision global H I profiles of isolated spiral galaxies. AJ *115*, 62.

[48] Hazard, C., M. B. Mackey, and A. J. Shimmins (1963, March). Investigation of the radio source 3C 273 by the method of lunar occultations. Nature *197*, 1037–1039.

[49] Hellings, R. W. and G. S. Downs (1983). Upper limits on the isotropic gravitational radiation background from pulsar timing analysis. ApJ *265*, L39.

[50] Hewish, A., S. J. Bell, J. D. H. Pilkington, P. F. Scott, and R. A. Collins (1968, February). Observation of a rapidly pulsating radio source. Nature *217*, 709–713.

[51] Hey, J. S. (1973). *The Evolution of Radio Astronomy*. London: Elek Science.

[52] Ho, P. T. P. and C. H. Townes (1983). Interstellar ammonia. ARA&A *21*, 239–270.

[53] Hogg, D. W. (1999, May). Distance measures in cosmology. *ArXiv Astrophysics e-prints*.

[54] Jackson, J. D. (1975). *Classical Electrodynamics* (2nd ed.). New York: Wiley.

[55] Jackson, J. D. (1998, July). *Classical Electrodynamics* (third ed.). New York: Wiley.

[56] Jansky, K. J. (1933, October). Electrical disturbances of apparently extraterrestrial origin. *Proceedings of the Institute of Radio Engineers 21*, 1387–1398.

[57] Jenkins, E. B. and T. M. Tripp (2011, June). The distribution of thermal pressures in the diffuse, cold neutral medium of our galaxy. II. An expanded survey of interstellar C I fine-structure excitations. ApJ *734*, 65.

[58] Johnson, K. E., R. Indebetouw, and D. J. Pisano (2003, July). Searching for embedded super-star clusters in IC 4662, NGC 1705, and NGC 5398. AJ *126*, 101–112.

[59] Kaspi, V. M. and D. J. Helfand (2002). Constraining the birth events of neutron stars. In P. O. Slane and B. M. Gaensler (Eds.), *Neutron Stars in Supernova Remnants*, Volume 271 of *Astronomical Society of the Pacific Conference Series*, pp. 3.

[60] Konopelko, A., A. Mastichiadis, J. Kirk, O. C. de Jager, and F. W. Stecker (2003, November). Modeling the TeV gamma-ray spectra of two low-redshift active galactic nuclei: Markarian 501 and Markarian 421. ApJ *597*, 851–859.

[61] Kramer, M., I. H. Stairs, R. N. Manchester, M. A. McLaughlin, A. G. Lyne, R. D. Ferdman, M. Burgay, D. R. Lorimer, A. Possenti, N. D'Amico, J. M. Sarkissian, G. B. Hobbs, J. E. Reynolds, P. C. C. Freire, and F. Camilo (2006, October). Tests of general relativity from timing the double pulsar. *Science 314*, 97–102.

[62] Kraus, J. D. (1966). *Radio Astronomy*. New York: McGraw Hill.

[63] Kraus, J. D., M. Tiuri, A. V. Raisanen, and T. D. Carr (1986). *Radio Astronomy Receivers.* Powell, Ohio: Cygnus-Quasar Books.

[64] Laing, R. A., A. H. Bridle, P. Parma, L. Feretti, G. Giovannini, M. Murgia, and R. A. Perley (2008, May). Multifrequency VLA observations of the FR I radio galaxy 3C 31: Morphology, spectrum and magnetic field. MNRAS *386*, 657–672.

[65] Liebe, H. J. (1985, October). An updated model for millimeter wave propagation in moist air. *Radio Science 20*, 1069–1089.

[66] Lockman, F. J. and R. L. Brown (1975, October). The radio recombination line spectrum of Orion A - Observations and analysis. ApJ *201*, 134–150.

[67] Loinard, L., R. M. Torres, A. J. Mioduszewski, L. F. Rodríguez, R. A. González-Lópezlira, R. Lachaume, V. Vázquez, and E. González (2007, December). VLBA determination of the distance to nearby star-forming regions. I. The distance to T Tauri with 0.4% accuracy. ApJ *671*, 546–554.

[68] Longair, M. S. (1992, March). *High Energy Astrophysics. Vol.1: Particles, Photons and their Detection.* Cambridge, UK: Cambridge University Press.

[69] Lorimer, D. R. and M. Kramer (2005). *Handbook of Pulsar Astronomy.* Cambridge, UK: Cambridge University Press.

[70] Lyne, A. G. and F. Graham-Smith (2006). *Pulsar Astronomy.* Cambridge, UK: Cambridge University Press.

[71] McNamara, B. R., P. E. J. Nulsen, M. W. Wise, D. A. Rafferty, C. Carilli, C. L. Sarazin, and E. L. Blanton (2005, January). The heating of gas in a galaxy cluster by X-ray cavities and large-scale shock fronts. Nature *433*, 45–47.

[72] Mezger, P. G. and A. P. Henderson (1967, February). Galactic H II regions. I. Observations of their continuum radiation at the frequency 5 GHz. ApJ *147*, 471.

[73] Moran, J. M. (2008, August). The black-hole accretion disk in NGC 4258: One of nature's most beautiful dynamical systems. In A. H. Bridle, J. J. Condon, and G. C. Hunt (Eds.), *Frontiers of Astrophysics: A Celebration of NRAO's 50th Anniversary*, Volume 395 of *Astronomical Society of the Pacific Conference Series*, pp. 87.

[74] Nyquist, H. (1928, July). Thermal agitation of electric charge in conductors. *Physical Review 32*, 110–113.

[75] Oster, L. (1961, October). Emission, absorption, and conductivity of a fully ionized gas at radio frequencies. *Reviews of Modern Physics 33*, 525–543.

[76] Osterbrock, D. E. (1974). *Astrophysics of Gaseous Nebulae.* San Francisco: W. H. Freeman.

[77] Pacholczyk, A. G. (1970). *Radio Astrophysics. Nonthermal Processes in Galactic and Extragalactic Sources.* San Francisco: Freeman.

[78] Parsons, A. R., A. Liu, J. E. Aguirre, Z. S. Ali, R. F. Bradley, C. L. Carilli, D. R. DeBoer, M. R. Dexter, N. E. Gugliucci, D. C. Jacobs, P. Klima, D. H. E. MacMahon, J. R. Manley, D. F. Moore, J. C. Pober, I. I. Stefan, and W. P. Walbrugh (2014, June). New limits on 21 cm epoch of reionization from PAPER-32 consistent with an X-ray heated intergalactic medium at z = 7.7. ApJ *788*, 106.

[79] Pennucci, T. T., A. Possenti, P. Esposito, N. Rea, D. Haggard, F. K. Baganoff, M. Burgay, F. Coti Zelati, G. L. Israel, and A. Minter (2015, May). Simultaneous multi-band radio and X-ray observations of the galactic center magnetar SGR 1745–2900. *ArXiv e-prints*.

[80] Penzias, A. A. and R. W. Wilson (1965, July). A measurement of excess antenna temperature at 4080 Mc/s. ApJ *142*, 419–421.

[81] Piner, B. G., S. C. Unwin, A. E. Wehrle, A. C. Zook, C. M. Urry, and D. M. Gilmore (2003, May). The speed and orientation of the parsec-scale jet in 3C 279. ApJ *588*, 716–730.

[82]   Planck Collaboration, P. Ade, N. Aghanim, M. Arnaud, M. Ashdown, J. Aumont, C. Baccigalupi, A. J. Banday, R. B. Barreiro, J. G. Bartlett, et al. (2015, February). Planck 2015 results. XIII. Cosmological parameters. *ArXiv e-prints*.

[83]   Planck Collaboration, N. Aghanim, M. Arnaud, M. Ashdown, J. Aumont, C. Baccigalupi, A. J. Banday, R. B. Barreiro, J. G. Bartlett, N. Bartolo, et al. (2015, July). Planck 2015 results. XI. CMB power spectra, likelihoods, and robustness of parameters. *ArXiv e-prints*.

[84]   Quireza, C., R. T. Rood, D. S. Balser, and T. M. Bania (2006, July). Radio recombination lines in galactic H II regions. ApJS *165*, 338–359.

[85]   Quireza, C., R. T. Rood, T. M. Bania, D. S. Balser, and W. J. Maciel (2006, December). The electron temperature gradient in the Galactic Disk. ApJ *653*, 1226–1240.

[86]   Reber, G. (1940, June). Notes: Cosmic static. ApJ *91*, 621–624.

[87]   Reber, G. (1958, January). Early radio astronomy at Wheaton, Illinois. *Proceedings of the Institute of Radio Engineers 46*, 15–23.

[88]   Reid, M. J. (1993). The distance to the center of the Galaxy. ARA&A *31*, 345–372.

[89]   Reid, M. J. and M. Honma (2014, August). Microarcsecond radio astrometry. ARA&A *52*, 339–372.

[90]   Rindler, W. (1980). *Essential Relativity*. Berlin: Springer.

[91]   Rindler, W. (1991). *Introduction to Special Relativity* (second ed.). USA: Oxford University Press.

[92]   Roberts, M. S. and R. N. Whitehurst (1975, October). The rotation curve and geometry of M31 at large galactocentric distances. ApJ *201*, 327–346.

[93]   Rodriguez-Rico, C. A., F. Viallefond, J.-H. Zhao, W. M. Goss, and K. R. Anantharamaiah (2004, December). Very Large Array H92α and H53α radio recombination line observations of M82. ApJ *616*, 783–803.

[94]   Rubin, R. H. (1968, October). A discussion of the sizes and excitation of H II regions. ApJ *154*, 391.

[95]   Ruze, J. (1965, September). Lateral feed displacement in a paraboloid. *IEEE Transactions on Antennas and Propagation 13*, 660–665.

[96]   Ruze, J. (1966, April). Antenna tolerance theory – A review. *IEEE Proceedings 54*, 633–642.

[97]   Rybicki, G. B. and A. P. Lightman (1979). *Radiative processes in astrophysics*. New York: Wiley-Interscience.

[98]   Ryle, M. and R. W. Clarke (1961). An examination of the steady-state model in the light of some recent observations of radio sources. MNRAS *122*, 349.

[99]   Ryle, M. and P. A. G. Scheuer (1955, July). The spatial distribution and the nature of radio stars. *Royal Society of London Proceedings Series A 230*, 448–462.

[100] Scheuer, P. A. G. and A. C. S. Readhead (1979, January). Superluminally expanding radio sources and the radio-quiet QSOs. Nature *277*, 182–185.

[101] Schmidt, M. (1963, March). 3C 273 : A star-like object with large red-shift. Nature *197*, 1040.

[102] Sullivan, III, W. T. (2009). *Cosmic Noise: A History of Early Radio Astronomy*. Cambridge University Press.

[103] Swaters, R. A., T. S. van Albada, J. M. van der Hulst, and R. Sancisi (2002, August). The Westerbork HI survey of spiral and irregular galaxies. I. HI imaging of late-type dwarf galaxies. A&A *390*, 829–861.

[104] Taylor, J. H., L. A. Fowler, and P. M. McCulloch (1979, February). Measurements of general relativistic effects in the binary pulsar PSR 1913+16. Nature *277*, 437–440.

[105] Tegmark, M., A. de Oliveira-Costa, and A. J. S. Hamilton (2003, Dec). High resolution foreground cleaned CMB map from WMAP. *Phys. Rev. D 68*, 123523.

[106] Thompson, A. R., J. M. Moran, and G. W. Swenson, Jr. (2001). *Interferometry and Synthesis in Radio Astronomy* (2nd ed.). New York: Wiley.

[107] Tingay, S. J., R. Goeke, J. D. Bowman, D. Emrich, S. M. Ord, D. A. Mitchell, M. F. Morales, T. Booler, B. Crosse, R. B. Wayth, C. J. Lonsdale, S. Tremblay, D. Pallot, T. Colegate, A. Wicenec, N. Kudryavtseva, W. Arcus, D. Barnes, G. Bernardi, F. Briggs, S. Burns, J. D. Bunton, R. J. Cappallo, B. E. Corey, A. Deshpande, L. Desouza, B. M. Gaensler, L. J. Greenhill, P. J. Hall, B. J. Hazelton, D. Herne, J. N. Hewitt, M. Johnston-Hollitt, D. L. Kaplan, J. C. Kasper, B. B. Kincaid, R. Koenig, E. Kratzenberg, M. J. Lynch, B. Mckinley, S. R. Mcwhirter, E. Morgan, D. Oberoi, J. Pathikulangara, T. Prabu, R. A. Remillard, A. E. E. Rogers, A. Roshi, J. E. Salah, R. J. Sault, N. Udaya-Shankar, F. Schlagenhaufer, K. S. Srivani, J. Stevens, R. Subrahmanyan, M. Waterson, R. L. Webster, A. R. Whitney, A. Williams, C. L. Williams, and J. S. B. Wyithe (2013, 1). The Murchison widefield array: The square kilometre array precursor at low radio frequencies. *PASA - Publications of the Astronomical Society of Australia 30*, e007.

[108] Tully, R. B. and J. R. Fisher (1977, February). A new method of determining distances to galaxies. A&A *54*, 661–673.

[109] Urry, C. M. and P. Padovani (1995, September). Unified schemes for radio-loud active galactic nuclei. PASP *107*, 803.

[110] Verschuur, G. L. and K. I. Kellermann (1988a). *Galactic and Extra-Galactic Radio Astronomy.* Springer.

[111] Verschuur, G. L. and K. I. Kellermann (1988b). *Galactic and Extragalactic Radio Astronomy* (second ed.). Berlin: Springer.

[112] Voelk, H. J. (1989, July). The correlation between radio and far-infrared emission for disk galaxies - A calorimeter theory. A&A *218*, 67–70.

[113] Walsh, D., R. F. Carswell, and R. J. Weymann (1979, May). 0957 + 561 A, B - Twin quasistellar objects or gravitational lens. Nature *279*, 381–384.

[114] Weaver, H., D. R. W. Williams, N. H. Dieter, and W. T. Lum (1965, October). Observations of a strong unidentified microwave line and of emission from the OH molecule. Nature *208*, 29–31.

[115] Wilson, T. L., C. Henkel, S. Huttemeister, G. Dahmen, A. Linhart, C. Lemme, and J. Schmid-Burgk (1993, September). Hot ammonia emission - Kinetic temperature gradients in Orion-Kl. A&A *276*, L29.

[116] Wilson, T. L., K. Rohlfs, and S. Hüttemeister (2009). *Tools of Radio Astronomy.* Berlin: Springer.

[117] Wolszczan, A. and D. A. Frail (1992, January). A planetary system around the millisecond pulsar PSR1257 + 12. Nature *355*, 145–147.

[118] Yun, M. S., P. T. P. Ho, and K. Y. Lo (1994, December). A high-resolution image of atomic hydrogen in the M81 group of galaxies. Nature *372*, 530–532.

[119] Ziurys, L. M. and the ARO Staff (2006). Initial observations with an ALMA Band 6 Mixer-Preamp. *NRAO Newsletter 109*, 11–13.

Index